D0235757

HUMBERSIDE LIBRARIES
Loan
Collection
CENTRAL

Climate and building in Britain

Building Research Establishment Report

Climate and building in Britain

A review of meteorological information suitable for use in the planning, design, construction and operation of buildings

R E Lacy, BSc (Eng), FRMetS
Building Research Station

Department of the Environment
Building Research Establishment
London, Her Majesty's Stationery Office

Full details of all new BRE publications
are published quarterly in BRE NEWS.
Requests for BRE NEWS or placing on
the mailing list should be addressed to:

Distribution Unit
Application Services Division
Building Research Establishment
Garston, Watford WD2 7JR

ISBN 0 11 670544 2

First published 1977

Foreword

Meteorological information in a form suitable for the building and construction industry is required at all stages of a job, from town planning to the operation of the completed structure. There is an increasing awareness in all sections of the industry that the climate must be taken into account; this is shown by the many enquiries received both by the Building Research Station and by the Meteorological Office. The latter receives annually some thousands of enquiries, mainly for design purposes, while the Weather Centres and the various local forecast offices have tens of thousands of requests for forecasts or information on current weather of concern to the industry.

There are no doubt several reasons for this increased interest in the weather. Some designers want to use the information in order to provide more economical buildings or ones better suited to particular climates. Others merely wish to avoid mistakes which have been made in the past. Constructors may wish to make a more accurate estimate of the time likely to be lost by bad weather during a contract. Whatever the query, or the reason for making it, the need is for suitable data on the weather. If suitable data are to be provided, the way in which the weather affects the problem in question must be known. This work sets down what is known of the interactions between the climate and a building, including some consideration of town planning and of the construction processes. Much remains to be done, but it is hoped that this will be a useful introduction to the subject.

J B Dick

Director, Building Research Establishment, 1976

Author's preface

Some years ago it was planned to produce a Code of Practice on Climate, which was to be in effect a climatic atlas incorporating meteorological information of use to the building industry. This was never completed, but the first part, prepared largely by H C Shellard of the Meteorological Office, with certain sections by R E Lacy and Professor J K Page, has been used as the basis of Chapter 1 of the present work, although considerably re-written and enlarged.

Of the remainder of the book, the section on wind loading consists mainly of the text of a Building Research Station Digest written by C W Newberry and that on ice loading was largely prepared by N C Helliwell. Some of the data used in the book refer to Garston, Hertfordshire, and are derived from the measurements made at the climatological station maintained by the Building Research Establishment. Most of the remainder come from the Meteorological Office, either directly (usually from specially commissioned studies), or from various published handbooks, memoranda or articles in journals, these being individually acknowledged. It is hoped that this general acknowledgement will serve to mark the debt we owe to the Meteorological Office for the effective way in which it collects, stores, analyses and publishes such a quantity of climatological data.

The author thanks all those individuals who have helped to provide material, and apart from his colleagues at the Building Research Establishment, would like to name especially Messrs H C Shellard, S F G Farmer and the late N C Helliwell of the Meteorological Office, and Mr O Griffiths of the then Tredegar Urban District Council.

Contents

Editorial Note

The following mathematical signs have been used to aid the layout of certain tables and maps:

$>$ 'greater than'
$<$ 'less than'
$\geqslant$ 'greater than'or 'equal to'
$\leqslant$ 'less than' or 'equal to'

List of figures

Chapter 1

Chapter 2

Chapter 3

Chapter 4

Chapter 4—*continued*

Acknowledgements

Most of the maps showing the distribution of climatic elements over the whole country are taken from Meteorological Office originals and are the latest available at the time of writing. Memoranda with updated analyses are published from time to time by the Meteorological Office (see lists at end of Chapter 1). Many other maps and diagrams were derived by the author from Meteorological Office analyses, often previously unpublished, or from Meteorological Office publications of various kinds. Diagrams incorporating Garston data are from analyses of the measurements made at the climatological station maintained by the Building Research Establishment. Figures taken from other sources are all separately acknowleged.

List of tables

Chapter 1

Chapter 2

Chapter 3

Chapter 4

Introduction

The way that climatic information is presented depends on the problem being considered. In each case the problem must be identified and described. Next, the way the climate (and any other external factors) affects the process or material must be defined. Finally the relevant climatic data may be extracted from the available climatological material. Until the first two steps have been completed, the last cannot be attempted. This is the so-called 'systems approach' to the problem, in which all the actions and reactions of the building problem upon the weather, and of the weather on the problem, are considered together. Too often, attempts have been made to solve the problems by starting from the climatological data, in the belief that summaries of data known to be important would be sufficient. For example, attempts have been made to relate the occurrence of frost damage to materials by forming a regression equation with climatic elements such as temperature and rainfall as variables. It is true that such an approach may be of initial help, but without the essential second, or linking, step, no complete solution is possible.

This is a major limitation in this work, because for one reason or another we do not always know the precise relations between the weather and a particular point of design or practice. When the problem is purely physical, for example the determination of the rate of heat loss through a wall, enough may be known to enable the heat losses to be calculated with adequate accuracy under any given set of weather conditions. But where human responses are involved we may be a long way from being able to make more than an inspired guess at a numerical relationship with the weather. From time to time attempts are made to produce a classification of climate which will serve as a guide to the best building type in a particular climate. This was done with some success many years ago for vegetation, by Köppen (Köppen and Geiger 1930) when he produced his famous world map of climatic zones, subdivided according to the vegetation natural to each zone. In such a case the zoning can be done with some confidence, for the naturally established vegetation does form an integrated index of the total climate of a place.

Buildings, on the other hand, are essentially artificial devices, the design of which, although to some extent conditioned by the local climate if they are of a long-standing indigenous pattern, may be influenced by many other ideas.

Moreover, although one may speak of the suitability of a building for a certain climate, in fact a complete building is normally made of many different parts. Each of these parts may behave differently with respect to the climate, either in the immediate or the long-term aspect. Thus some parts of the building may behave satisfactorily in one or more ways, while others may not – for example, some parts of the building may deteriorate rapidly and require renewing before others. Some materials or forms of construction may be sensitive to temperature, others to humidity, others again to wind-driven rain, and so on. It is clear that no overall index of climate can be devised that will serve as a measure of the behaviour of a complex thing like a building in all its aspects.

Any lack of balance in this contents of this book reflects the incomplete state of knowledge in many sections of the subject. Although the impact of weather or climate on some aspects of building has been realised for a very long time, it is only recently that attempts have been made to put the relationships into a numerical form. In the field of heating, for example, the relationship between the weather and heat loss is well understood (except perhaps in certain problems concerning moisture), and climatic data are in everyday use for heat loss calculations. In others, for example in the effects on the construction process, little is at present known.

Within these limitations, therefore, this work sets down that information which will be most useful to the building and construction industry in Britain. It was felt unnecessary to include the great mass of data needed for the design of heating and air-conditioning systems contained in the Guide of the Institution of Heating and Ventilating Engineers (1971).

In many of the problems which will be dealt with, climate is not the only factor to be considered. In some it may be the least important. However the balance may vary according to circumstances and a problem has not been omitted merely because at present the climatic influences seem to be of little moment.

The cost of weather

Although in this work stress is laid on the three steps or stages of each problem, beginning with the identification of the problem, followed by the determination of the linking relationships between the building factor and the weather, and finally producing the required climatic statistics, there is in reality a further stage. This fourth stage concerns the additional cost to the project, or to the running costs, caused by the weather. Estimates of the variations in the costs of heating a building caused by variations in temperature, wind speed and amount of sunshine can be made without undue difficulty.

The variations in the costs of the construction processes are much more difficult to assess. A simple approach is to say that if 1 per cent of the time is lost on a particular contract because of bad weather, then one adds 1 per cent to x million pounds to cover the cost of making good this lost time.
It is unlikely that the situation is really as simple as this, for it is probable that for at least part of the time, men laid off one task can be put on another (although they may carry it out less efficiently than those trained to do it).

So far little work has been done towards this estimation of the cost of weather. It must be regarded as a further stage in the work which has been started in the present book.

Chapter 1 General description of the climate of the United Kingdom

This Chapter consists mainly of a general description of the climate of the United Kingdom as it affects buildings and building operations of various kinds. No attempt has been made to give a physical explanation of the weather: for this the reader is referred to the books listed on page 54.

The Chapter provides information on the basic climatic elements, how these are measured and how they vary from one part of the country to another, together with some notes on local variations. The information is set out in maps, tables or diagrams as seems most appropriate, although the presentation in some cases depends on the information available.

It is not the purpose of this Chapter to draw attention to particular ways in which climate affects buildings or their design and construction (these are dealt with in subsequent chapters) but to give general background information on the climatic elements and their variations, with the applications of this knowledge to the building industry in mind.

Most climatological averages refer to a period of 30 years, although 35 years has been used for rainfall in this country and shorter periods are sometimes used for other elements. It is well known, however, that there may be large variations between one year and another in mean monthly or annual values. There are longer-term trends in the climate which may also be important. A study of the instrumental and other records of the past few hundred years reveals secular variations which are statistically significant, implying real climatic changes during this period. Probably the most conspicuous feature, at least so far as Britain is concerned, is a warming tendency from the late 19th century to about 1940, particularly in the winter months. Since about 1940 this trend has ceased or may even have been reversed. Such climatic trends do not appear to be sufficiently regular or consistently maintained to be extrapolated with any confidence into the future. Nevertheless, it is clear that the relatively mild period from about 1900 to 1940 cannot be regarded as representative of British winters, and that the long-term average of one winter in six or seven with mean temperature below 3°C (mean of December to February inclusive in London) is more representative than the one such winter in 15 over that period. Moreover, a return to the regime of the 1880s or of the early 19th century cannot be ruled out. The average winter temperature in London for the 17 winters 1878–79 to 1894–95 was 3·4°C with mean temperatures below 3°C once every three or four years. Clearly it will be prudent to take more precautions against frost and snow than was considered necessary before 1940.

Air temperature

Measurement of air temperature

The most important single climatic element for our purpose is the temperature of the air measured at a height of 1·2 m above ground, in the shallow layer in which human beings move. In some circumstances the variation of temperature with height above ground is important and this is discussed where appropriate. However, the statistics described in the following paragraphs deal, unless otherwise noted, with measurements taken at the standard height of 1·2 m.

The air temperature, often described as the 'screen' or 'shade' temperature, is measured by thermometers carefully shielded from the sun by a louvred box (usually of wood) the most common type being known as a 'Stevenson screen'. The

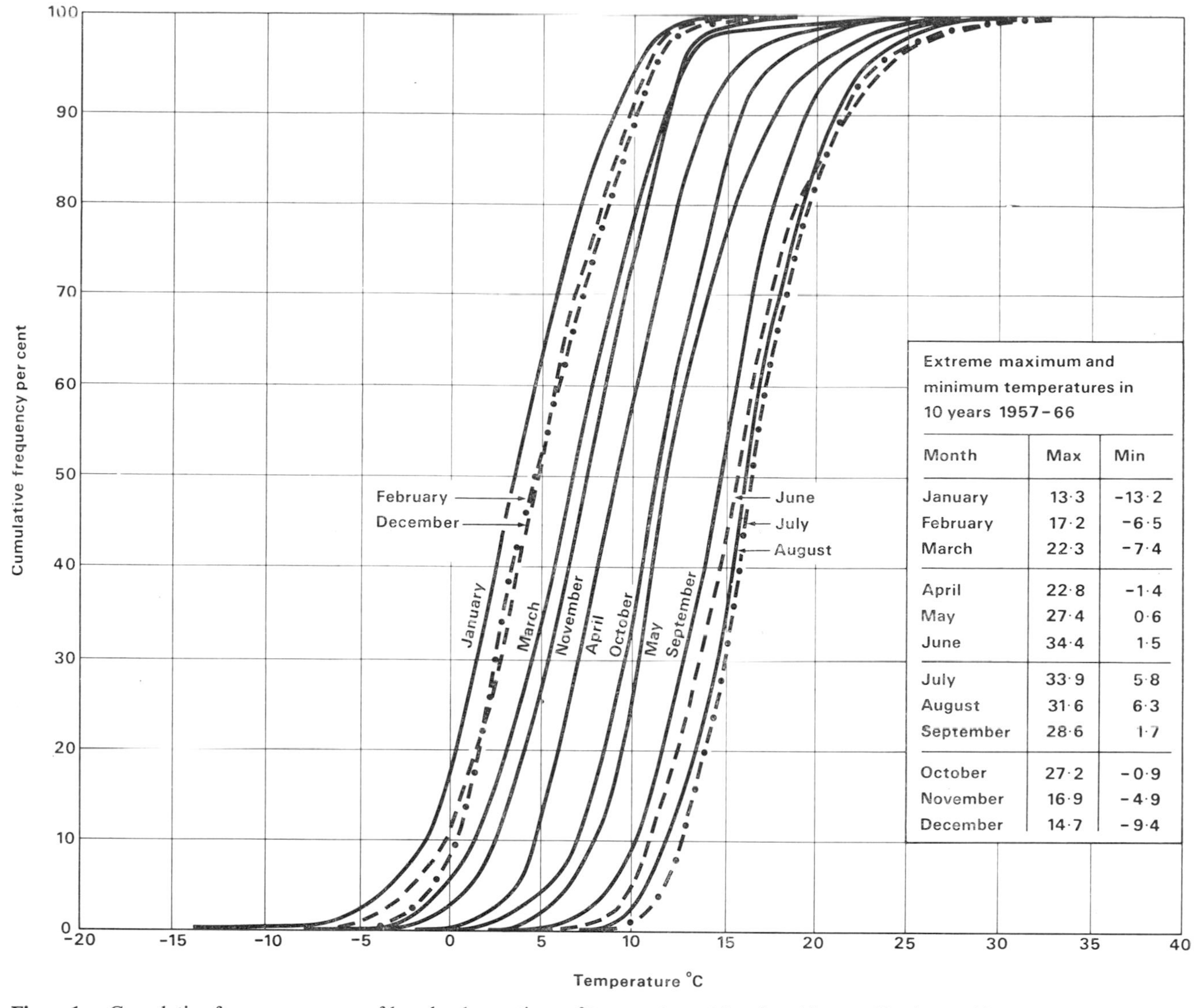

Month	Max	Min
January	13·3	-13·2
February	17·2	-6·5
March	22·3	-7·4
April	22·8	-1·4
May	27·4	0·6
June	34·4	1·5
July	33·9	5·8
August	31·6	6·3
September	28·6	1·7
October	27·2	-0·9
November	16·9	-4·9
December	14·7	-9·4

Figure 1 Cumulative frequency curves of hourly observations of temperature at London Airport, Heathrow, 1957–66

screen shields the thermometers from (i) radiation by day from the sun, the ground or neighbouring objects, (ii) loss of heat by radiation to the cold ground and sky at night, and (iii) precipitation, while at the same time allowing free passage of air. This is achieved by the use of double louvres in the sides and door, a double roof with a ventilated air-space and a floor consisting of three partially overlapping boards separated by an air-space. The screen is painted white and mounted on a stand so that the thermometer bulbs are about 1·2 m above the ground. Ideally the ground should be level, covered with short grass and well away from trees, buildings, walls and other obstructions. The screen normally contains four thermometers, the dry-bulb and wet-bulb thermometers and two self-registering thermometers, the maximum and minimum, designed to indicate the highest and lowest temperatures reached since the instruments were last set. The summarized temperature data for climatological stations are based mainly on the latter; the dry-bulb thermometer gives the air temperature at the time of observation, and is primarily an adjunct to the wet-bulb thermometer for the purpose of measuring the humidity (see page 15). The dry-bulb and wet-bulb thermometers are read at each observation hour and the maximum and minimum thermometers are reset immediately after they are read. At an ordinary climatological station the thermometers are read and reset daily at 09 00 GMT. More frequent observations are taken at some stations.

Air temperature statistics

The air temperature at any one place is continually varying. The variations over short time intervals, of a few seconds, or even minutes, may be ignored for most purposes, but variations from one hour to another will be significant in their effects on buildings. At a number of climatological stations temperatures are read every hour (at airfields, for example) or recorded continuously. The temperatures at each hour from such stations can be analysed to give cumulative frequency curves like those in Figure 1. This Figure shows 12 curves, one for each month of the year. Each is based on observations over a period of 10 years (1957–66 in this case) at one place. From these we can read off the proportion of the time for which the temperature was below or above any given value in a particular month. For example, on average during these 10 Januaries, the temperature at Heathrow was below 0°C for about 18 per cent of the time. The 50 per cent value is nearly the same as the mean temperature for the month. The wide spread of values in any month is noticeable – in the Figure a table shows the extreme temperatures recorded in each month during the 10 years. There is a considerable overlap of

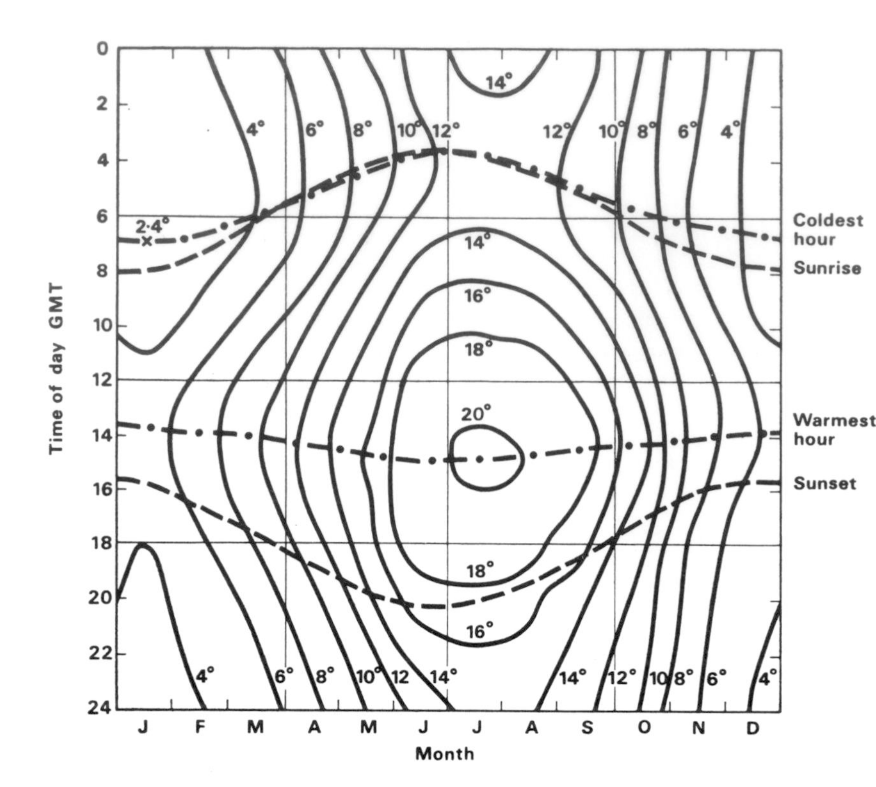

Figure 2 Diurnal and annual variation of air temperature at London Airport, Heathrow, from average hourly values for ten years 1957–66

the individual curves – and it can be seen, for example, that some 20 per cent of the hours in January are warmer than the coolest July hours in an average year.

Another way of presenting the data is shown in Figure 2. In this, the measurements at any given hour of the day in, say, each January of the 10 years are averaged and plotted. The same is done for each hour of each month, and isotherms drawn to show how, on average, the temperature varies with time of day and with time of year. It can be seen how the warmest part of the day is on average about two hours after noon in winter, and nearer three hours after noon in summer. The coldest part is a little before sunrise – as much as an hour before in winter, but almost coinciding in summer. The coldest part of the year is in January, about a month after the midwinter solstice, and the warmest in July, a month after Midsummer Day.

Figure 1 shows the variability of temperature within a month, taking all hours together. It is also of interest to know how the temperature varies from day to day at a given time of day. Cumulative frequency curves similar to those of Figure 1 could be drawn for each hour, but it would clearly be an inconvenient method of presentation, with 288 (ie 12 × 24) separate curves. An alternative method is to plot the diurnal and annual variation of the standard deviation of the hourly temperatures (Figure 3). The standard deviation is a measure of the dispersion of the individual values of the hourly readings about the mean value (see for example Conrad and Pollak 1950 for a description of the use of statistical methods in climatology). If the individual values are normally distributed about the mean, then about 68 per cent of the values will be within the limits of one standard deviation either side of the mean and about 95 per cent within two standard deviations either side. For example, from Figure 3, if the mean temperature at noon in June is θ°C, then on 68 per cent of occasions (two days in three) the temperature will lie between $(\theta - 3{\cdot}5)$°C and $(\theta + 3{\cdot}5)$°C. On 95 per cent of occasions it will lie between $(\theta - 7)$°C and $(\theta + 7)$°C.

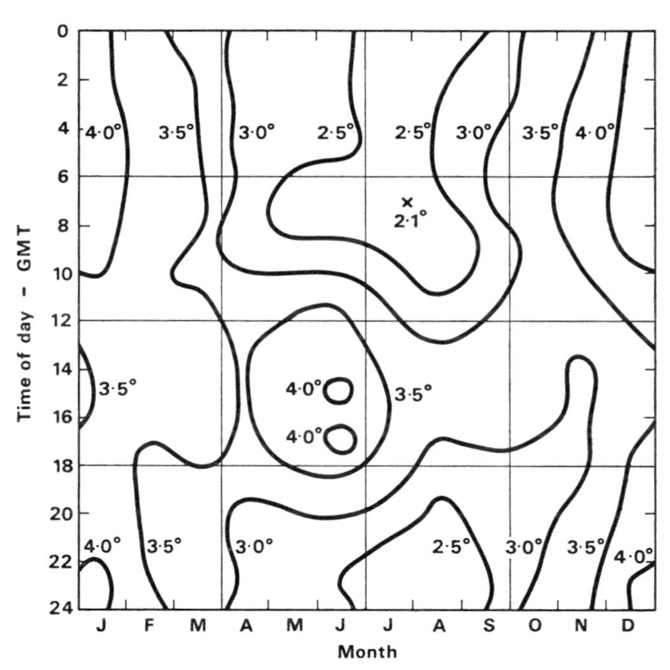

Figure 3 Diurnal and annual variation of standard deviation of temperature at London Airport, Heathrow, from hourly readings in ten years 1957–66

Figure 3 shows that at Heathrow in midwinter the greatest variation in temperature takes place at night, the standard deviation being then over 4 deg C, falling slightly to 3·5 deg C in the afternoon. In summer, on the other hand, the greatest variation is in the afternoon, and the daytime standard deviation is almost twice that at night. Considering days as a whole, it is clear from Figure 3 that the standard deviation of daily mean temperature at Heathrow is higher in winter than in summer. This is to say, there are larger, or more frequent, changes of temperature from one day to the next in winter than in summer. An analysis made by Hoblyn (1928) of air temperatures at Rothamsted (a country site in Hertfordshire, north of London) gave similar results for mean daily temperatures. Hoblyn also showed that maximum temperatures had their greatest standard deviation in spring and early summer; while minimum temperatures had their greatest standard deviation in early winter – October to December.

Diagrams of this kind can be drawn for a number of other places and will show generally similar patterns, though changing in detail according to latitude, distance from the sea, and altitude.

Variations in air temperature across the country

The variations in air temperature from one part of the country to another are best shown on maps. Figures 4 and 5 are maps of average means of daily mean temperature in a winter month (January) and a summer month (July) respectively. The averages refer to the standard 30-year period 1941–70. For this purpose, the average daily mean temperature is defined as half the sum of the average daily maximum and minimum temperatures. Before being plotted, the station readings were corrected to mean sea-level values. On average in Britain the air temperature decreases by about 0·6 deg C with every increase of 100 metres in altitude (Meteorological Office 1972, p. 165). Because of this, a plot of the original values would largely reflect the topography of the country, and would be so complicated that it would be difficult to observe regional differences. Therefore each reading was increased by an amount corresponding to the altitude of the station above sea-level, so that the maps show readings as though the stations were all at sea-level. This has the advantage, not only of making the regional differences apparent, but of making possible the estimation of temperatures at any place if its elevation above sea-level is known.

The winter map, Figure 4, shows that in winter the run of the isotherms is mainly from north to south, so that temperature is lower in the east than in the west. Indeed, the winter mean temperature in the East Midlands of England is much the same as that in the Orkney and Shetland Islands, although because of the stronger winds the weather of the Northern Isles is markedly more severe. In summer, on the other hand (Figure 5), there is a marked gradient of temperature from SSE to NNW, from about 17°C in south-east England to around 14° in north-west Scotland, while east and west coast areas are at much the same temperature at a given latitude.

Temperatures in coastal areas are generally more equable than those inland, so that the interior of the country is the coldest part in winter and the warmest in summer. Occasional exceptions to this rule occur with easterly winds, especially in winter when the extreme south-eastern part of England may suffer from very cold weather with a cold wind from the Continent, while further north its severity becomes moderated by the passage over the relatively warm sea (see Figure 69).

The temperature at any place on the maps may be determined by reading the temperature given on them for the required geographical location, and subtracting 0·6 × H/100 deg C, where H is the height of the place above sea-level in metres. For example, from Figure 5 the July mean temperature, reduced to sea-level, of Dunstable is about 17·2°C. The altitude is 150 metres above sea-level, so the correction is 0·6 × 150/100, that is 0·9 deg C. Thus the true mean temperature at Dunstable is 17·2 − 0·9 = 16·3 °C.

Extremes of air temperature

Figures 6 and 7 are maps respectively of the average maximum temperature for the 31 days of July and average minimum temperature for the 31 days of January, over the period 1941–70. It is clear from them that both the highest and the lowest temperatures are recorded inland, the area of highest maxima being in south-east England while the areas of lowest minima are in central areas of Scotland, northern England, Wales and the Midlands.

The highest and lowest air temperatures that have been recorded in each month at a selection of stations, mainly in or near the principal centres of population, are given in Table 1. The period of each record and the years in which the extremes occurred are given.

The months which have contributed high temperatures over large areas are August 1911, September 1906 and August 1947; the lowest temperatures over wide areas have been mainly contributed by February 1929, January 1940, February 1947 and January 1963. Outstandingly warm days were 9 August 1911 and 19 August 1932. Outstandingly cold days were 14 February 1929 and 21 January 1940; on the latter date the whole of the United Kingdom experienced air temperatures below 0°C and more than half the area temperatures below −6°C.

Figure 8 shows graphs of the diurnal variation of temperature on clear and overcast days, both in winter and summer, in open country. Each curve is derived from the averages of the hourly measurements on a number of days. The lowest temperature usually occurs at sunrise, and the highest about two hours after noon in summer, but less than an hour after noon in winter.

Frequency of cold weather. Days with minimum temperature 0°C or below, cold spells, ground frost, and frost hollows

Figure 9 is a map showing the average annual number of days with minimum air temperature 0°C or below, in the 10 years 1956–65. The frequency varies from 20 days or less in the extreme south-west to over 100 days in the hilly areas of Northern England and Scotland. The map was based on data from some 200 stations and was drawn with due consideration of topographical features, but it was clearly impossible to indicate all local variations. Frost frequency depends not only on altitude and distance from the coast, but also on other local topographical features. Thus at a station in a 'frost-hollow' near Rickmansworth the average frequency of air frosts over the period 1930–42 was 134 compared with only about 60 for the area deduced from Figure 9. This station recorded minimum air temperatures of 0°C or below in every month of the year, at some time during the 21 years during which records were taken (Hawke 1944).

Winter cold spells in this country normally result either from the arrival of easterly winds with a long track over the continent of Europe, or from northerly winds with a direct trajectory from the Arctic. Long cold spells, such as occurred in 1947 and 1962–63, are normally made up of a mixture of the two types, sometimes with intervening transient periods of slow thaw. During these warmer intervals, mild air from over the Atlantic penetrates to some parts of the country, or even to the whole of Britain for limited periods.

Table 2 gives mean temperatures at Kew Observatory, based on hourly readings, for the coldest spells since 1879 of lengths ranging from 1 day to 75 days, arranged in order of severity. It will be noted that a considerable number of the spells occurred before 1896. Winters in the present century (in bold type) are 1917, 1929, 1940, 1947 and 1963, the last three predominating especially for the longer spells. The length of a spell has an important bearing on its rank; thus the exceptionally severe winter of 1962–63 did not produce an outstanding cold spell of less than 3 days in length. Its coldest 3-day spell ranked seventh, for 10-day spells it ranked fifth, for 15 to 45-day spells it ranked fourth or third, for 50 to 60-day spells it was second while it took the lead for spells of 65 days and over.

A day with 'ground frost' is one on which the reading of a grass-minimum thermometer is 0·0°C or less.

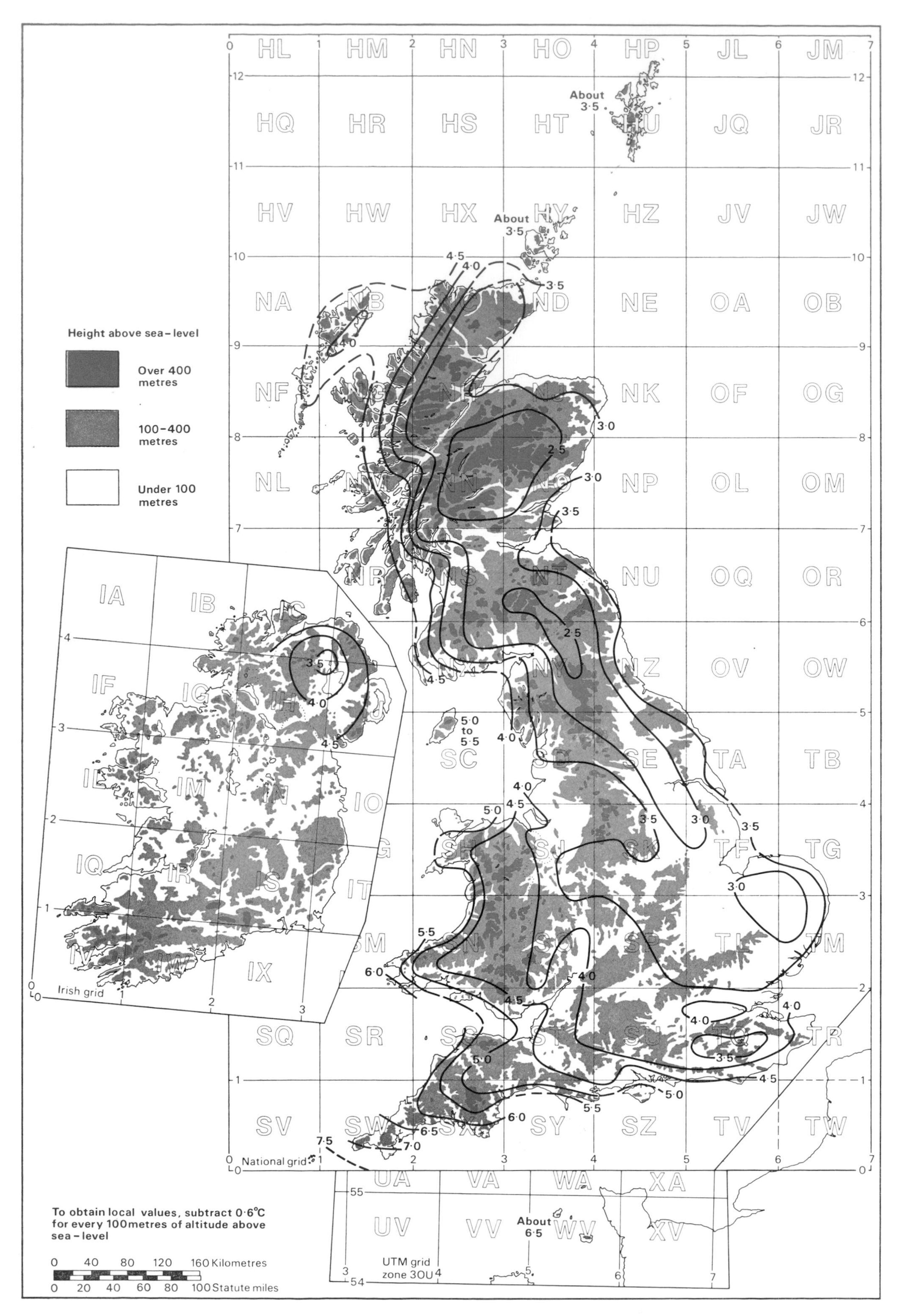

Figure 4 Mean monthly air temperatures in the British Isles in January, 1941–70 averages, °C, reduced to mean sea-level. To obtain local values subtract 0·6°C for every 100 m of height

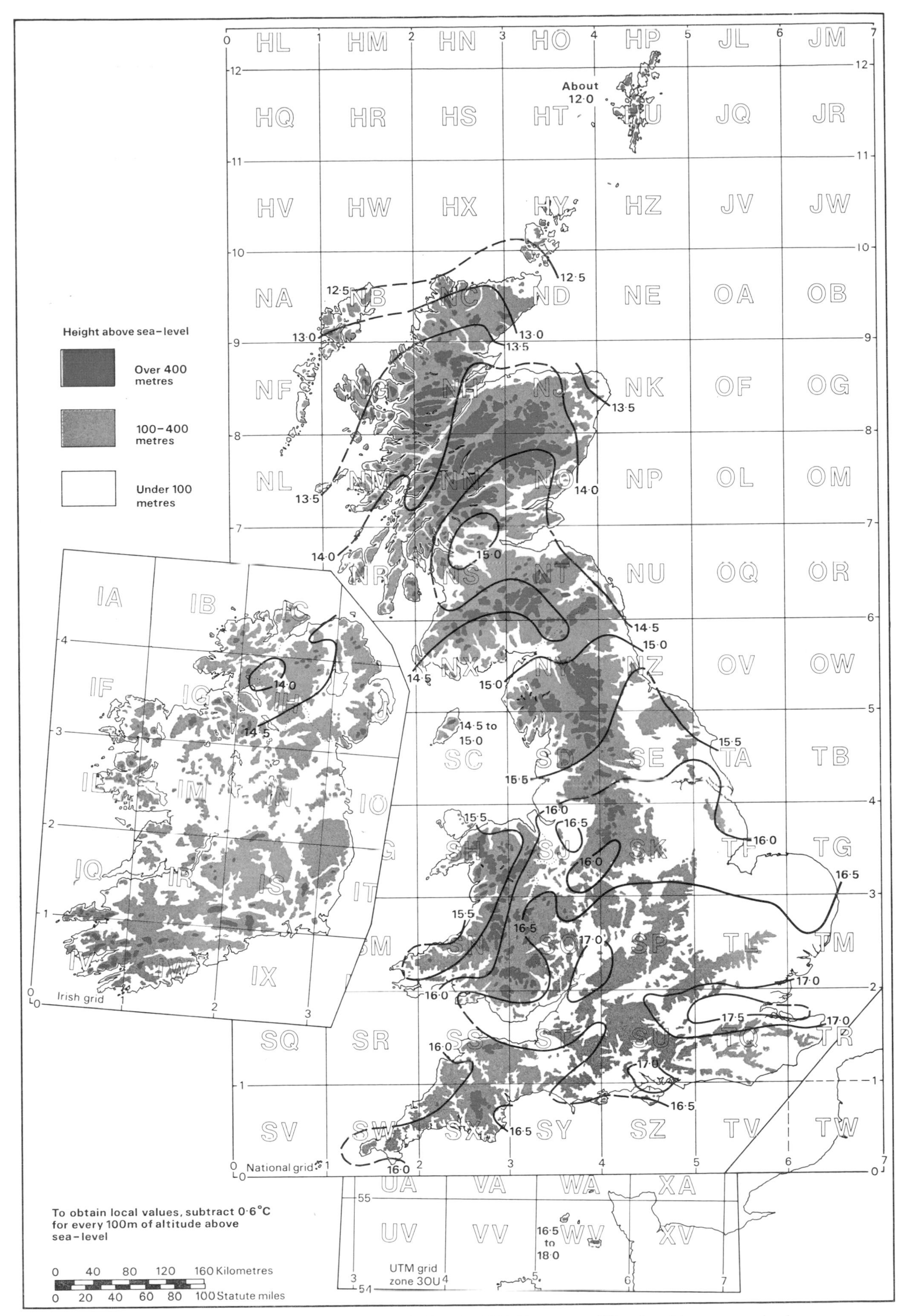

Figure 5 Mean monthly air temperatures in the British Isles in July, 1941–70 averages, °C, reduced to mean sea-level. To obtain local values subtract 0·6 deg C for every 100 m of height

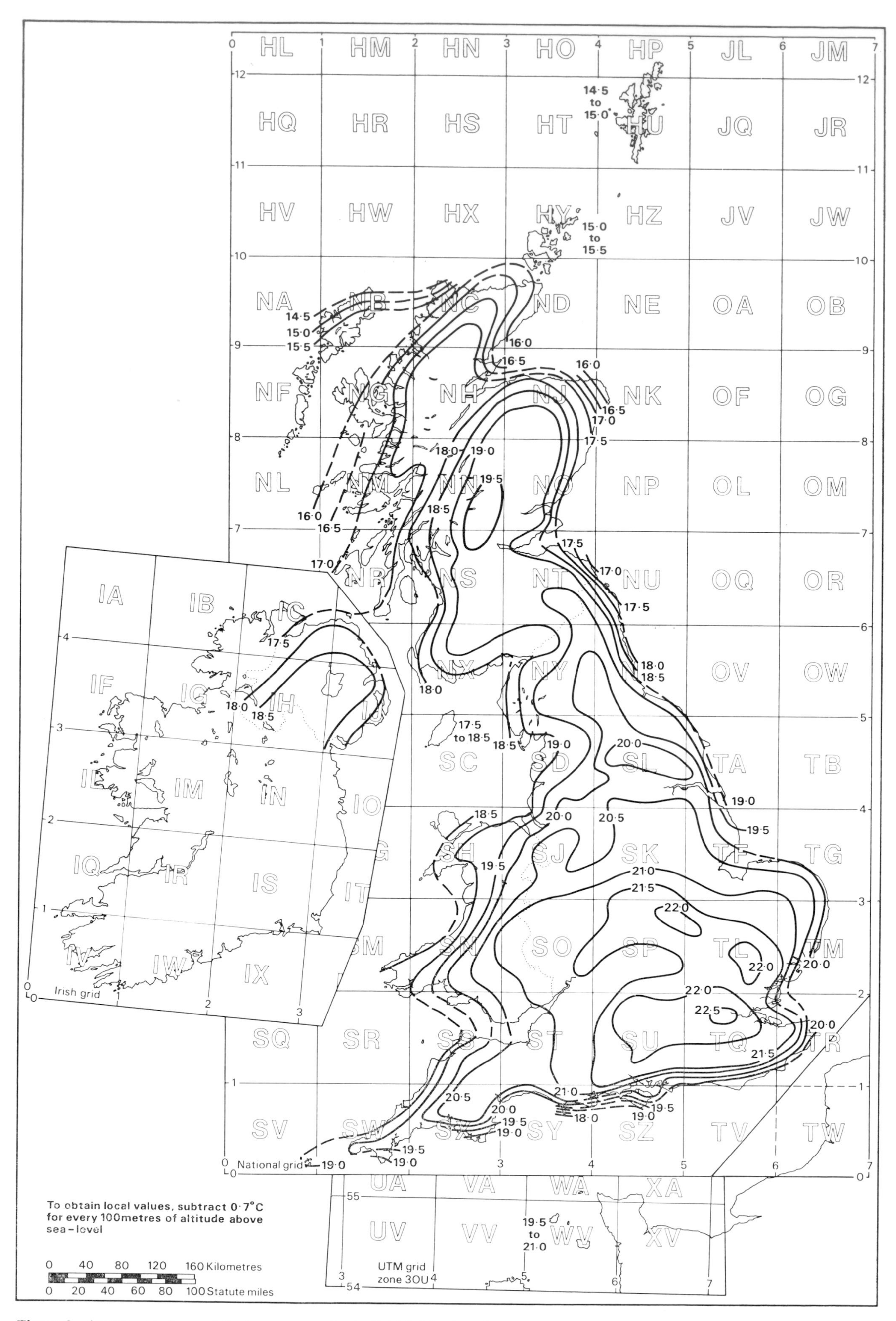

Figure 6 Average maximum daily temperatures in the British Isles in July, 1941–70 averages, °C, reduced to mean sea-level. To obtain local values subtract 0·7 deg C for every 100 m altitude

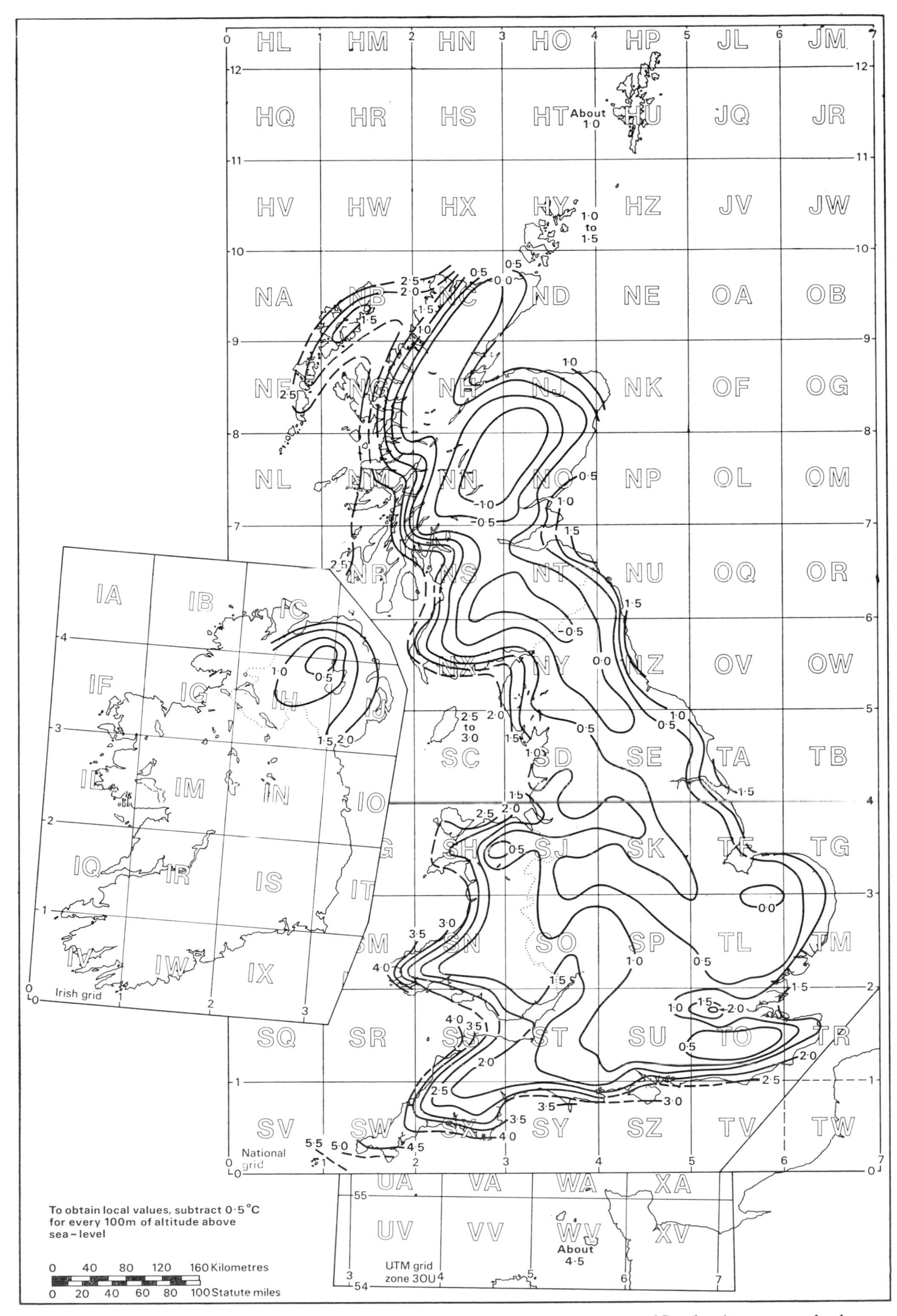

Figure 7 Average minimum daily temperatures in the British Isles in January, 1941–70 averages, °C, reduced to mean sea-level. To obtain local values subtract 0·5 deg C for every 100 m altitude.

Table 1 Highest and lowest air temperatures recorded at selected stations (°C). Highest and lowest temperatures experienced at each station during the period given are shown in bold type

Station and height above mean sea-level	Period of record		January	February	March	April	May	June	July	August	September	October	November	December
London (Kensington Palace)	1919–1965	Highest	15·6	18·3	23·3	28·9	32·2	34·4	33·9	**35·0**	31·1	28·9	20·0	17·2
25 m		*Year*	*1922*	*1959*	*1948/65*	*1949*	*1944*	*1947/57*	*1948/59*	*1932*	*1926/49*	*1921*	*1938*	*1953*
		Lowest	**−10·0**	**−10·0**	−6·7	−2·8	−0·6	2·8	6·1	5·6	0·6	−2·8	−4·4	−8·9
		Year	*1947*	*1929/47*	*1931*	*1922*	*1935*	*1962*	*1919*	*1920*	*1919*	*1931*	*1923/56*	*1920*
Cambridge (Botanic Garden)	1904–1965	Highest	14·4	17·8	23·3	26·1	31·1	33·9	34·4	**35·6**	33·9	26·7	21·1	15·6
13 m		*Year*	*1954**	*1945/59*	*1965*	*1934*	*1922*	*1947*	*1923/33*	*1911/32*	*1906/11*	*1921/59*	*1938*	*1931*
		Lowest	−16·1	**−17·2**	−11·7	−6·1	−4·4	−0·6	2·2	3·3	−2·2	−6·1	−13·3	−15·6
		Year	*1963*	*1947*	*1947*	*1910*	*1938/41*	*1923/62*	*1920*	*1964**	*1919*	*1931*	*1904*	*1920*
Birmingham (Edgbaston)	1901–1965	Highest	13·9	15·6	22·3	23·9	29·4	30·6	33·3	**34·4**	32·8	26·1	19·4	14·4
163 m		*Year*	*1930*	*1920/59*	*1965*	*1946/49*	*1944*	*1941/47*	*1923*	*1911*	*1906*	*1921*	*1938*	*1931*
		Lowest	**−11·7**	−11·1	−7·4	−3·3	−1·1	2·8	5·6	5·6	0·6	−2·2	−6·7	−10·0
		Year	*1940*	*1917*	*1965*	*1917*	*1941*	*1936*	*1907*	*1922**	*1919*	*1926/31*	*1904*	*1908*
Bristol (Whitchurch)	1938–1949	Highest	13·9	17·2	21·7	23·9	28·9	30·6	**32·3**	30·6	28·4	25·6	18·9	15·0
64 m		*Year*	*1946/48*	*1948*	*1965*	*1946**	*1944*	*1950*	*1948/49*	*1942/47*	*1961*	*1959*	*1946*	*1948*
(Filton)	1950–1965	Lowest	−14·5	**−15·6**	−11·7	−4·4	−2·8	1·4	4·4	4·2	0·0	−5·0	−5·6	−9·4
59 m		*Year*	*1962*	*1947*	*1947*	*1938*	*1945*	*1962*	*1938*	*1964*	*1948*	*1951*	*1965*	*1948**
Southampton	1901–1965	Highest	14·4	17·2	22·2	26·1	31·7	31·7	33·3	**33·9**	30·6	26·1	17·8	15·0
20 m		*Year*	*1956*	*1959*	*1929*	*1945/49*	*1944*	*1947*	*1952*	*1947*	*1929*	*1921/49*	*1946/63*	*1954**
		Lowest	−10·6	−11·1	**−11·7**	−3·3	−1·7	3·3	6·1	4·4	0·0	−6·7	−7·2	−8·9
		Year	*1901/63*	*1929*	*1909*	*1917*	*1945*	*1923/62*	*1954*	*1920*	*1919*	*1933*	*1923*	*1908*
Plymouth (The Hoe)	1901–1965	Highest	13·9	15·0	20·6	23·9	26·7	28·9	**30·6**	**30·6**	28·9	25·6	16·7	15·0
36 m		*Year*	*1908*	*1948*	*1929*	*1945*	*1944*	*1950*	*1923*	*1947*	*1911*	*1908*	*1946**	*1953*
		Lowest	**−8·9**	**−8·9**	−6·7	−2·2	−0·6	2.8	6.1	5.6	2.2	−1·7	−5·0	−6·1
		Year	*1940/47*	*1947*	*1965*	*1958**	*1945*	*1916*	*1902*	*1912/20*	*1919*	*1926*	*1944*	*1946*
Cardiff	1904–1965	Highest	15·0	16·1	21·1	23·9	28·9	30·6	**32·8**	**32·8**	31·1	24·4	18·3	15·0
62 m		*Year*	*1938/64*	*1960*	*1929/65*	*1945*	*1944*	*1950*	*1911*	*1911/32*	*1911*	*1921*	*1946*	*1948*
		Lowest	**−16·7**	−11·1	−8·9	−3·3	−1·1	3·3	4·4	3·3	1·1	−4·9	−7·8	−8·3
		Year	*1945*	*1929*	*1965*	*1906/08*	*1906*	*1962*	*1907*	*1912*	*1919*	*1905*	*1904*	*1964*
Liverpool (Bidston)	1901–1965	Highest	14·4	15·6	21·1	22·8	28·3	30·6	**31·1**	30·6	26·7	23·3	17·8	14·4
60 m		*Year*	*1916*	*1959*	*1965*	*1949**	*1918*	*1950*	*1901*	*1911*	*1961*	*1959*	*1946*	*1931/54*
		Lowest	**−9·4**	**−9·4**	−7·2	−5·0	0·0	4·4	7·2	5·0	1·1	−1·7	−3·9	−7·2
		Year	*1940*	*1929*	*1965*	*1917*	*1907/35*	*1926**	*1954**	*1923*	*1919*	*1926*	*1952**	*1908*

Table 1—*continued*

Station and height above mean sea-level	Period of record		January	February	March	April	May	June	July	August	September	October	November	December
Manchester (Whitworth Park)	1901–1957	Highest	14·4	16·1	22·2	24·4	29·4	**32·8**	32·2	32·2	**32·8**	26·7	20·0	14·4
38 m		*Year*	*1957*	*1922*	*1946*	*1946/49*	*1947*	*1950*	*1943*	*1930/53*	*1906*	*1908*	*1946*	*1948/54*
		Lowest	**−13·3**	−12·2	−8·9	−6·1	−1·1	1·1	4·4	3·3	0·0	−5·0	−8·9	−9·4
		Year	*1940*	*1917*	*1909/47*	*1917*	*1921/45*	*1903*	*1902/29*	*1919*	*1903/19*	*1931*	*1904*	*1950*
Sheffield	1901–1965	Highest	14·4	16·1	23·3	23·3	28·9	31·7	31·7	**33·3**	32·8	25·6	18·9	15·0
131 m		*Year*	*1910/32*	*1961*	*1955*	*1901/46*	*1947*	*1941*	*1921/43*	*1911*	*1906*	*1906*	*1927*	*1918*
		Lowest	**−11·7**	−9·4	−9·4	−7·8	−0·6	1·7	3·9	5·0	1·7	−2·8	−5·6	−10·0
		Year	*1940*	*1929*	*1947*	*1917*	*1941**	*1962*	*1961*	*1964*	*1919/32*	*1909*	*1952*	*1908*
Bradford	1908–1965	Highest	13·9	15·6	21·7	23·9	28·3	**31·1**	30·6	**31·1**	27·2	25·0	18·9	14·4
134 m		*Year*	*1916/37*	*1960*	*1929/65*	*1949*	*1947*	*1941*	*1935/43*	*1911/53*	*1949**	*1921*	*1946*	*1953*
		Lowest	**−13·9**	−13·3	−11·1	−10·6	−2·2	0·6	2·8	3·3	−0·6	−5·6	−7·8	−9·4
		Year	*1940*	*1929*	*1947*	*1917*	*1927*	*1962*	*1929*	*1931*	*1919*	*1931*	*1919*	*1908*
Durham	1901–1965	Highest	15·0	15·0	21·7	22·8	27·8	30·0	**30·6**	30·0	30·0	25·0	18·9	14·4
102 m		*Year*	*1916/44*	*1926/45*	*1953*	*1946*	*1922*	*1940**	*1921/43*	*1911/53*	*1906*	*1908*	*1927/46*	*1921/54*
		Lowest	**−16·1**	−12·2	−15·0	−11·1	−4·4	−1·1	1·1	1·7	−1·7	−5·6	−8·9	−9·4
		Year	*1940*	*1955**	*1947*	*1917*	*1927/29*	*1927*	*1929*	*1938**	*1954*	*1926*	*1919*	*1964**
Edinburgh (Blackford Hill)	1901–1965	Highest	13·9	15·6	20·0	22·2	24·4	28·3	28·9	27·8	**29·4**	24·4	19·4	14·4
134 m		*Year*	*1957**	*1920*	*1945/65*	*1946*	*1922/39*	*1939*	*1911*	*1930/55*	*1906*	*1908*	*1946*	*1948*
		Lowest	**−10·0**	−9·4	−9·4	−6·7	−1·7	2·8	5·0	4·4	0·6	−2·2	−7·8	−7·2
		Year	*1918*	*1956**	*1965*	*1917*	*1909*	*1935**	*1905/65*	*1956**	*1943/54*	*1926/31*	*1919*	*1908/61*
Aberdeen (Observatory)	1901–1947	Highest	14·4	15·6	20·0	20·6	23·9	**28·3**	**28·3**	27·2	27·8	22·8	16·7	16·1
24 m		*Year*	*1953*	*1953*	*1965*	*1946*	*1902*	*1939/50*	*1921/32*	*1959**	*1906*	*1908*	*1938*	*1931*
(Mannofield)	1950–1965	Lowest	−13·9	**−15·0**	−12·8	−5·6	−2·2	0·6	1·7	1·7	−1·1	−3·9	−10·6	−12·8
52 m		*Year*	*1910*	*1955*	*1958*	*1908*	*1957*	*1951/52*	*1965*	*1964**	*1954*	*1949*	*1919*	*1912*
Glasgow (Renfrew)	1921–1965	Highest	13·3	13·9	21·1	21·7	26·1	**30·0**	28·9	29·4	26·7	23·9	16·7	14·4
6 m		*Year*	*1958**	*1953/60*	*1945*	*1921*	*1941/48*	*1940/50*	*1934/43*	*1955*	*1959*	*1959*	*1927/46*	*1948*
		Lowest	**−17·8**	−15·0	−13·9	−7·2	−5·0	−1·7	1·7	0·6	−3·9	−8·3	−8·9	−11·1
		Year	*1940*	*1960*	*1947*	*1922*	*1927*	*1928*	*1922*	*1932*	*1928/32*	*1926*	*1952**	*1935/37*
Belfast (Aldergrove)	1926–1965	Highest	13·3	13·9	20·2	20·6	26·1	28·3	**29·4**	27·8	25·6	21·1	16·1	14·4
90 m		*Year*	*1957*	*1945*	*1965*	*1946*	*1952*	*1949/50*	*1934*	*1955*	*1959*	*1926/59*	*1927/46*	*1948*
		Lowest	**−12·8**	−11·7	−12·2	−4·4	−3·3	−1·2	2·2	1·1	−2·2	−5·0	−6·6	−10·7
		Year	*1958*	*1956*	*1947*	*1958**	*1938/54*	*1962*	*1929/63*	*1938*	*1942*	*1926*	*1962*	*1961*

*Most recent date of several with temperature given.

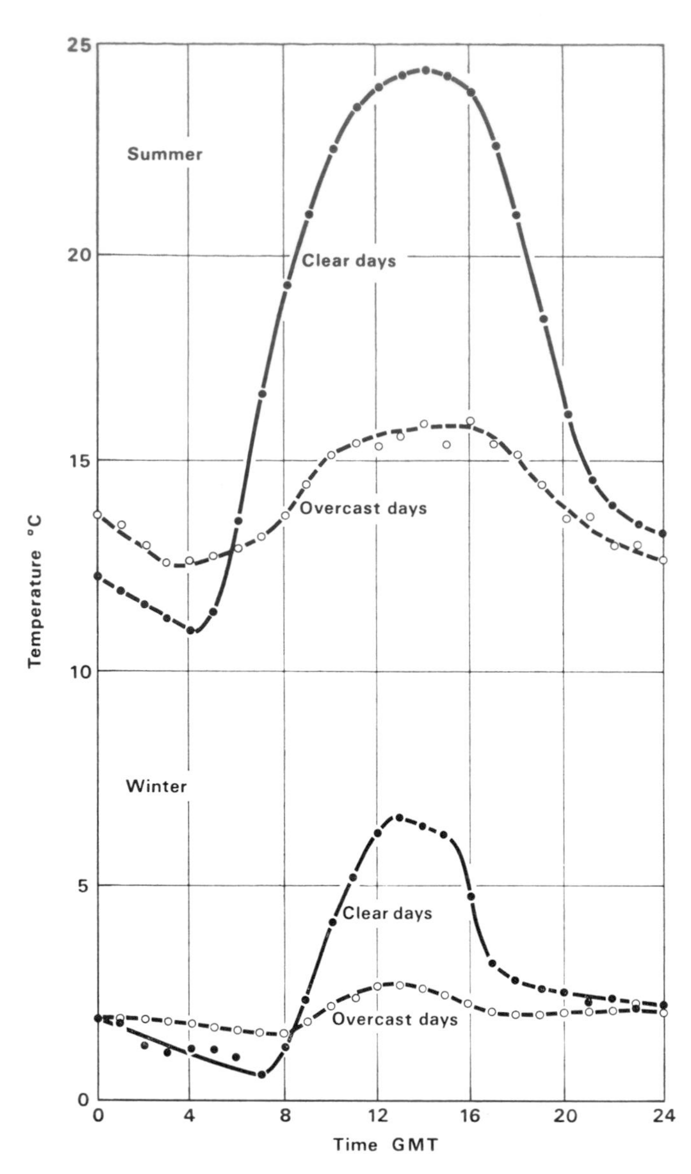

Figure 8 Diurnal variation of temperature on clear and overcast days in winter and summer at Rye, Sussex, 1·1 m above ground. (Best *et al* 1952.)

Table 2 Coldest spells of lengths from 1 to 75 days at Kew Observatory since 1879, in ranking order, °C (dates denote first day of spell, years of 20th century in bold type)

1 day	°C	2 days	°C	3 days	°C	5 days	°C
22.12.90	−8·3	6. 2.95	−7.5	7. 2.95	−7·6	6. 2.95	−6·8
5. 1.94	−8·1	8. 2.95	−7·3	14. 1.81	−6·9	13. 1.81	−6·0
7. 2.95	−7·9	14. 1.81	−7·3	13. 2.**29**	−6·1	11. 2.**29**	−5·2
15. 1.81	−7·8	4. 1.93	−6·6	4. 1.94	−5·6	1. 1.93	−4·8
9. 2.95	−7·8	10. 1.91	−6·5	3. 1.93	−5·6	3. 1.94	−4·8
6. 2.95	−7·1	4. 1.94	−6·3	9. 1.91	−5·3	20.12.91	−4·5
4. 1.93	−6·9	14. 2.**29**	−6·1	23. 1.**63**	−5·3	21. 1.**63**	−4·3
8. 2.95	−6·9	16. 1.81	−6·1	26. 1.80	−5·3	25. 1.80	−4·1
14. 1.81	−6·8	13.12.90	−5·8	13.12.90	−4.9	7. 1.91	−3·9
14.12.90	−6·7	1. 1.87	−5·7	21.12.91	−4·8	18. 1.81	−3·9
10 days		15 days		20 days		30 days	
5. 2.95	−5·2	12. 1.81	−4.1	26. 1.95	−3.5	13.12.90	−2·4
13. 1.81	−4·9	4. 2.95	−3·9	7. 1.81	−2·8	22. 1.95	−2·3
27.12.92	−3.9	24.12.92	−3·1	12.12.90	−2·6	8. 1.**63**	−1·6
11. 2.**29**	−3.3	11. 1.**63**	−3·0	7. 1.**63**	−2·3	27. 1.**47**	−1·5
16. 1.**63**	−3·3	12.12.90	−2·9	24.12.92	−1·8	21.12.92	−0·7
13.12.90	−3·2	11. 2.**47**	−2·2	6. 2.**47**	−1·7	30.12.80	−0·7
30.12.93	−2·7	29.12.90	−2·1	11. 1.**40**	−1.4	28.11.79	−0·5
16. 2.**47**	−2.6	10. 1.**40**	−1·9	11. 2.**29**	−1·3	14. 1.**17**	−0·3
14. 1.**40**	−2·5	11.12.78	−1·7	6.12.78	−1·3	1. 1.**40**	−0·3
2. 1.91	−2·4	26. 1.**17**	−1·6	1. 1.91	−1·2	4. 2.**29**	−0·2
40 days		50 days		60 days		75 days	
11.12.90	−2·0	1.12.90	−1·2	25.11.90	−0·9	19.12.**62**	−0·1
23. 1.**47**	−1·2	23.12.**62**	−0·8	23.12.**62**	−0·5	25.11.90	−0·4
25. 1.95	−1·1	22. 1.**47**	−0·7	30.12.94	−0·1	27.12.94	−0·6
29.12.**62**	−1.1	2. 1.95	−0·5	17. 1.**47**	0·2	31.12.**46**	0·9
22.12.**39**	−0·2	28.12.**39**	−0·2	21.12.**39**	0·3	10.12.**39**	1·1

Note that in this Table the date refers to a day running from midnight to midnight, not from 09 00 to 09 00 GMT as is more usual in climatological summaries.

The grass-minimum thermometer is freely exposed with its bulb just in contact with the tips of the grass-blades on a level area of short grass. The thermometer is set out in the evening and read the following morning so that a 'grass-minimum' temperature is obtained. The conditions affecting a grass-minimum thermometer are complex; they include the rate of radiation from the surface of the earth to the sky, the radiation reflected back by clouds and the heat radiated towards the earth by these clouds, the upward conduction of heat through the soil, the run-off of the chilled air and its replacement by warmer air from a higher level, the latent heat released when the temperature falls to the dew-point and condensation takes place, and the degree of turbulence in the air. The reading may be regarded as a measure of temperature conditions on the grass surface on which it is exposed, this being the surface which is radiating heat to the sky. Although it does not accurately represent the temperature of anything else exposed to the night sky it may be regarded as a convenient index of the night conditions with regard to temperature and radiation. In general the grass-minimum thermometer reads lower than the screen-minimum thermometer, the difference being greatest on clear calm nights. On such nights the temperature may increase with height up to about 100 m above ground by the end of the night. Such a situation is called a 'temperature inversion' since it is the reverse of that on windy nights and in the daytime, when temperature decreases with height.

The annual frequency of ground frosts varies widely from less than 30 in the Channel Islands to over 150 at Balmoral. It is in the neighbourhood of 100 in places as widely separated as London, Glasgow and Birmingham. Coastal places generally show low frequencies while high frequencies occur at inland places. The fact that there are often large discrepancies between the figures for neighbouring stations makes it very difficult to present the available information in the form of a map. On a calm clear night the grass-minimum temperature may be more than 5 deg C lower than the air temperature measured in a screen at the height of about 1·2 m. This low temperature is largely due to the insulating effect of the mat of grass, with still air trapped between the grass blades. Thus heat cannot readily travel up from the warm soil below to replace that radiated to the cold sky. Minimum temperatures over bare soil or other materials having a greater conductivity are usually higher than those over grass. Indeed, in the summer half-year the minimum surface temperature of a substantial area of concrete laid on the ground will usually be at least as high as the minimum air temperature, and may be several degrees higher in June-August (Figure 10).

On clear nights with light winds the ground is cooled by radiation and in turn cools the air in contact with it. Cold air being denser than warm air begins to flow down slopes and along the bottoms of valleys. Such flows of cold air are known

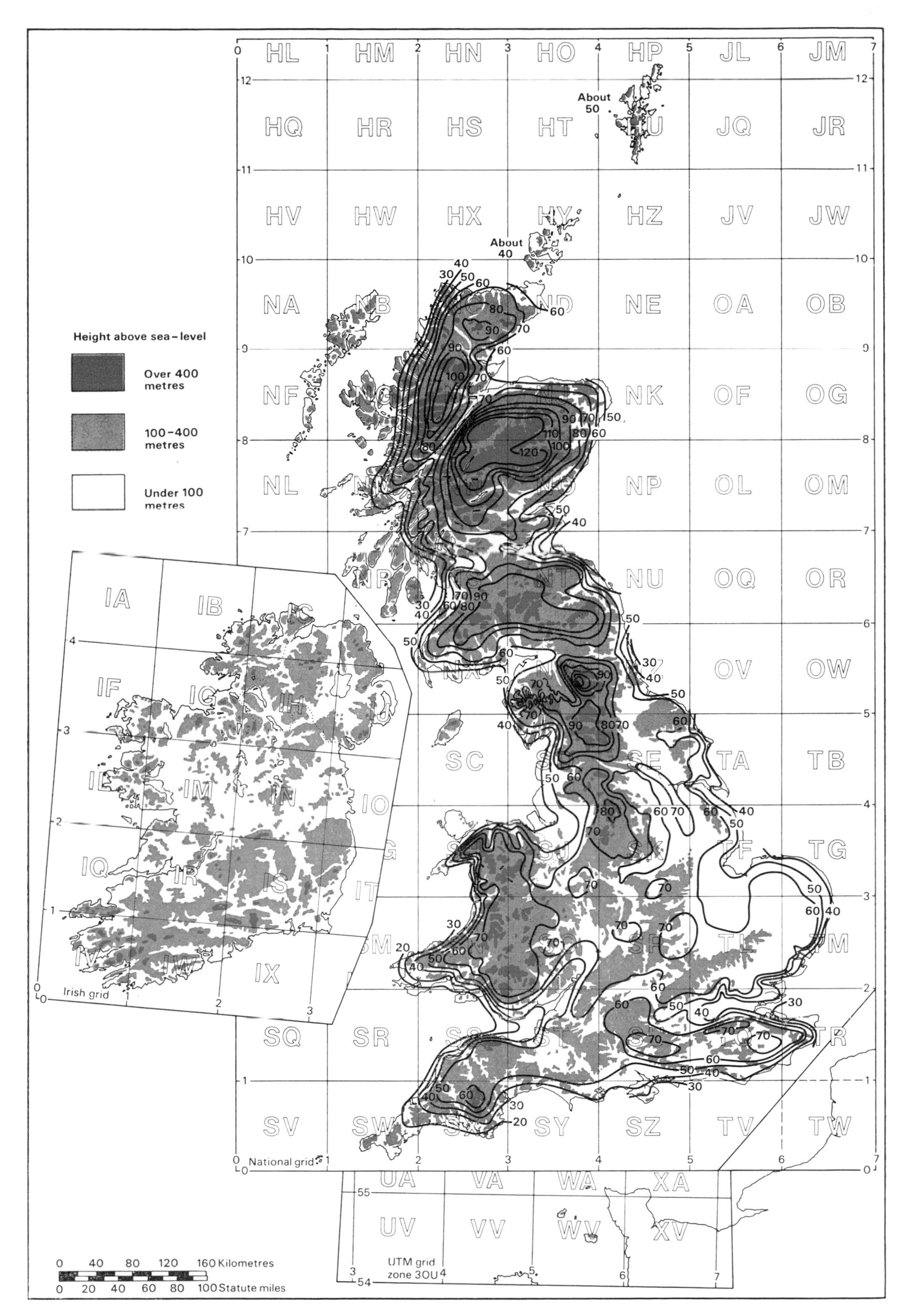

Figure 9 Average annual number of days in Great Britain with minimum temperatures of 0°C or less, 1956–65

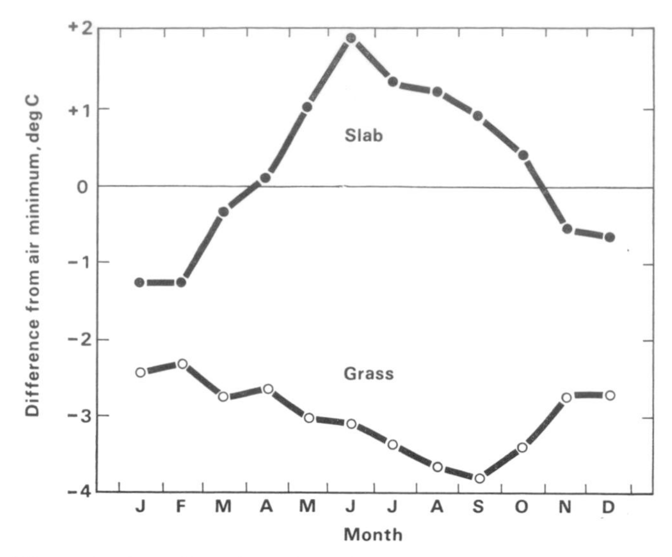

Figure 10 Monthly values of the difference between daily screen minimum temperature and corresponding minima recorded by thermometers exposed immediately above grass and on a concrete slab. Garston, averages for 1962–64

as 'katabatic winds'. If a valley is tortuous or obstructed, however, the flow is restricted and a pool of cold air accumulates. Such valleys or hollows are known as 'frost hollows' because they are particularly liable to frosts. The classic example in Britain, already quoted (Hawke 1944), of a frost hollow is the small valley in the foothills of the Chilterns near Rickmansworth. The conditions in this valley are probably typical of those of many hillside valleys in this country. The screen minimum there has fallen to 0°C even in July and reached the very low figure of −20°C in January 1947. The lowest grass minimum was −24°C in the same month. By contrast the days are hot, reaching 35°C one day in July 1943. In comparison with Rothamsted, a climatological station 20 km to NNE with no special site peculiarities, mean daily maxima at Rickmansworth are about 2 deg C higher in the summer months and about 1 deg C higher in the winter months. Mean daily minima on the other hand are about 2 to 3 deg lower in all months. For the whole year the mean daily range at Rickmansworth exceeds 11 deg C compared with a value of 7 deg C at Rothamsted. To find a station with similar values of minimum temperature we have to go to high-level stations in Scotland such as Braemar.

Earth temperature

Earth temperature is measured with a mercury thermometer which has its bulb embedded in paraffin wax. The thermometer is suspended in a steel tube and the wax keeps the reading constant while the instrument is hauled up to be read each morning. The usual depths of measurement are 0·3 m, 1·22 m and occasionally 3·05 m (1 ft, 4 ft and 10 ft). For depths up to 100 mm, mercury thermometers with their stems bent at right angles to lie flat on the ground are used. Records of the temperatures measured at a number of stations have been published by the Meteorological Office in *Averages of earth temperature at depths of 30 cm and 122 cm for the United Kingdom 1931–60*. These records and the problem of frost in the ground are discussed in Chapter 3, page 126.

Long-wave radiation

In the previous Section, where the readings of the grass-minimum thermometer were discussed, the factors mentioned included the radiation exchange between the ground and the sky and clouds. Short-wave solar radiation is discussed below on pages 45–50, but it is convenient to deal with long-wave (infra-red) radiation now, along with air temperature. It is conventional to speak of the radiation emitted by bodies at temperatures near to that of the atmosphere as 'long-wave' radiation. The wavelengths are in the range 4 to 100 micrometres (μm).

All bodies (including gases) are continually gaining and losing heat by radiation (radiation being a transfer of energy by electro-magnetic waves), the amount and quality of the radiation loss depending only on the temperature of the body and (except in the case of gases) on the nature of its surface. Most bodies radiate at about the same rate as the so-called 'black body'*; except polished metals, which emit and absorb radiation at a much lower rate. In consequence, the temperature of the surface of any body depends not only on the temperature of the air with which it is in contact, but also on whether it is absorbing more or less radiation from its surroundings than it is giving to them. If it is losing more heat by radiation than it is gaining, its temperature may fall below that of the air. When this happens, the surface will pick up heat directly by conduction from the air; the amount of heat it can gain from the air depends on the speed of movement of the air – the stiller the air, the smaller the rate of conduction by convection heat-transfer to the surface, and the lower the temperature of the surface.

This exchange of long-wave radiation goes on continually, by day and by night. Normally, at night, and particularly when the sky is clear, the long-wave radiation received from the atmosphere is less than that emitted by the ground and other surfaces. In other words, there is usually a net loss of long-wave radiation by these surfaces, so that they are tending to cool. (A thin sheet of material with low heat-capacity, exposed horizontally and with good insulation below it, may cool at night to a temperature of some 5 deg C below that of the air, providing dew is not forming on it.)

However, during the day the quantity of incoming solar (short-wave) radiation is usually so much more than the net long-wave loss that surface temperatures rise above air temperature. Sometimes, on a very clear day, a north-facing surface shielded from direct sunshine may lose more heat to the sky than it gains from scattered solar radiation, so that it will be a little colder than the air.

The loss of radiation to the sky is, at night, dependent on the amount and type of cloud. With a completely clear sky the net loss from a horizontal surface which is at 5 deg C below air temperature is, in our climate, normally in the range of 55 to 65 W/m². Cloud reduces this by an amount which depends on the cloudiness C ($C = 0$ is cloudless, $C = 1$ is overcast) in the following way:

$$R_c = R_o(1 - kC) \quad \ldots\ldots(1)$$

where R_o is the radiation loss with a cloudless sky. The constant k depends on a number of variables – for thick clouds a value of 0·76 is suitable, decreasing to 0·26 with thin high cirrus. It should be noted that this rate of exchange of long-wave radiation is almost independent of the temperature of the air.

The same relationship should hold for daytime, but the radiation exchange is of course complicated by the incoming short-wave radiation. For a more detailed study of long-wave exchange over Britain see Monteith (1961) and Monteith and Sceicz (1962).

*A perfect 'black body' is one which absorbs all the radiation falling on it and which emits, at any temperature, the maximum amount of radiant energy. The term arises from a correlation between darkness of colour and proportion of visible light absorbed. A body which appears white because it scatters the visible light falling on it may, however, act nearly as a black body to radiation of a different wavelength. Snow is an example, being effectively a black body for wavelengths greater than 1.5 micrometres. (Meteorological Office 1972).

Humidity

Measurement

The humidity of the air is usually measured by a pair of thermometers exposed side by side. One is an ordinary, perfectly dry, thermometer (known as the dry-bulb thermometer) while the other has its bulb covered by a thin muslin sleeve kept moist with distilled water and is known as the wet-bulb thermometer. The wet-bulb temperature is usually lower than the dry-bulb temperature because of evaporation of water from the wet-bulb and the difference is greater the drier the air. The humidity of the air can be deduced from the two readings. The thermometers may be exposed in a thermometer screen with their bulbs 1·2 m above the ground (see page 2). Alternatively they may be mounted in a frame which can be whirled by hand or in one on which there is a fan driven by clockwork or electricity to ventilate the bulbs. It is important to distinguish between thermometers that have forced ventilation and those which have not, because the depression of the wet-bulb reading below the dry-bulb reading depends on the speed of the air past the bulb up to a speed of about 4 m/s. The various humidity parameters (see below) can be obtained for any combination of dry-bulb and wet-bulb temperature, with the aid of tables (Meteorological Office 1964) or of a humidity slide-rule (Meteorological Office 1956b).

Humidity parameters

It is important to note that the humidity of the air can be expressed in a number of different ways. This frequently leads to confusion and care must be taken to select the appropriate information when dealing with any particular design problem depending on humidity. All air in its natural state contains water vapour, but a volume of air at a given temperature can hold only a certain maximum amount of water vapour. If it does contain this maximum amount the air is said to be saturated. The amount of water vapour needed to saturate a given volume of air increases with temperature. The humidity may first of all be expressed as a *vapour pressure*, that is as the partial gaseous pressure exerted by the water vapour. Vapour pressure is normally measured in millibars, or less frequently in millimetres of mercury. It may also be expressed on a weight basis, thus heating and ventilating engineers often express the *moisture content* of the air in kg of water vapour per kg of *dry* air; this is equivalent to the *mixing ratio,* ie the ratio of the mass of water vapour to the mass of dry air with which it is associated, sometimes used by meteorologists. Another way of expressing the humidity is as *vapour concentration, vapour density,* or *absolute humidity* which are alternative names for the mass of vapour per unit volume of *moist* air, for example in grams per cubic metre. Normally the air will not be saturated and its degree of saturation may be stated in terms of its *relative humidity,* which is the actual vapour pressure expressed as the percentage of the saturation vapour pressure at the same temperature. If unsaturated air is cooled it will eventually reach saturation and condensation will occur. The temperature at which saturation is reached is known as the *dew-point.* The dew-point of the air is therefore the temperature of saturated air having the same vapour pressure as the air under consideration.

Relative humidity, or wet-bulb temperature, standing on its own, without an indication of the dry-bulb temperature recorded *simultaneously* with it, gives no information about the absolute moisture content of the air. It is preferable therefore to record values of absolute humidity expressed as vapour pressure, dew-point or moisture content rather than relative humidity or wet-bulb temperature. Some problems which may arise in building construction are linked with relative humidity, however. For example, the equilibrium moisture contents of many organic substances like timber depend mainly on the relative humidity of the air and not on its absolute humidity. While it may therefore be necessary to use relative humidity in considering such problems, it should be realized that it can quite easily be obtained from the dry-bulb temperature and absolute humidity or vapour pressure.

For further information see British Standard 1339: 1965, *Definitions, formulae and constants relating to the humidity of the air.*

Variations in vapour pressure and relative humidity across the country

Figures 12 and 13 are maps of average means of vapour pressure, reduced to sea-level by adding 1 mb for each 600 m of elevation, at 13 00 GMT in January and July. They indicate the variations in the amount of water vapour in the air, that is the absolute humidity, over the country. In general, values are greatest in the south-west and least in the north and also diminish with distance from the sea. There is nearly twice as much water vapour in the air in July as there is in January. The diurnal variation is on the average small and of the order of 5 per cent, the vapour pressure being slightly higher by day because of evaporation from various surfaces (Figure 11). The fact that the maps refer to 13 00 GMT is therefore unimportant. Corresponding maps of average means of relative humidity at sea-level at 13 00 GMT are shown in Figures 14 and 15. At all times of the year the relative humidity is lowest in the early afternoon and drying can be expected to be most rapid at this time (Figure 16). It will be noted that the January map (Figure 14) is rather featureless; this is because the areas with the higher values of absolute humidity, for example south-west England, also tend to have the higher temperatures in winter. Nevertheless the higher relative humidities are to be found on the coasts and the lower ones inland, especially in the lee of hill masses. In summer this variation is very much more marked; the July map (Figure 15) in fact somewhat resembles that of average maximum temperature, indicating that in summer the major factor in relative humidity variation is the change in temperature. Distance from the sea is thus the dominant factor and the lowest humidities are found in the Home Counties and behind the Welsh hills. The maps may be used

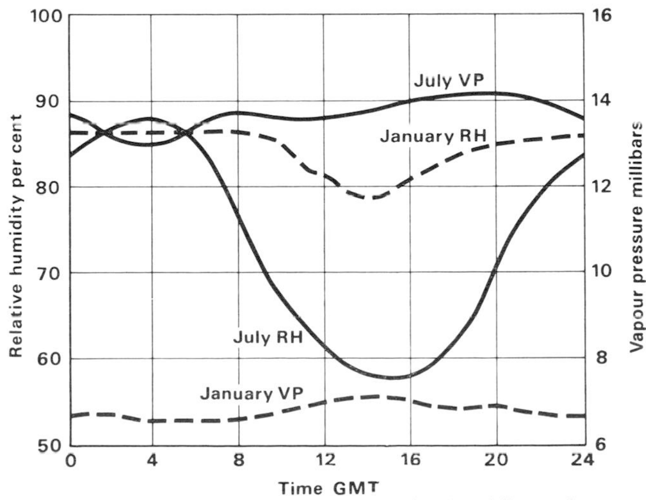

Figure 11 Mean diurnal variation of relative humidity and vapour pressure at Kew in January and July. Averages for 30 years 1886–1915.

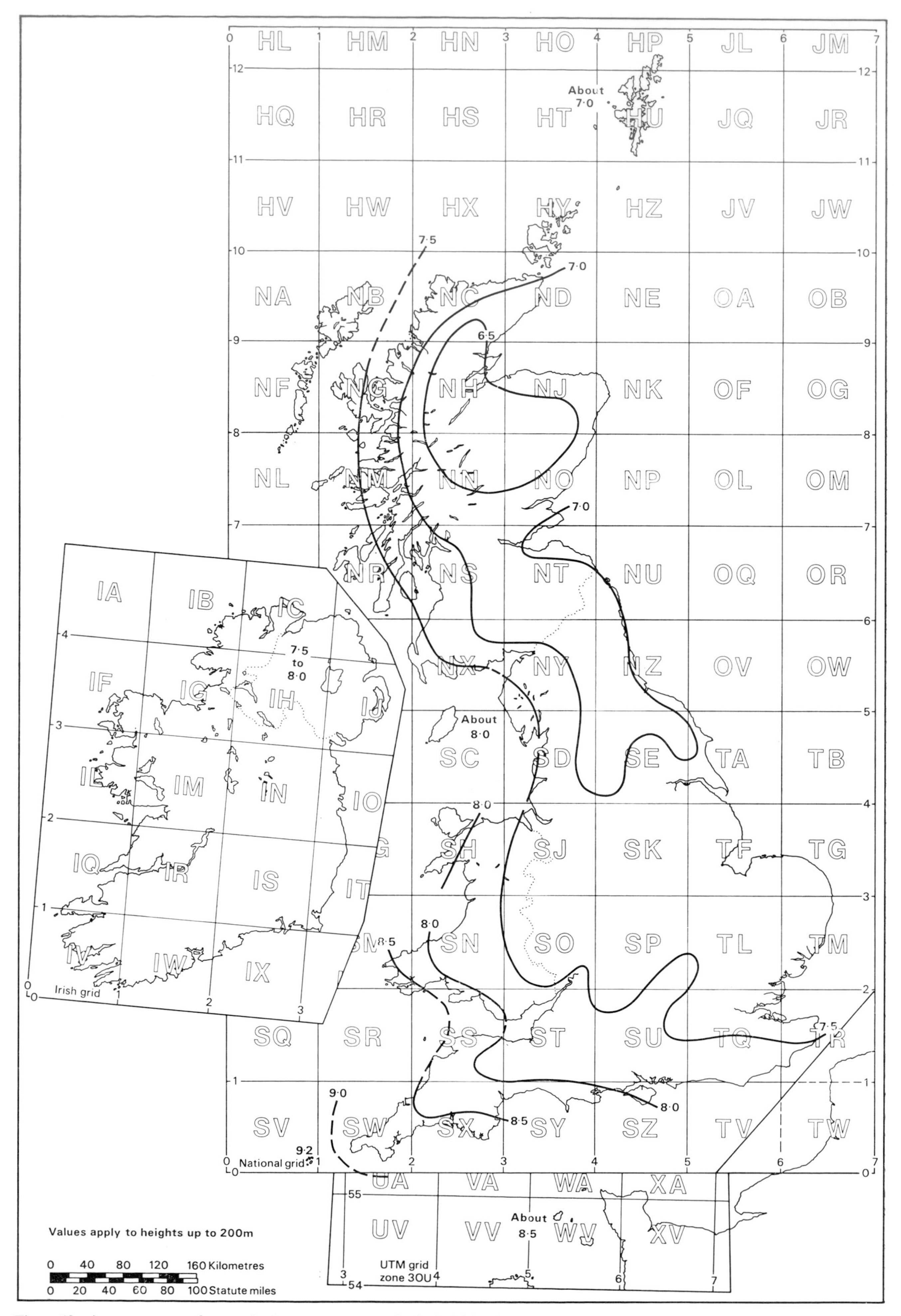

Figure 12 Average means of atmospheric vapour pressure in the British Isles at 13 00 GMT during January, 1921–35 (millibars)

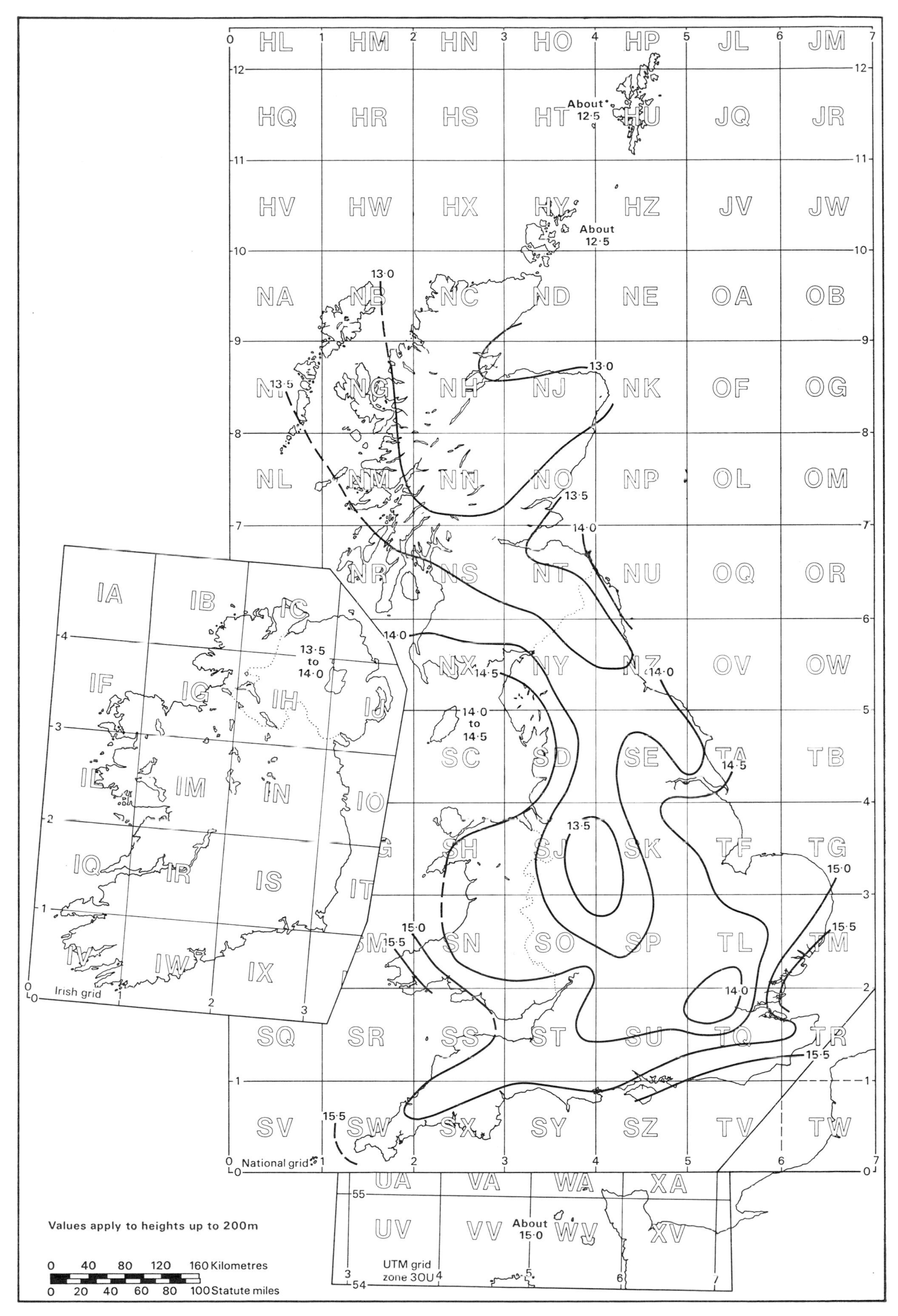

Figure 13 Average means of atmospheric vapour pressure in Britain, at 13 00 GMT during July, 1921–35 (millibars)

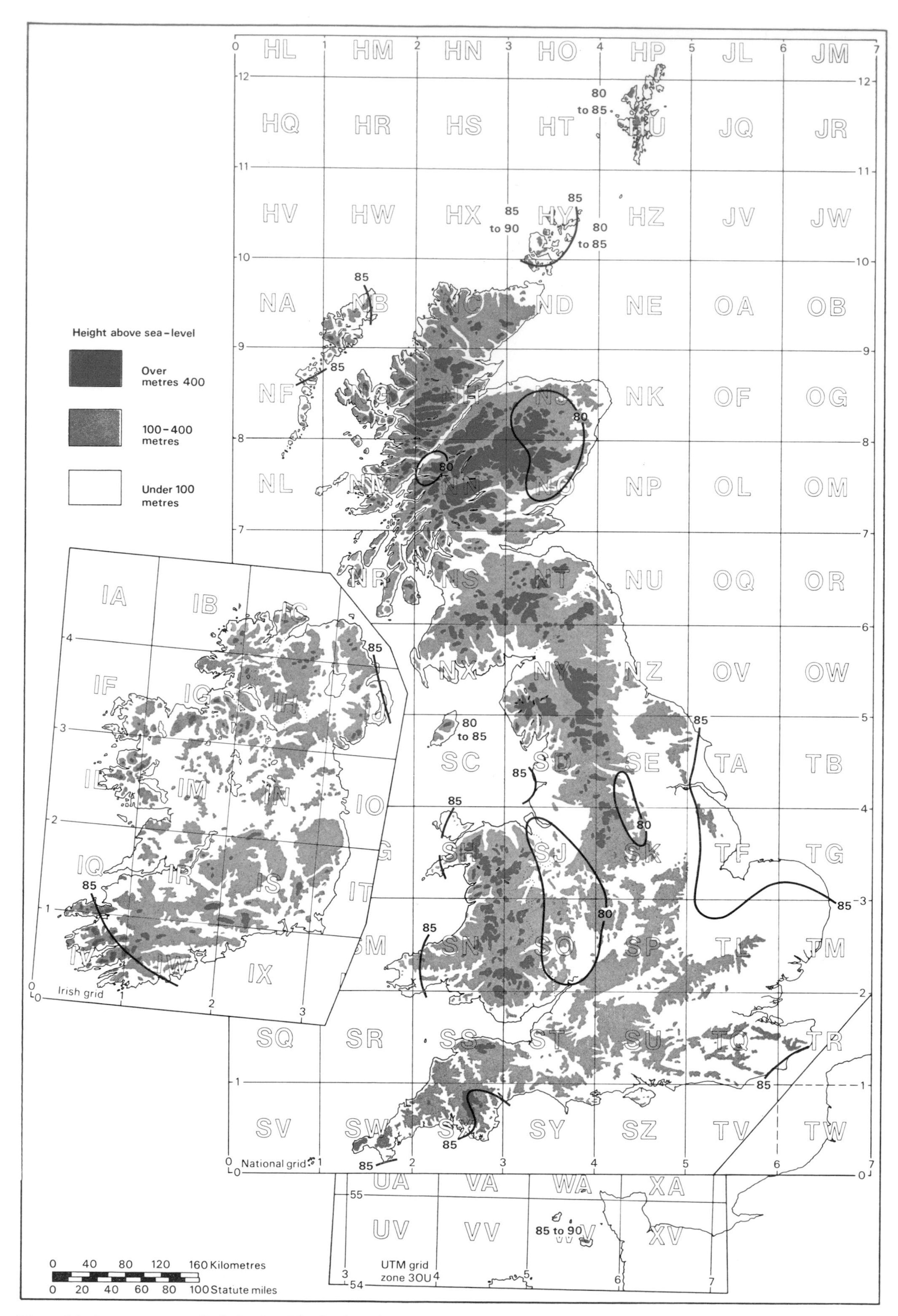

Figure 14 Average means of relative humidity in the British Isles, at 13 00 GMT during January, 1921–35 (per cent)

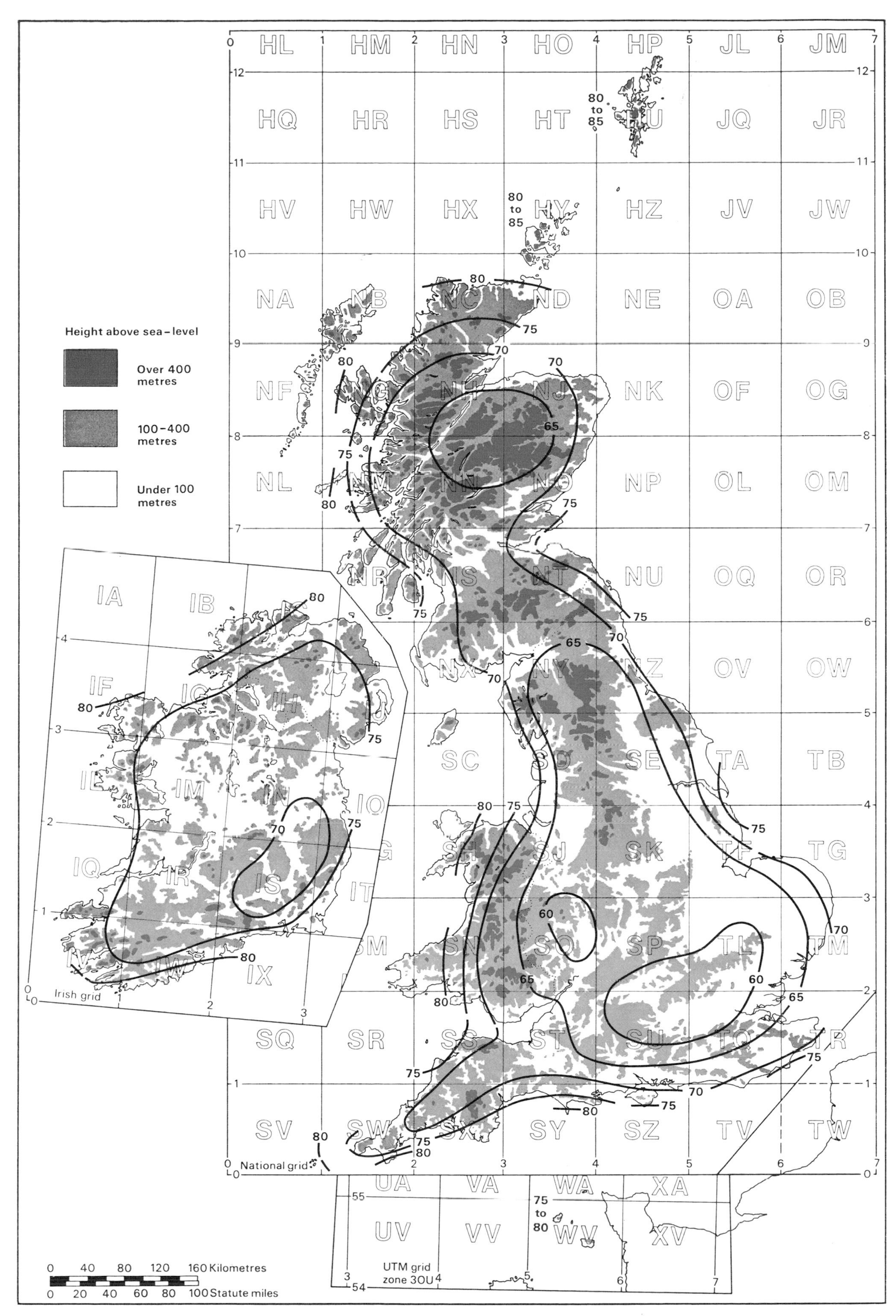

Figure 15 Average means of relative humidity in the British Isles, at 13 00 GMT during July, 1921–35 (per cent)

Table 3 Corrections to relative humidity at mean sea-level at 13 00 GMT for obtaining relative humidity at 300 metres (for any other height *h* metres multiply correction by *h*/300). Positive corrections to be added

Temperature at Station °C	Relative humidity (per cent at mean sea-level) 90	80	70	60	50
−5	2.2	0.8	−0·7	−2·1	−4·0
0	4·6	3·2	1·7	0·4	−1·0
5	6·6	5·1	3·8	2·5	1·1
10	7·7	6·3	5·0	3·7	2·5
15	8·3	7·0	5·7	4·5	3·2
20	8·6	7·4	6·2	5·0	3·8
25	8·7	7·6	6·5	5·3	4·1
30	8·7	7·6	6·6	5·4	4·3

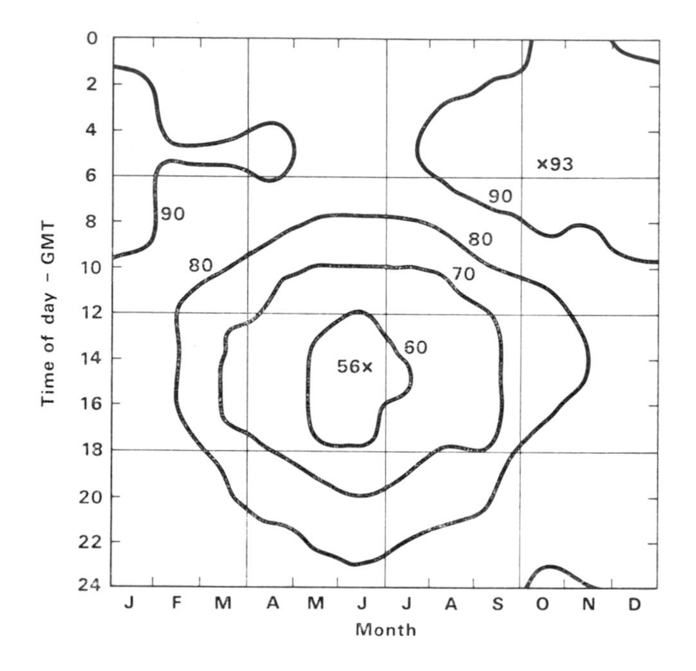

Figure 16 Diurnal and annual variation of monthly values of hourly mean relative humidity at London Airport, Heathrow. Mean for 1957–66

to give the appropriate average relative humidity at any place, for its own level, by using the corrections given in Table 3.

More detailed data for over 40 individual stations are given in the Meteorological Office publication MO 421, *Averages of humidity for the British Isles.*

Day-to-day variations in vapour pressure and dew-point
Figures 12, 13, 14 and 15, showing monthly mean values of vapour-pressure and relative humidity, indicate variations in humidity over the country, but not how this varies from day to day. Figure 17 consists of frequency curves showing cumulative percentage frequencies of occurrence of vapour pressures (and the corresponding dew-points) each month at London Airport (Heathrow). The data used to construct the curves are the vapour pressures measured each hour of the day, during the 10 years 1957–66.

It is notable that a wide range of vapour pressure may be experienced in any month, and that there is a big overlap of the summer and the winter curves. Thus even in midwinter there may be days when the atmospheric moisture-content is much greater than that on some days in midsummer, and vice versa.

Frequency of occurrence of high relative humidities
Drying of wet surfaces will be very slow when the relative humidity of the air is high and the average values discussed above may be of limited use when considering problems of drying-out of materials. It has been found experimentally, however, that the duration of wetness of an exposed unheated surface is highly correlated with the duration of relative humidity of 90 per cent or more (Shellard 1962). Figure 18 shows the average annual number of hours with a relative humidity of 90 per cent or more over England, Wales and Scotland. It can be seen that areas on the east side of high ground have noticeably shorter periods of time with high relative humidity, the most favourable areas being that around Liverpool which is in the lee of both the Welsh hills and the Pennine Range, according to wind direction, and the shores of the Moray Firth. They have respectively about one-quarter and one-third of all hours with relative humidity of 90 per cent or above. On higher land the duration of such humid conditions exceeds 46 per cent and is no doubt well above 50 per cent in many upland regions.

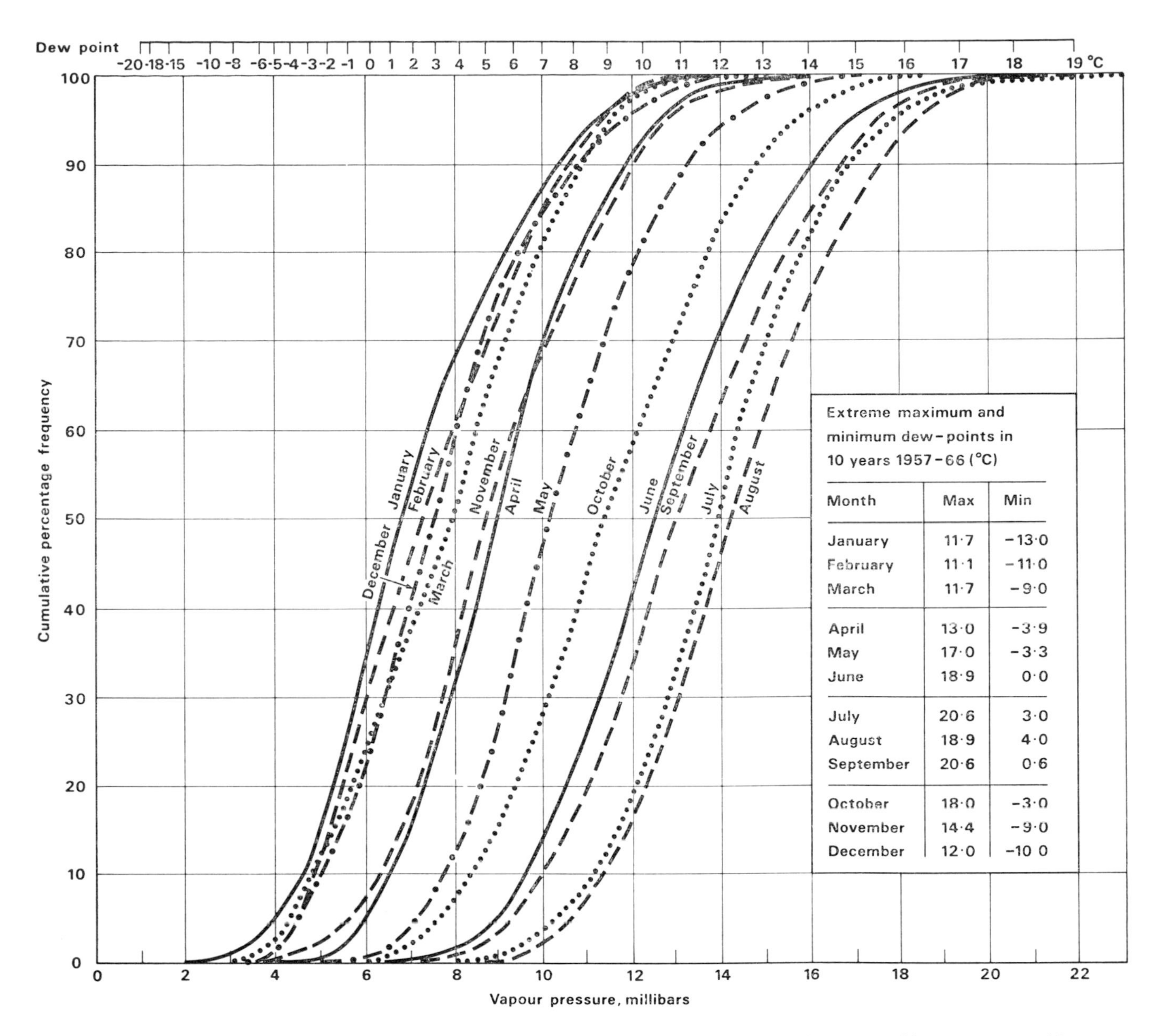

Month	Max	Min
January	11·7	−13·0
February	11·1	−11·0
March	11·7	−9·0
April	13·0	−3·9
May	17·0	−3·3
June	18·9	0·0
July	20·6	3·0
August	18·9	4·0
September	20·6	0·6
October	18·0	−3·0
November	14·4	−9·0
December	12·0	−10·0

Figure 17 Cumulative frequency curves of atmospheric vapour pressure, London Airport, Heathrow. Monthly averages over 10 years from hourly readings, 1957–66

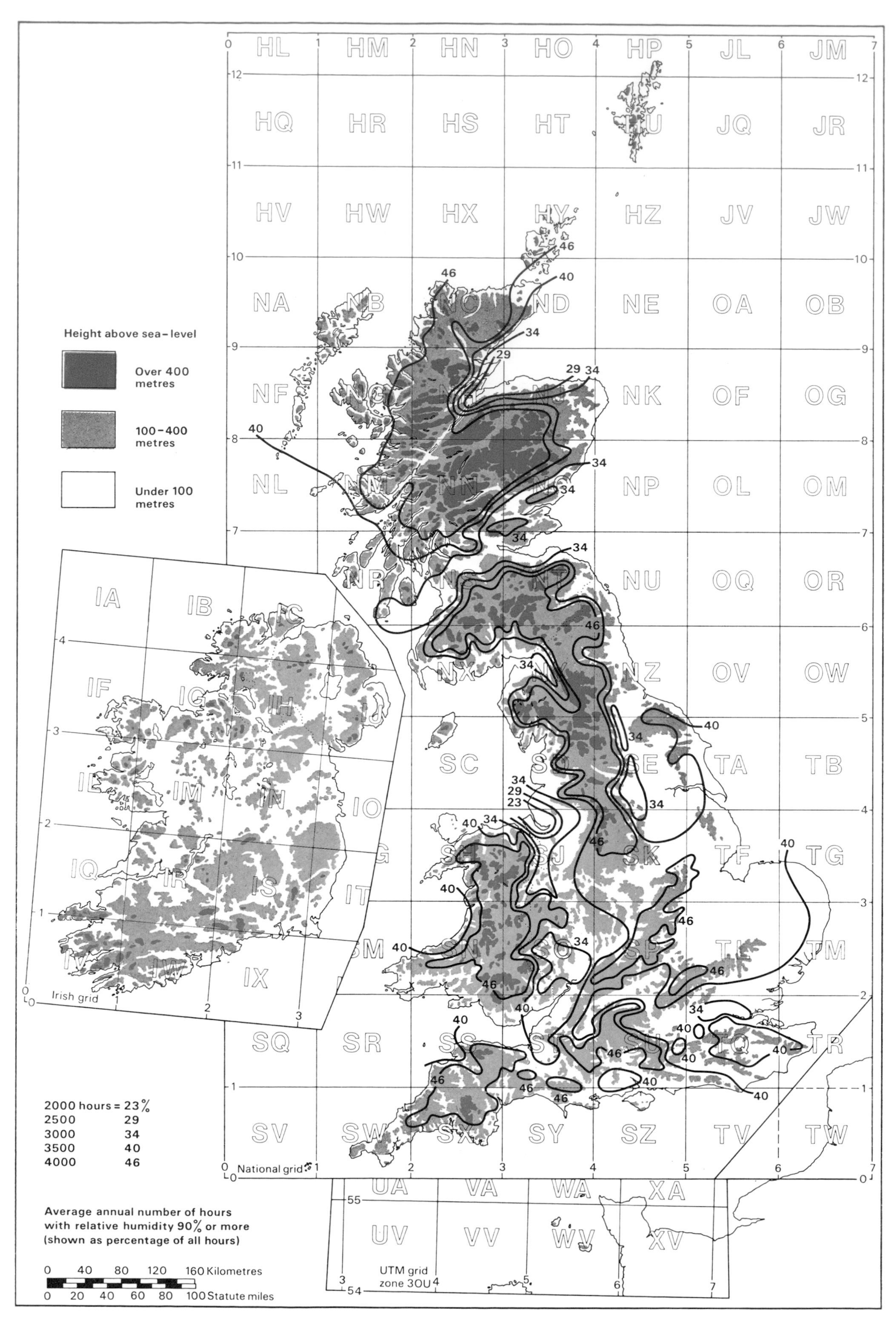

Figure 18 Average annual number of hours in Great Britain with relative humidity 90 per cent or more

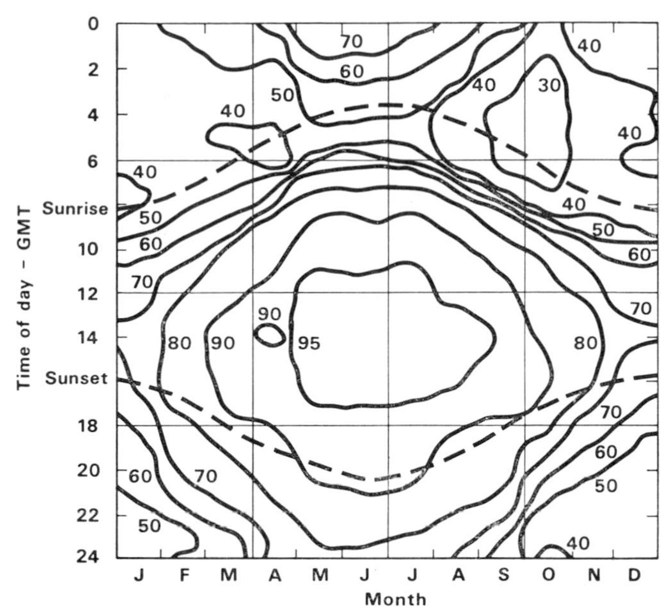

Figure 19 Diurnal and annual variation of percentage of time during which the relative humidity is below 90 per cent at Heathrow. From hourly observations during the 10 years 1957–66

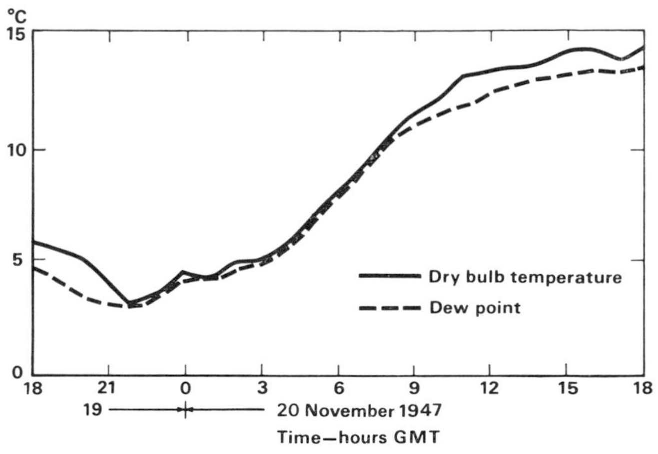

Figure 20 Rapid rise of temperature and dew-point at Kew, 19–20 November 1947

An alternative way of displaying information on humidity is shown in Figure 19, which gives the annual and diurnal variation of the percentage of the time for which the relative humidity is less than 90 per cent, at a single station. The diagram is derived from hourly data of humidity at London Airport (Heathrow) in the 10 years 1957–66, and is fairly typical of the lower part of the Thames Valley, although it may be expected that in the built-up area of London the frequencies will be rather lower because of the generally higher temperatures in London in the night-time (see also Chapter 3, page 127).

Rapid changes in humidity

With the passage of depressions and anticyclones, the British Isles may be successively covered by airstreams of different origin. In any locality and season each of these airstreams has its own special characteristics of temperature, humidity and so on. Thus in winter, low absolute humidities are normally associated with cold Continental air coming from central Europe, while relatively high absolute humidities are normally associated with rather mild maritime air coming from the Atlantic. Rapid changes in the moisture content of the air can thus occur over short periods of time when one type of air-stream is replaced by another. Figure 20 shows graphically a rapid rise in temperature and dew-point that occurred at Kew Observatory on 19–20 November 1947. Such changes may lead to heavy temporary condensation on structures, especially those of heavyweight materials (see also Chapter 4, page 164). It may often be possible to forecast such events some 24 hours ahead: readers are referred to Appendix 2 for details of the weather forecasts that are available from the Meteorological Office.

Opposite page

Note: This map, prepared by the Meteorological Office, is based on data for a limited number of stations, mainly for the period 1957–66

The map is considered satisfactory in broad outline, but caution is needed, in interpreting the map in detail: relative humidity can vary considerably over short distances in areas of varied topography, but such details cannot be adequately shown on a map of the present scale

Precipitation (rainfall, snow, hail)

Measurement

Rainfall is measured by collecting it in a standard raingauge on a selected site. The standard raingauge consists essentially of a funnel of special design supported upon an outer cylindrical vessel, partly sunk into the ground, within which is placed a receiver to collect the rain entering the funnel. In the older gauges, the aperture of the funnel is a brass ring usually 5 inches (127 mm) in diameter and bevelled to a fine edge but is of plastic material in the latest ones. To minimize loss from splashing the sides of the funnel are extended to form a cylinder about 100 mm deep. The outer vessel is firmly fixed into the ground so that the rim is horizontal and 0·3 m above ground-level. To make the measurement, usually daily at 09 00 GMT, the collected water is poured into a rain measure graduated to give readings directly in millimetres and tenths. The amount of rainfall is thus expressed in terms of the depth, in millimetres, of the layer of water which has fallen into the aperture of the gauge since the previous measurement. The gauge must be so exposed that the amount recorded is a good sample of the rain falling on the ground in the vicinity. This involves following certain rules about siting the gauge sufficiently clear of obstacles such as buildings, trees, etc, while at the same time providing adequate shelter from the full force of the wind. In the British Isles the rainfall measurements include rain, drizzle, snow, sleet and hail and also small amounts from dew, hoar frost, rime and wet fog collecting in the funnel. Frozen precipitation is thawed and then measured as equivalent rainfall.

There are outstanding problems of rainfall measurement in areas where it is difficult to find a suitable site, especially where there is over-exposure to strong winds as in mountain and moorland terrain. There are also special problems in the measurement of moderate to heavy snowfall and of hail. These problems are receiving attention, but owing to their difficult nature it is unlikely that satisfactory solutions will be found in the near future.

There are some 6000 raingauges of the simple type in use in Britain, giving daily amounts of rainfall. In addition there are about 400 recording gauges, which produce a trace on a paper chart. From the chart, the times at which it has rained,

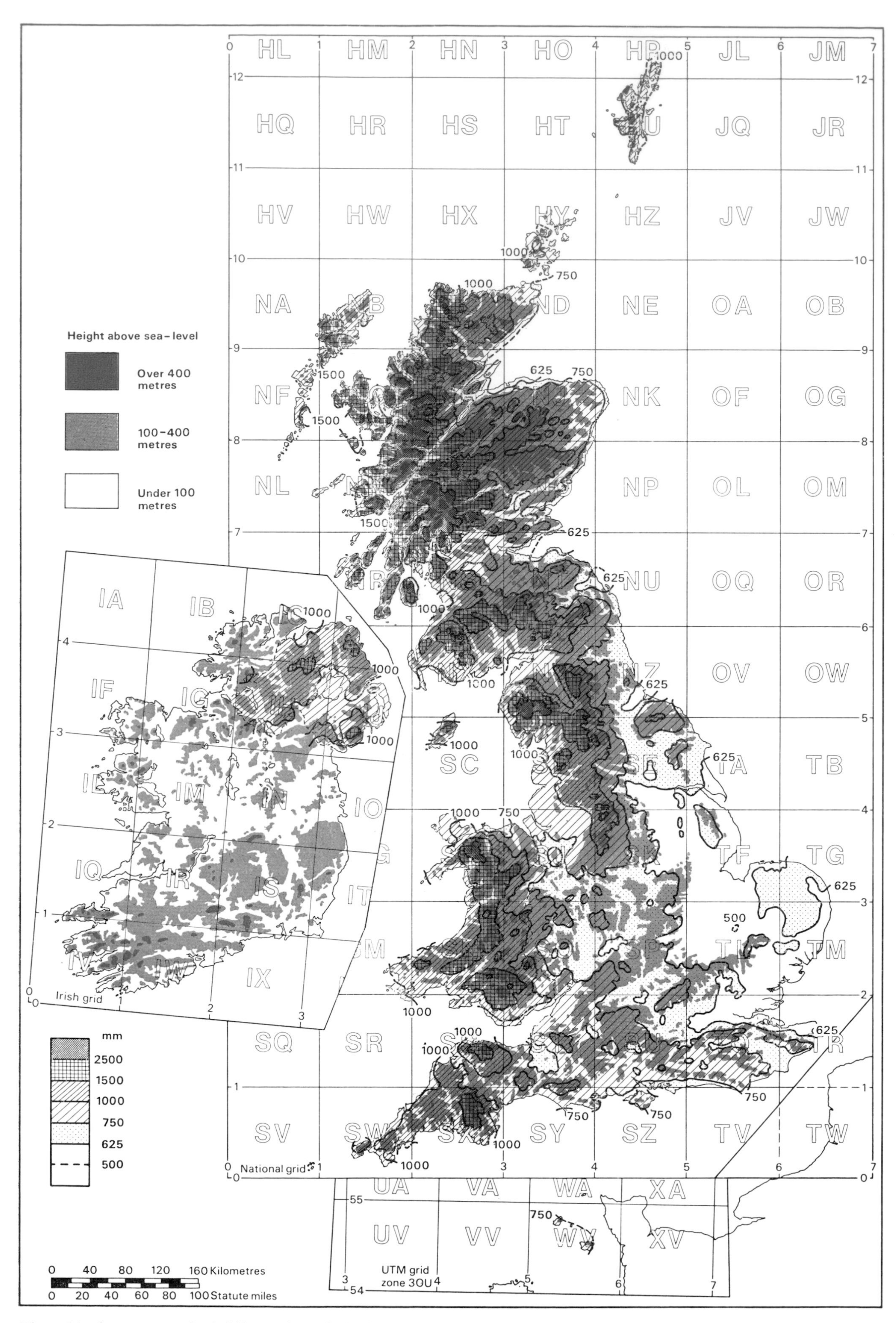

Figure 21 Average annual rainfall over the United Kingdom, 1916–50

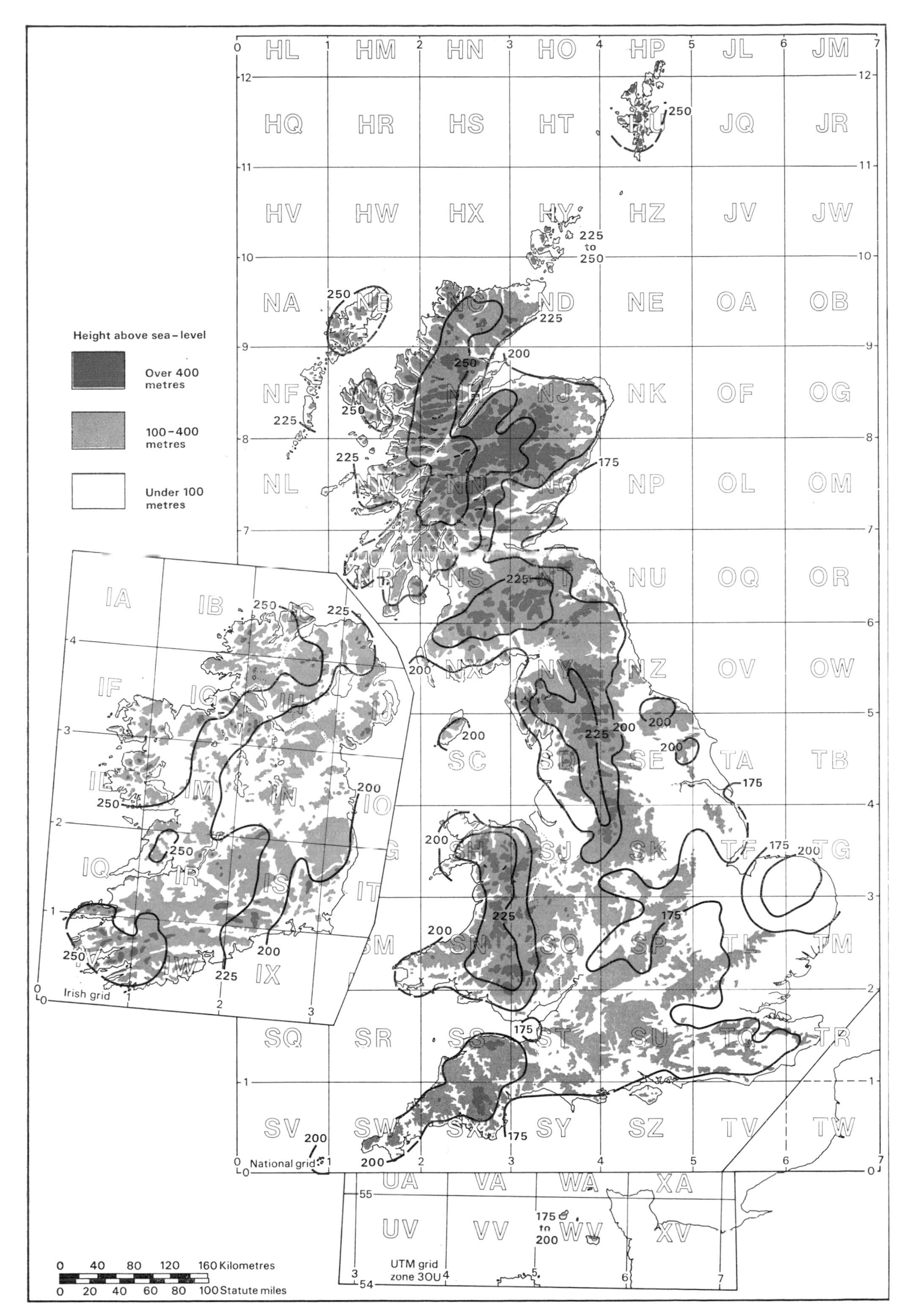

Figure 22 Average annual number of rain-days in the British Isles, 1901–30

the duration of rainfall, and the rate of rainfall at any instant, can be obtained. In addition, at a few places rate-of-rainfall recorders have been installed. These give a better estimate of the rate than an ordinary rainfall-recorder, especially at high intensities (although some of the latter have been modified with a different chart mechanism, so that high rates of rainfall can more easily be determined).

It is from gauges such as these that the information on rainfall duration and on the frequency of high rates of rainfall have been determined (see Chapter 3, page 140 and Chapter 4, page 154).

Mean annual rainfall across the United Kingdom: the influence of topography
Figure 21 is a map of average annual rainfall during the years 1916–50. This makes it clear that average rainfall is very largely controlled by the configuration of the ground and the distance from the Atlantic Ocean. The mountainous areas of the west and north are areas of high rainfall and the relatively flat plains of the south and east are areas of low rainfall. The relationship between rainfall and altitude is not as close, however, as appears at first sight. In general, for similar altitudes rainfall is greater in the west than in the east. The local topography is also important; thus the rainfall may be higher on a mountain range at right-angles to the south-westerly winds which bring most rain, than on one running parallel to them. There may be great local variations in rainfall in regions of rugged topography, over distances of a few hundred metres. This was demonstrated strikingly by Balchin and Pye (1947) in their study of the climate around the city of Bath.

A large proportion of eastern England and some narrow strips on the east coast of Scotland have average annual totals of less than 750 mm. On the other hand there are considerable areas in the west of Scotland, Cumbria and North Wales where the average fall exceeds 2500 mm.

It is important to note that the annual rainfall of any one place shows very considerable fluctuations from year to year. Such fluctuations are of great importance to those responsible for water supplies and drainage.

Average annual number of rain-days
The distribution over the country of the average annual number of rain-days is shown in Figure 22. A rain-day is defined as a period of 24 hours commencing at 09 00 GMT during which 0·2 mm or more of rain is recorded. The distribution somewhat resembles that of average rainfall in Figure 21, the number of rain-days being large in the west and north and small in the south and east. The range is from under 175 in the south-east to over 250 in the north-west. On the whole June has the least number of rain-days and October or December the greatest.

Figure 22 provides only a rough guide as to the frequency with which building operations are likely to be interrupted. It is doubtful, for example, whether a fall of only a millimetre or so spread over several hours would be of any importance in this respect. Allowance also needs to be made for the fact that building operations are mainly carried out during the daytime, so that those days when the rainfall is at night will not be relevant. A special analysis of hourly rainfall data for the hours 06 00 – 18 00 GMT indicates that in the London area the average monthly percentages of time likely to be lost due to appreciable precipitation (defined for this purpose as a rate of 0·5 mm/h or more) range from under 3 per cent in April and June to nearly 6 per cent in February. (See also Chapter 4, page 154.)

Intensity of heavy rainfall and effect of state of ground
The occurrence of high-intensity rain, that is heavy falls in short periods, is of importance in the design of the drainage for roofs and yards. Here we are concerned with falls in which the rate rather than the total amount is exceptional and with durations lasting from a matter of minutes up to an hour or two. Examination of rainfall records has shown that the frequency of the more intense rains is not related to the mean annual rainfall although it varies from place to place. Taking all available information various formulae have been derived relating frequency to rate of rainfall and duration. Of these the best known is that introduced by Bilham in 1936. This has been used for many years but has now been modified following the study of the great amount of information accumulated since then. An average relationship is shown by the curves in Figure 23, and by the data in Tables 52 and 53 in Chapter 3. Since these were compiled in 1967 by Holland (1967a, 1968) further work by the Meteorological Office has resulted in the production of relationships which permit the computation of data for individual places (see page 140).

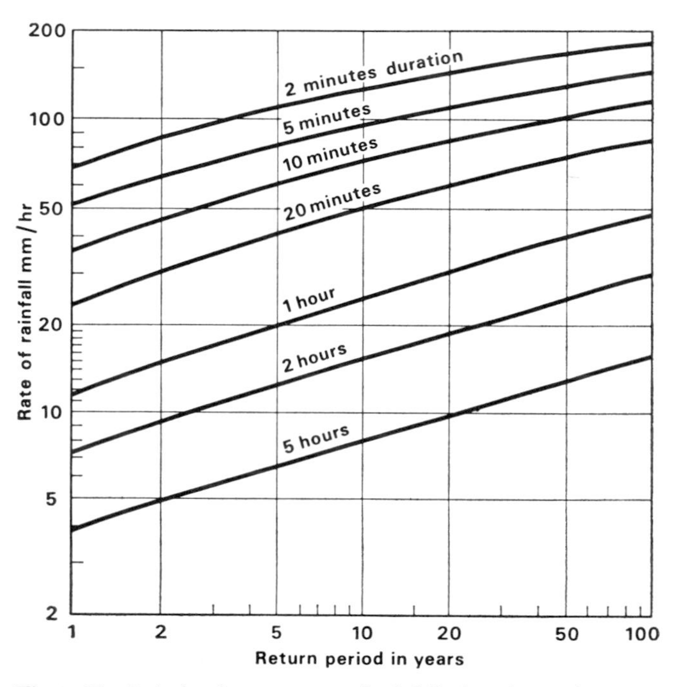

Figure 23 Relation between rate of rainfall, duration and frequency of occurrence. A return period of *n* years means that there is a likelihood of the event recurring not more than once in that number of years. (After Holland 1967)

The state of the ground has an important bearing on what happens to a fall of rain after reaching the ground. Most of the rain which falls on roofs and roads will run off down gulleys and drains into streams or rivers. What happens to that portion which falls on soil, however, depends on the state of the soil, on the soil type and on the rate of evaporation (see page 29). There are many complications but, broadly speaking, if the soil is dry and unfrozen the upper layers only will be wetted and the moisture will soon be re-evaporated or transpired by vegetation; if it is wet but not saturated water will penetrate the upper layers to be stored and drained slowly by rivers in dry weather. If there is a large proportion of impervious rock, run-off will be rapid during heavy rains, subsiding quickly afterwards. If on the other hand there are deep pervious rocks, such as the chalk of southern England, there will be little surface run-off and most of the rainfall will either evaporate or be absorbed by the chalk.

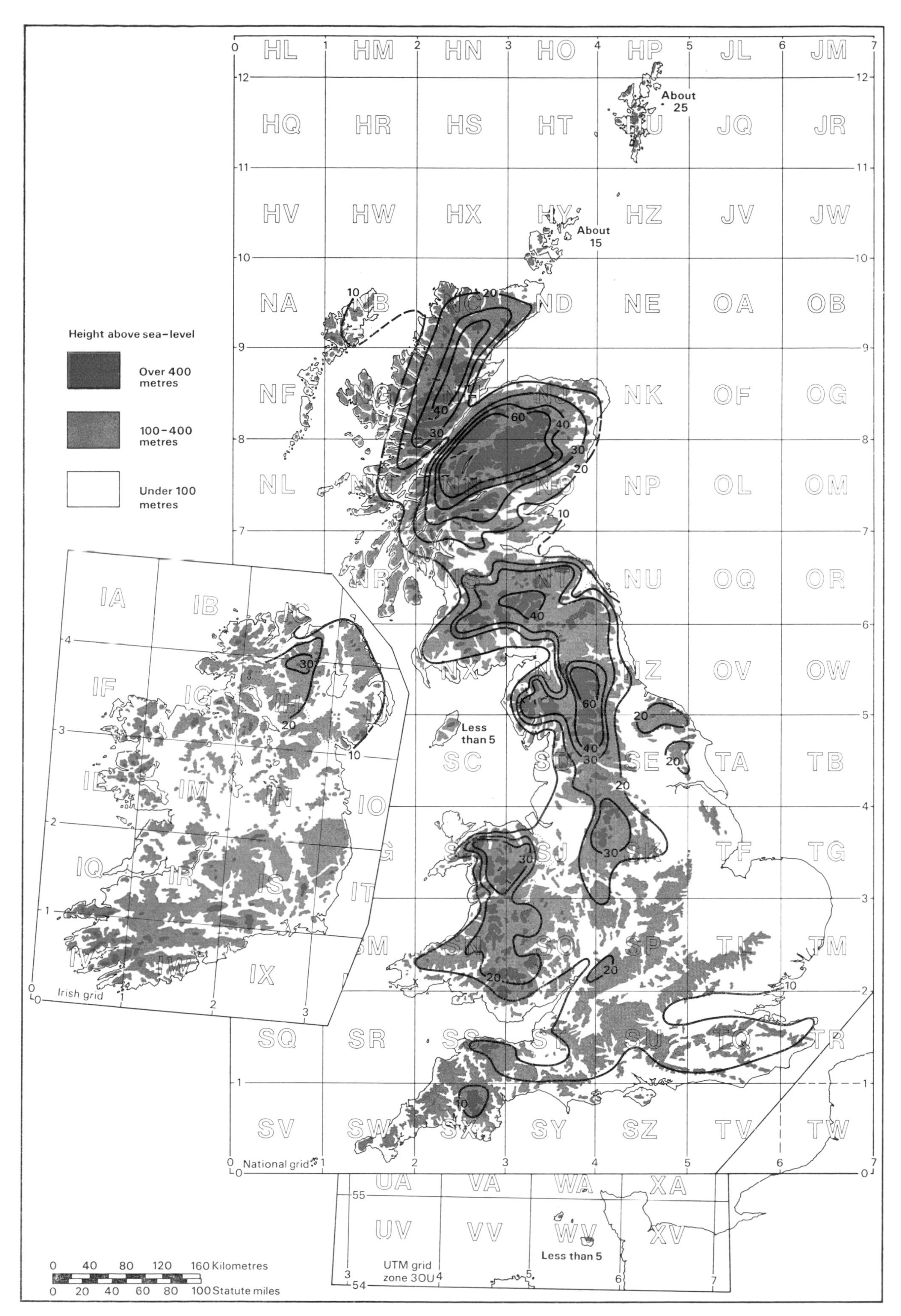

Figure 24 Average annual number of mornings with snow lying in the British Isles, 1941–70

Table 4 Direction of wind when snow is falling
Percentage distribution of surface wind for occasions with snow falling* at London (Gatwick) Airport during 1959 to 1970

Wind directions	Wind speed (m/s) 0	0.3–1.5	1.6–3.3	3.4–5.4	5.5–7.9	8.0–10.7	10.8–13.8	13.9–17.1	17.2 or more	All speeds
Calm	6·6									6.6
350°–010°		0·6	2·0	2·9	1·7					7.2
020°–040°		1·7	4·9	6·2	7·4	0·6		0·3		21·1
050°–070°		1·7	4·0	8·6	8·3	3·1	1·1	0·3		27·1
080°–100°		1·7	2·0	2·3	3·4	0·6				10·0
110°–130°		0·3	0·3	0·9	1·9		0·3			3·7
140°–160°		0·3	0·6	0·6	1·7	0·2				3·4
170°–190°		0·3	0·3	1·4	0·6					2·6
200°–220°		0·9	1·1	0·3	0·3					2·6
230°–250°		1·1								1·1
260°–280°		0·6	0·3	0·5						1·4
290°–310°		0·3	1·1	0·6	0·3					2·3
320°–340°		1·4	2·3	4·0	2·3	0·9				10·9
All directions	6·6	10·9	18·9	28·3	27·9	5·4	1·4	0·6		100·0

* From observations at 3-hourly intervals – 350 observations of snow in all.

Lamb (1969) has suggested that because of the reduced vigour of the westerly wind circulation over Britain in the past few years, depressions may sometimes move more slowly than usual across the country. In consequence there is a greater risk of an individual depression producing more rain at a point on the ground below than if the depression moves rapidly. Certainly there has been an increased frequency of floods in the upper Severn valley, for example, but it is not possible to say whether this is more than a chance variation.

Frequency of occurrence of snow

The amounts of snow which fall over the British Isles are measured as equivalent rainfall and are included in the rainfall statistics already discussed. There are three other aspects of snowfall for which separate records are available: (i) number of days with snow falling on low ground, (ii) number of days with snow lying, and (iii) frequencies of different snow-depths. The average annual number of days with snow falling on low ground increases from south-west to north-east across the country, from less than 5 days in southern Cornwall to more than 35 days in north-east Scotland. At places above about 60 metres the average number of days increases by about 1 day for every 15 metres of elevation, with a greater rate of increase above 300 metres. If half of the ground representative of a station is covered with snow at the morning observation hour, the day is counted as a day with snow lying. Figure 24 shows the average annual number of mornings with snow lying, 1941–70. The frequency of snow lying is one of the most variable of meteorological elements over the British Isles. In some winter months the duration has been almost negligible, whereas during severe winters snow has lain on the ground for long periods, for instance for more than 60 days over a wide area in 1962–63. The duration of snow cover depends greatly on altitude but also shows a marked increase from south-west to north-east. This is largely due to the influence of the Atlantic Ocean on the south-west regions, but also to the greater frequency of snow and on the whole the greater quantity of snowfall in northern and eastern districts (see also Chapter 2, page 74). It is thought that over the past few years, since about 1940, the frequency of snowfall has increased somewhat.

Depth of snow in heavy snowfalls

The depth of undrifted snow does not often exceed 0·15 m on level ground at low altitudes, but in our worst storms depths of 0·3 to 0·6 m may fall over a wide area. About 1·5 m are stated to have fallen in the Isle of Wight and Hampshire in January 1881. In March 1947 level snow 0·5 m deep was measured at Cranfield, Bedfordshire and in February 1955 level snow was 0·6 m deep at Braemar. In the severe 1962–63 winter depths of 1·6 m and 0·75 m were reported from Tredegar in South Wales and Bellingham in Northumberland respectively.

When depths exceed 0·15 m or so in conjunction with strong winds serious drifting can occur. Drifts may reach some metres in depth in the lee of buildings and other obstacles to the wind, and in hollows and cuttings deep accumulations 5 to 8 m or more in depth may occur, especially in hilly areas. The drifting of snow can be controlled to some extent by the use of snow fences specially designed to form drifts in places where they are relatively harmless and so to prevent wind-blown snow from accumulating on roads or footpaths.

When snow is falling, the direction of the wind may be quite different from that from which it blows during rain. Most rain falls with wind directions from between south through west to north-west but, as shown in Table 4 and Figure 25, when snow falls the most common directions lie between 020° and 070°, that is about north-east. These statistics are from London Airport, Gatwick, 40 km south of central London, and are probably typical for south-eastern England. The wind speeds during snow may be quite high even at this low-level place, up to about 15 m/s.

Data on maximum rates of snowfall in short periods are almost entirely lacking, but 0·125 m in an hour has been recorded and falls of around twice this rate are entirely possible. The density of freshly fallen snow varies over a wide range, the snow/water depth-ratio ranging from about 2 to almost 30 in this country (Thomson 1963). Hence the density of freshly fallen snow varies widely, but an average figure is around 100 kg/m^3. Thus 0·61 m of snow on a roof,

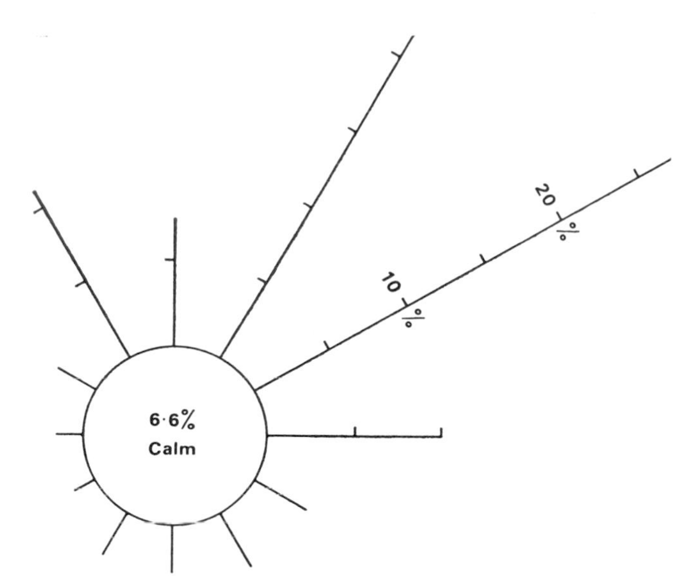

Figure 25 Wind-rose showing wind directions at London Airport, Gatwick, for hours during which snow is falling. Annual mean for period 1959–70. Lengths of lines proportional to percentage of time when snow is falling. For a more detailed analysis see Table 4

with an additional allowance for drifting in the case of a large flat roof, may produce an imposed load of around 720 N/m² (see *BS CP3*, Chapter V, Loading). At a number of meteorological stations, the density of the snow lying on the ground is now measured regularly. In a series of measurements made at Garston, Hertfordshire, during the severe winter of 1962–63, it was found that over a period of about 2 months the density of settled snow tended to rise to and remain at about 300 kg/m³ (Lacy 1964). This figure is probably typical of lowland Britain. In the high Alps settled snow has a density of about double this (then being known as *firn*).

Hail and thunderstorms

Hail falls from time to time in Britain, but normally the hailstones are not more than about 10 mm in diameter and cause little or no damage. Damaging hailstorms – those which caused sufficient damage to growing crops or to glasshouses for them to be noticed and recorded by weather observers or in newspapers – occurred somewhere in Britain on 169 days in the 50 years 1906–55, a little over 3 times a year on average (Rowsell 1956). All but 20 of these days fell in the months May to September. June, with 44 days, had the highest frequency, July (41) being next. The frequency of damaging hailstorms varied considerably from one part of the country to another. It was greatest in the London region (one, or a little more, occasions per 100 years, per 100 sq km) with lesser peaks in the Bristol Channel – Severn valley, in the north-east Midlands and in the lower Clyde valley. The latter area had a frequency about one-third of that in the London region.

Much less commonly, perhaps in about 1 year in 5, very large hailstones, variously described as being as large as walnuts, eggs, or even tennis balls or grapefruit, may fall. Such stones, with a diameter of 75 mm or even more, and weighing 100 g or more, are produced by 'severe local storms' (Figure 26). These storms are a special form of thunderstorm, with exceptionally strong up-draught, and usually associated with violent squally winds and even tornadoes (Browning and Ludlam 1962; Carlson and Ludlam 1968). Thus even the giant hailstones may be blown against windows with destructive effect.

This type of storm can occur only in summer with a particular weather situation which is fortunately quite rare in Britain, and has never been reported north of about latitude 53°N. The most recent cases were on 22 September 1935 and 11 May 1945 in Northamptonshire, on 5 September 1958 in Sussex, on 9 July 1959 in Hampshire and Berkshire, on 13 July 1967 in Wiltshire, on 21 April 1968 in Warwickshire and on 1 July 1968 in Glamorgan (Stevenson 1969). The storms usually travel in a northerly or north-easterly direction at a speed of around 10 m/s. They may be active for some hours, travelling perhaps some tens of kilometres in this time.

Damage to buildings by lightning strikes is much commoner than that from hail, but no statistics are available. It is thought that in much of England there is on average about one lightning strike to earth to the square kilometre in a year, and appreciably less than this over much of Scotland, Wales and south-west England (see also Chapter 3, page 140).

Evaporation and potential evaporation

The rate of evaporation from a water surface, or from a horizontal porous surface saturated with water, varies from a

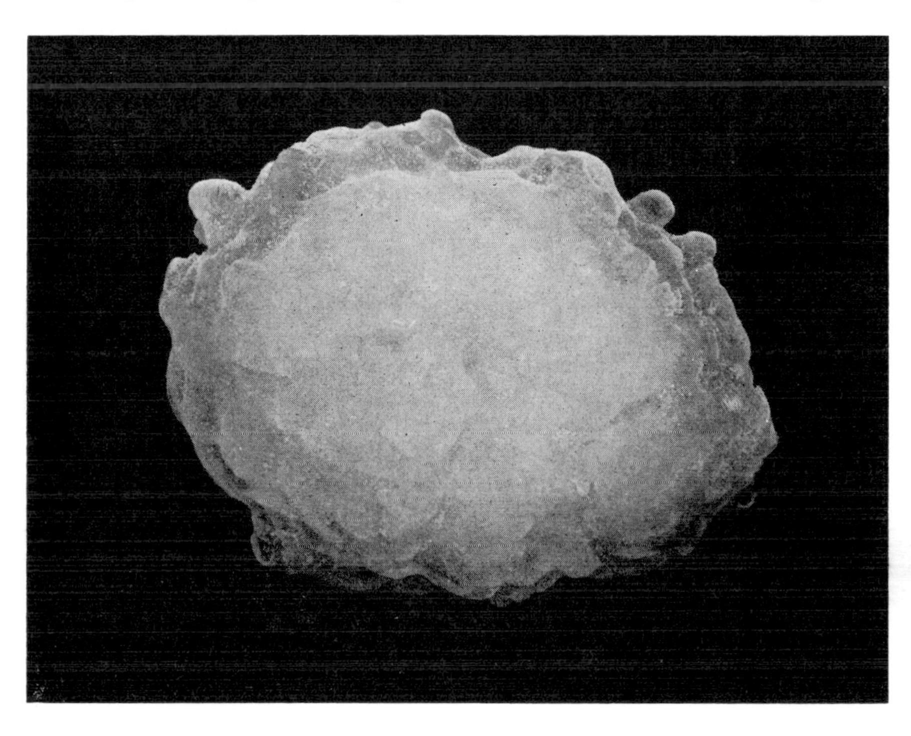

Figure 26 Photograph of a large hailstone which fell on 1 July 1968 at Cardiff (full size). (From *Weather*, *24* (*4*), *1969*)

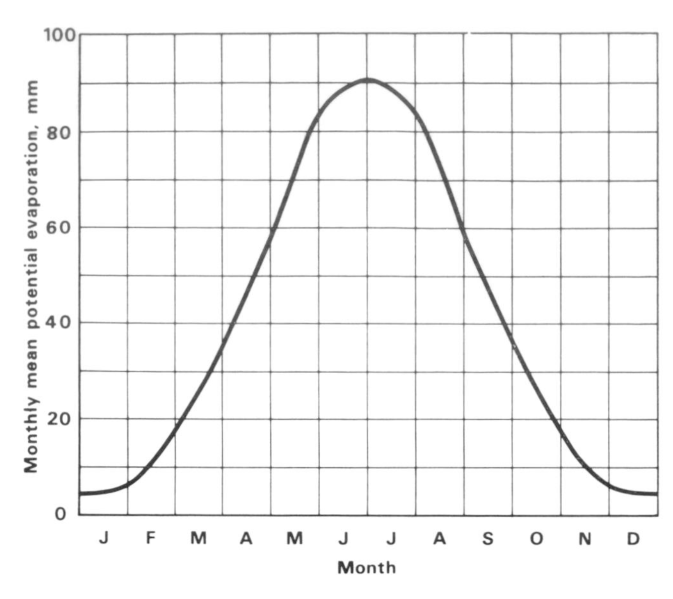

Figure 27 Annual march of monthly mean potential evaporation in Britain. (After Holland 1967b and Monteith 1966)

negligible amount each day in midwinter, to perhaps 7 mm on a sunny day in the summer. Evaporation from water tanks such as are used to measure rates of evaporation (Meteorological Office 1956b) varies from one site to another. Annual mean amounts of water evaporated are in the range from about 400 to 600 mm (Holland 1967b). It is thought that most of this variation is caused by differences in the degree of shelter from wind at the individual sites.

The rate of evaporation from a water surface is generally thought to be the same (or nearly so) as that from soil covered with growing vegetation, provided that the plants can get all the water that they need from the soil. This theoretical rate is known as the potential evaporation (PE), and is usually greater than the actual loss from plants in summer, when there is often a deficit of moisture in the soil. The potential evaporation can be calculated from a formula due to Penman, the latest results from which, using revised radiation data (Monteith 1966), indicate that annual mean PE varies from about 450 mm in the far north of Scotland to 500 mm in southern England. Figure 27 shows the annual march of monthly mean PE, based on an annual total of 500 mm, and on the standardised apportioning to months suggested by Holland (1967b). It appears that evaporation from water tanks, sunk in the ground so that the surface of the water is level with the soil, may be some 10 to 20 per cent higher.

It must be emphasised that these results apply only to horizontal surfaces at ground-level. Elevated ones, such as roofs, may be more exposed to wind, moreover they may be artificially heated from below. Both these factors will tend to increase the rate of evaporation. Vertical surfaces will have a still more complicated exposure, for rain will only reach them when driven by wind, while the incident angle of solar radiation will be quite different from that on a horizontal surface. Thus walls may absorb much more solar radiation than the ground in winter if the orientation is suitable, which with artificial heating will evaporate appreciably more water from the wall than can be evaporated from the ground or vegetation. Even so, and especially for unheated buildings, for example those under construction, Figure 27 does give a guide as to relative rates of evaporation from materials at different seasons. Evaporation from masonry walls is discussed on pages 114–6, and from swimming pools on page 148.

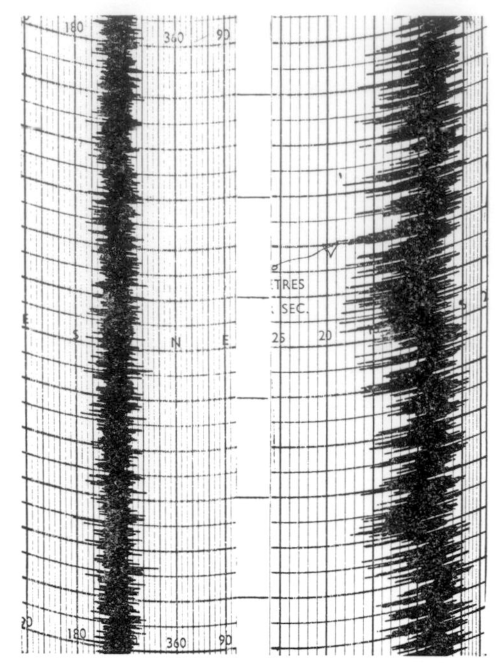

Figure 28 Copy of the trace from an anemograph, showing the very variable nature of the speed of the wind. The duration of the spikes on the trace, which are caused by gusts, may be as short as three seconds. The direction in which the wind blows varies in a similar manner (l.h. trace). From an anemograph at Garston, 15 m above ground, 11 30 to 17 45 GMT, 13 March 1969

Wind

The wind

The wind is never steady, because as the air moves over the surface of the ground, obstructions interfere with the free flow of the air, deflecting it one way or another. The degree of interference depends on the size of the obstructions, large ones having more effect than small ones (the departure of the surface texture from smoothness is described as its 'roughness'). An extensive flat field covered with short grass would be regarded as aerodynamically fairly smooth, while a town would be extremely rough.

The air acts as though there is friction between it and the surface, and the lower layer of the air is slowed down relative to that above. Each layer of air in its turn slows down the one above, and the result is normally (unless there is a temperature inversion as described below) a vertical gradient of wind speed, the speed being zero at the surface, rising rapidly in the first few metres, the rate of change falling gradually to zero at about 300 to 500 metres above the ground. Above this height, the speed of the wind depends upon the general weather situation. The rougher the underlying surface, the greater the extent to which the wind is slowed down in the lower layers, and the more disturbed its motion becomes.

The short-period variations in wind speed and direction caused by the surface roughness are called gusts. Figure 28 is reproduced from the record of a wind-recording instrument (anemograph) and shows how variable the speed and direction are at about 10 metres above the ground. The shortest gusts recorded last for some three seconds and consist of fast-moving parcels of air a few tens of metres across. Gusts lasting a longer time represent larger masses of air, usually moving more slowly than the short-period gusts.

This simple picture applies when the temperature falls with height above the ground, which is the usual state of affairs in the daytime, and when the wind is strong. When there is an inversion of temperature with height, as often happens over open country on clear nights (see page 78), the inversion layer, which may be 100 metres or more deep, will usually be calm, or with only a gentle drift of air. Thus under these circumstances, the variation of wind with height may be quite complex and may depend upon local topographical influences. Over towns there may be still further complexity, as the heat released during the night delays the formation of a temperature inversion.

Because of the short-period fluctuations of the wind, the instrument used to measure wind speed must be chosen with care – one with a slow response will not be able to measure the high-frequency gusts that influence structures. The instrument must also be sited carefully, because local variations in the roughness of the ground make a big difference to the wind speed in a short distance. Both these aspects of the measurement of wind are discussed briefly in the following paragraphs.

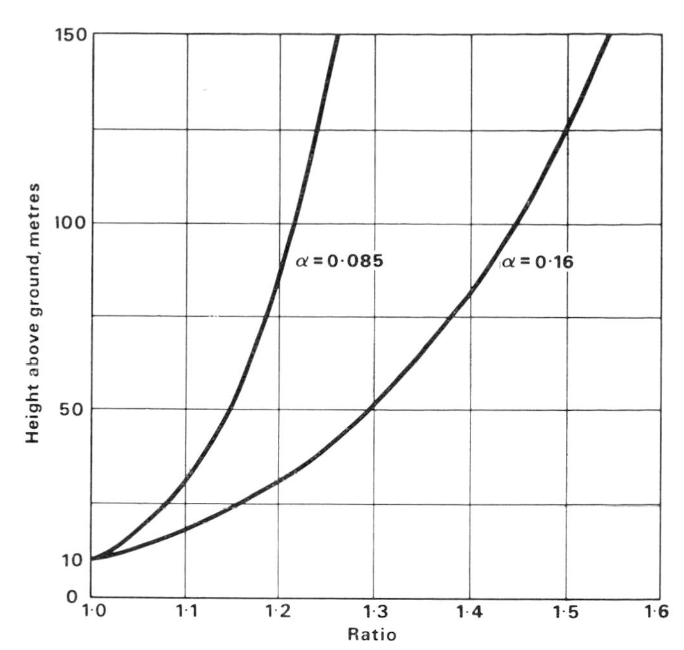

Figure 29 Variation with height above ground in open country of gust speed (α = 0·085) and hourly mean wind speed (α = 0·16), expressed as ratio to the speed at a height of 10 m, when the wind is strong

Measurement of wind

Wind speed is usually measured by means of some form of anemometer, a wind vane being used to measure direction. Anemometers may be of the indicating or of the recording type. Non-recording anemometers usually depend upon the rotation about a vertical axis of a system of cups, designed in such a way that the rate of rotation is directly proportional to the wind speed. The anemometers used in routine work are of two main patterns. 'Contact' anemometers are fitted with electrical contacts or other devices so that when connected in series with a power supply signals are produced with a frequency proportional to the wind speed. Counters add up the pulses from the anemometer and indicate the 'run of wind' on an indicator, so that the mean speed over a given time-interval, usually a day, can be computed from the change in reading over that interval. 'Generator' anemometers generate a voltage which is indicated as a speed on a dial graduated directly in knots or metres per second. Until recently the most common type of recording anemometer was the pressure-tube anemograph, but this is now gradually being superseded by the electrical cup-anemograph (Meteorological Office 1956b). Both these instruments give continuous records of speed and direction on a moving chart and have response times of one or two seconds. Thus gusts with durations of three seconds or longer are fully recorded. In tabulating the records, mean wind speeds and directions over each hour are measured by means of suitable scales and details of the highest gust each hour are also noted. Most of our detailed knowledge about wind over the United Kingdom has been derived from anemograph records.

When measuring the wind the exposure of the instrument is a matter of the greatest importance. In order that the records of an anemograph may be representative of the wind passing over the immediately surrounding region, and also comparable with the records obtained from other anemographs, every effort is made to set up the instrument in a standard exposure or, if this is not possible, to make appropriate allowances for the actual exposure. The standard exposure is obtained when the head of the anemograph is 10 m above level ground with no appreciable obstacles within a radius of at least 200 m. Such an exposure is often difficult to achieve in practice because even small buildings, trees, hedges and so on interfere with the free flow of the wind. The head or cups must often be raised above the standard height of 10 m therefore, so as to be clear of the more violent eddies produced by such obstacles. Since wind speed increases with height above the ground (except when a temperature inversion is present), this is a compromise solution and if records from different anemographs are to be comparable there must be a means of

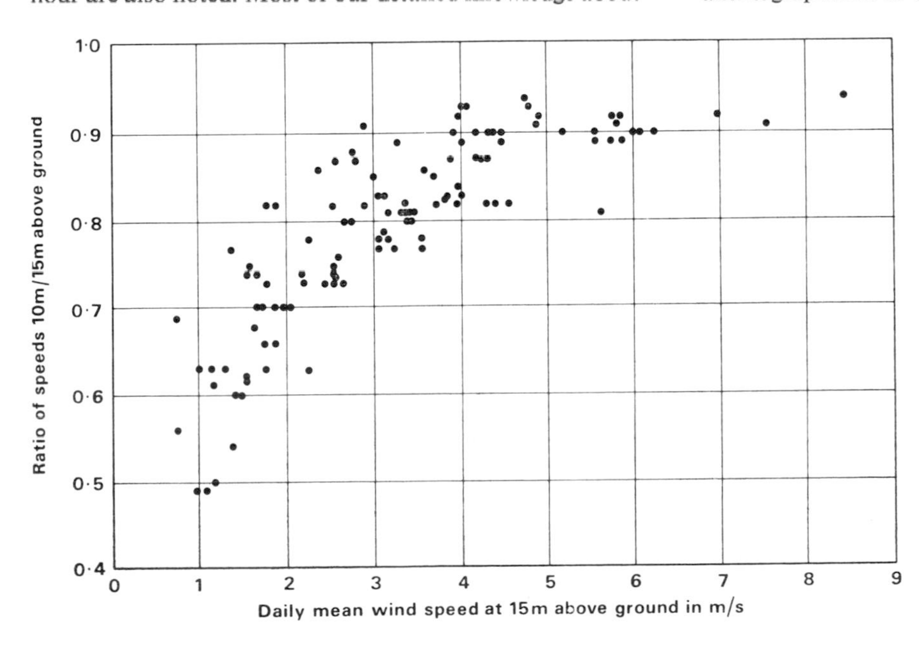

Figure 30 Variation of wind speed with height. Diagram shows how ratio of daily mean speeds at two heights varies with the speed of the wind. Garston, 1 January to 30 April 1972

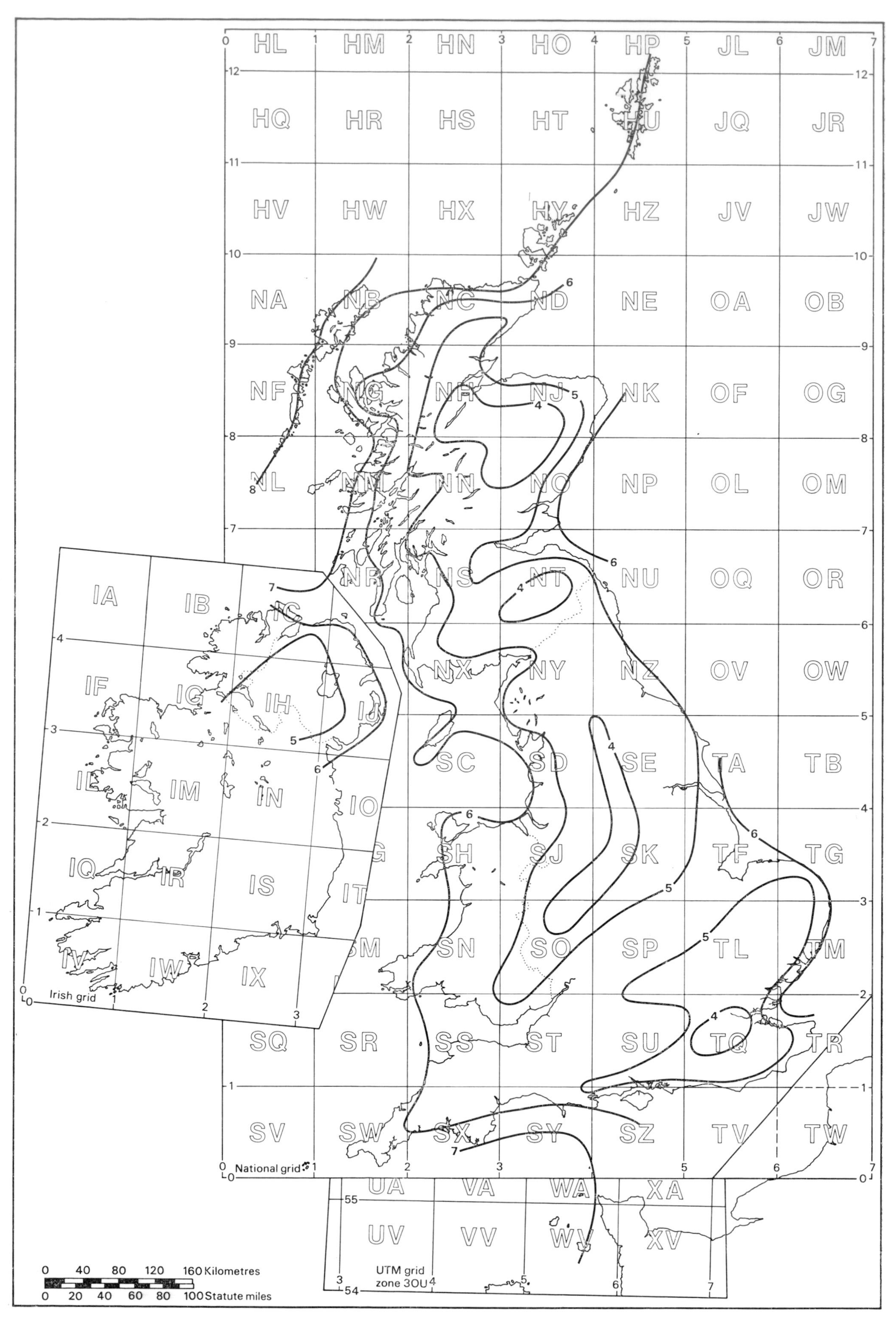

Figure 31 Annual average hourly mean wind speed (metres per second) in the United Kingdom at 10 m above ground in open country, based on data for 1965–69. Applies at altitudes up to 250 m above sea-level.

correcting them to the standard height. Such corrections are made by estimating an 'effective height' for each instrument, the effective height being the height over level unobstructed terrain in the vicinity of the anemograph which it is estimated would experience the same mean wind speeds as those actually recorded. If the effective height is h metres and a recorded hourly or 10-minute mean wind speed is V_h then the corresponding corrected speed V_{10} at the standard height of 10 m is given approximately (for strong winds) by the formula

$$V_{10} = V_h \times \left(\frac{10}{h}\right)^{0 \cdot 16} \qquad (2)$$

Gust speeds increase less rapidly with height and the correction is smaller, the corrected speed being given by a similar 'power-law' formula in which the index 0·16 is replaced by 0·085 (see Figure 29).

The difficulty of estimating the mean speed of the wind at a standard height at sites other than airfields with exceptionally clear exposure is illustrated by Figure 30. This is based on the daily mean wind speeds measured by two cup anemometers on the same mast, one 10 m above ground, the other 15 m. In the figure, the ratio of the daily mean speeds at the two heights is plotted against the corresponding mean speed at 15 m. If the power-law formula applied at all speeds with an exponent of 0·16, the ratio of the speeds would always be about 0·93. In fact this ratio is only rarely achieved at this site (gently rolling country, with open fields and hedgerow trees in some directions, woodland and buildings in others). The ratio falls off sharply as the daily mean speed falls, especially below about 3 m/s (which is the average wind speed at this site).

A more detailed examination of the data shows that the wind-speed gradient varies not only with the wind speed, but also with wind direction (reflecting the different ground cover in the various directions) and with time of year (because in summer the trees are clad with leaves and the grass is longer, causing more aerodynamic drag).

In towns the average wind speed is still further reduced among the buildings. As the speed at, say, 500 m above ground is much the same over the town as over the country, the change in speed with height is greater over the town. The rate of change can be roughly represented by changing the index in the equation from 0·16 to about 0·3 or even 0·4, according to the density of the buildings. However, as is explained in Chapter 2, page 66, some measurements of wind speeds over London suggest that above the general level of the roofs the wind speed increases as though there was a 'zero-plane' at or a little above the roof-tops. In the streets of a town with buildings all of generally similar height, except where particular configurations of buildings affect the wind and may sharply increase the speed, the speed is on average about half that at the same distance above ground in open country. There is further discussion on the local disturbances to the wind field by buildings in Chapter 2, with examples of the serious effects caused by certain layouts of tall buildings.

Wind-speed statistics

Statistics of wind speed may be presented in several ways. It may be most convenient to have a general view of the way wind speed varies on average over the country. Figure 31 demonstrates how the annual mean wind speed varies over the UK. This is based on the annual mean values of the speed measured each hour at about 100 places, adjusted where necessary to be representative of a common effective height of 10 m above ground in open country. It applies only to heights up to about 250 m above sea-level in inhabited localities: thus it should not be used to estimate speeds on high ground.

The map shows clearly that average speeds are highest around the coasts, and especially in western Britain. Even over the lower ground, local topography may often cause both the speed and direction of the wind at a place to differ materially from that which prevails over the surrounding country. Because wind speed generally increases with height, mean speeds at the top of an isolated hill some 120 m high may be about 50 per cent greater than over the surrounding plain. At the crest of a long ridge across the wind the increase is likely to be even greater especially if the windward slope is gradual and fairly smooth. A strong wind tends to blow even more strongly along a valley if the valley runs in the general direction of the wind. This so-called 'funnel effect' may be marked in valleys orientated in the direction of the strongest winds especially if the valley leads to or from a pass through a range of hills. But equally, a sheltered valley may have a higher proportion of calms than open country. Hill ridges and mountain ranges, when wind and temperature conditions in the lower atmosphere are favourable, may induce 'lee waves' in the atmosphere and when these are of considerable amplitude and at a low level, the surface wind speeds may be very considerably increased over a narrow band parallel to and leeward of the high ground. Such conditions are believed to have been operating during the damaging Sheffield gale of 16 February 1962 (see page 70).

The highest mean wind speeds generally occur in January and the lowest in July although at some inland stations in England there seems to be a tendency for March to be the windiest month and August or even September the quietest. Figure 32 shows the frequency distribution of hourly wind speeds at London Airport, Heathrow, averaged over the period 1957–66. The middle curve is for the whole year, the others for January (the windiest month) and September (the

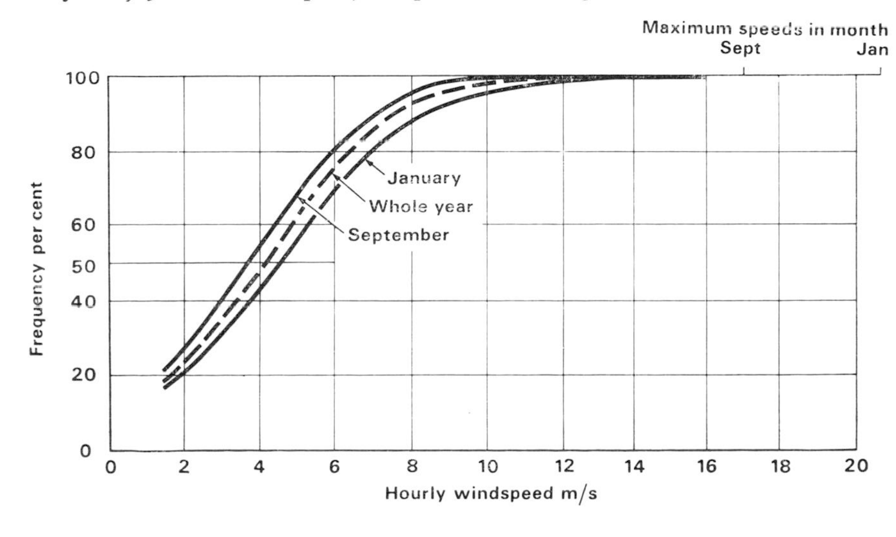

Figure 32 Frequency distributions of hourly wind speeds (spot readings) at London Airport, Heathrow, averages for 10 years 1957–66. Three curves: for whole year, for windiest month (January) and quietest month (September)

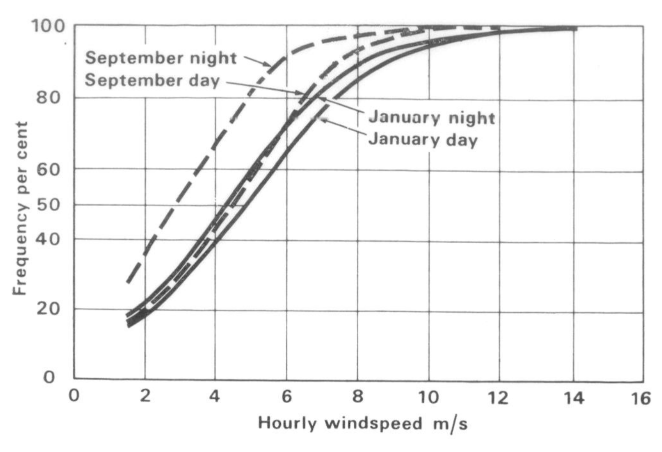

Figure 33 Frequency distributions of hourly wind speeds at London Airport, Heathrow, in daylight and at night, during January and September. Averages for 10 years 1957–66.

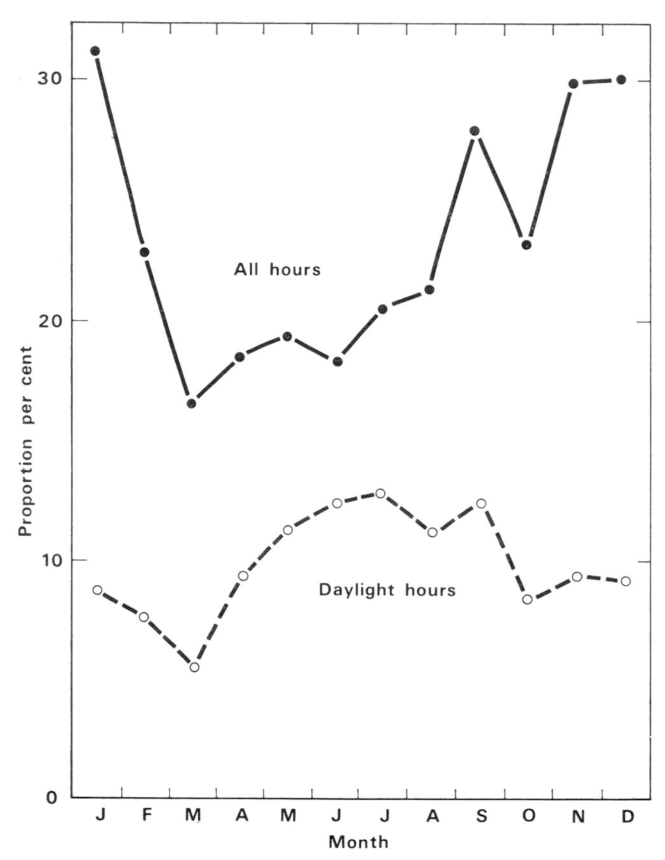

Figure 34 Annual variation of the proportion of time at Glasgow Airport, Renfrew, during which wind speed was 1·5 m/s or less. Average for 10 years 1956–65.

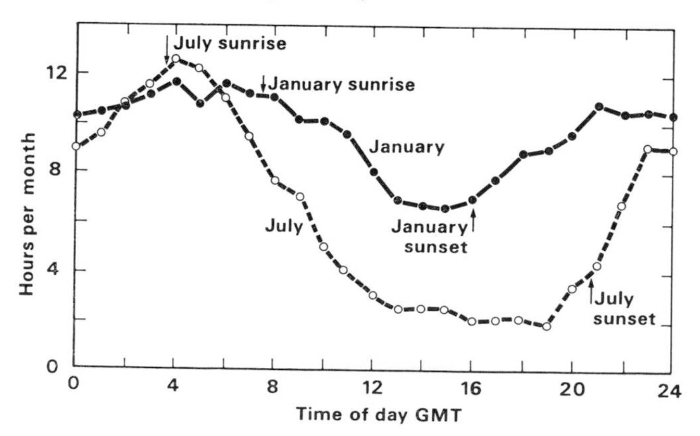

Figure 35 Diurnal variation of proportion of time at Glasgow Airport, Renfrew, during which wind speed was 1·5 m/s or less in January and July, in hours per month at each hour. Average for 1956–65

Table 5 **Frequency (per cent) of hours with wind speed less than 0·5 m/s at Filton Aerodrome, Bristol, in period 1957–70**

Season	Frequencies % Daytime 07 00 – 18 00 GMT	Night-time 19 00 – 06 00 GMT
Winter (December, January, February)	6·2	10·6
Spring (March, April, May)	2·1	12·6
Summer (June, July, August)	2·2	15·7
Autumn (September, October, November)	4·7	16·7

quietest month). From these curves can be read the median wind speed (that which is exceeded for 50 per cent of the time), or the speed exceeded for any desired proportion of the time, or the proportion of time for which a given speed is exceeded (the average speed over 24 hours is about 5 per cent higher than the *median* value). For example in this case the median speeds were respectively 4·2, 4·6 and 3·7 m/s, while a speed of 9 m/s was exceeded for 4, 7·5 and 2 per cent of the time respectively.

On average the wind speed is higher in the day than at night. Figure 33 shows the frequency distribution (at London, Heathrow as before) of hourly wind speeds separately during daylight hours and at night, during both January and September. In both months the wind is stronger by day, with median values of 4·9 and 4·5 m/s respectively, than by night, when the median values are 4·3 and 2·9 m/s respectively. The increased frequency in the daytime occurs all through the range of speeds. No data on calms are available from this particular analysis, but they are given in an analysis of wind speeds measured at Filton Aerodrome, Bristol, during the 14 years 1957–70 (Table 5). The average duration of winds of less than 0·5 m/s was greatest at night in all seasons (note that in this analysis the 'night' and 'day' are each 12 hours long whatever the season).

In summer, there were seven times as many hours with calm or very light winds at night as by day. Even in winter there were nearly twice as many. Calms were most frequent during autumn nights (when fog forms most commonly), with nearly one-sixth of all hours recording less than 0·5 m/s. On the other hand, at Bristol very light winds were rather more common on winter days than on autumn days. Thus at Bristol the annual frequency of very light winds of less than 0·5 m/s is about 9 per cent. At Glasgow Airport, Renfrew, the frequency is twice as high, about 18 per cent, while the frequency of all winds less than 1·5 m is as high as 23 per cent (1956–65 average). As shown in Figure 34, the frequency of light winds (1·5 m/s or less) varies markedly with time of year, from about 30 per cent of all hours in November, December and January, to about 17 per cent in March, and around 20 per cent in the summer months. The diurnal variation of light winds during January and July at Renfrew is shown in Figure 35. Light winds are always most frequent at night, occurring on about 30 per cent of occasions in both January and July during the hours around sunrise. The speed of the wind usually rises after sunrise, but even in July the lowest frequency of light winds in the afternoon is around 5 per cent – probably accounted for by anticyclonic weather situations which give rise to generally light winds. In January on the other hand the average frequency of light winds in the afternoon is as high as 20 per cent at Renfrew. This is because of the situation in a broad but sheltered valley. On many nights a deep temperature-inversion forms, and it appears that on the following day the wind tends to be kept above the inversion-level by the hills on each side of the valley, so that the pool of stagnant air in the valley bottom is not easily disturbed. Because of the resulting high frequency of calms and light winds, the annual mean wind speed at Renfrew is less than that at Heathrow, in spite of a greater frequency of strong winds at Renfrew. Indeed, there is a marked difference between the shapes of the frequency distribution curves from various places (Figure 36).

A study of the maximum wind speeds measured at the various anemograph stations in Britain has shown that:

(1) On exposed western coasts the highest hourly mean wind speed experienced on average not more than once in 50 years reached about 36 m/s with maximum gust (approximately 3-second mean) speeds of 54 to 56 m/s (Figures 123 and 86).

(2) On exposed eastern coasts the highest hourly mean wind speed experienced on average not more than once in 50 years is in the range 30 to 34 m/s with gusts of about 45 to 50 m/s.

Further information on extreme speeds and the method of using the data for the design of buildings is given in Chapter 3, page 86.

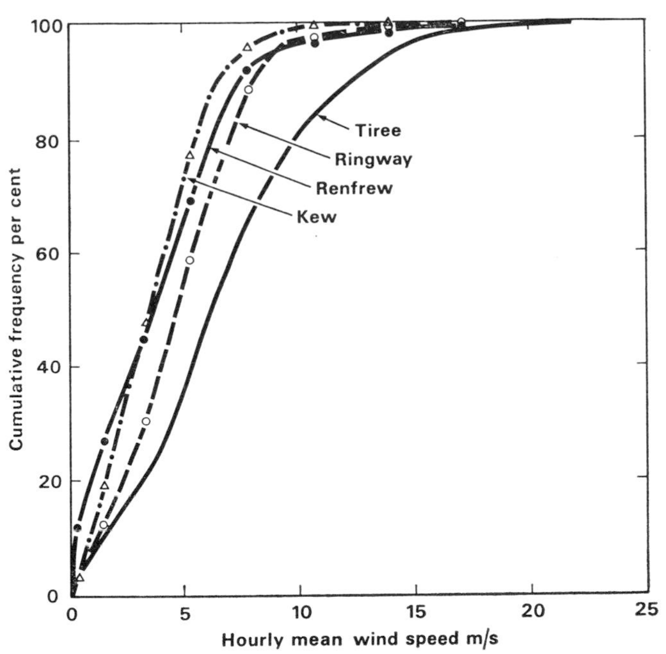

Figure 36 Annual frequency distribution of hourly mean wind speeds at: Tiree (effective height of anemograph, 13 m), Renfrew (12 m), Ringway (10–12 m) and Kew (15 m).

Wind direction and effects of topography

Statistics of the frequency, speed and direction of the wind at 35 places in Britain are given in *Tables of surface wind speed and direction over the UK* (Meteorological Office 1968). It is found that at most stations and over the year as a whole the dominant directions, are south, south-west or west. However the wind can blow from any quarter, and the wind-rose in Figure 37 shows that at Birmingham Airport, Elmdon, although on average throughout the period 1957–66 wind blew from directions 200–220° (approximately SSW)* for 14 per cent of all hours, it blew for at least 5 per cent of the time from each of the other directions. Furthermore, the speed was less than 0·5 m/s, or it was completely calm, for over 9 per cent of the time. Thus the expression 'prevailing wind' is a considerable over-simplification when applied to such a wind regime, which is typical of an open site in inland Britain, unaffected by prominent orographic features such as hills or deep valleys.

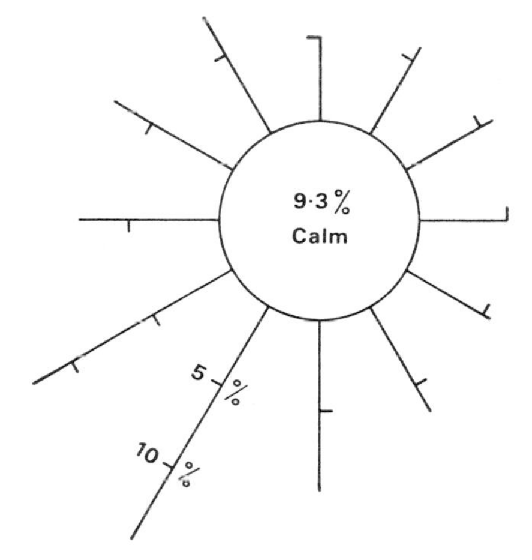

Figure 37 Wind-rose showing percentage frequency of hours during which wind was blowing from each of 12 directions (in 30-degree classes) at Birmingham Airport, Elmdon. Average for 10 years 1957–66.

Examples of other annual mean wind distributions are shown in Figure 38, which is an example of a method of displaying the combined frequency distribution of both wind speed and direction. It will be noted that at Glasgow Airport, Renfrew, during the 10 years 1956–65 there was on average 18 per cent of hours with light winds or calm.

Figure 38 shows clearly how much the dominant directions of airflow may be affected by local topography. At Birmingham, Elmdon, which is believed to be little influenced by local topography, the most common wind direction is about 210° (SSW) with a minor peak at 030–060°. At London, Heathrow, the peak around 210° is less pronounced than at Birmingham, while winds from between north and east are more frequent. This may indicate some degree of channelling up and down the Thames Valley. At Glasgow, Renfrew, the channelling effect of the Clyde valley seems to be quite important, with a well-defined concentration of wind directions centring on 240° (WSW) and 060° (ESE). The distribution of directions at Manchester, Ringway, on the other hand, is markedly different from the others. By far the most important wind direction is 180° (S), with lesser peaks at 270° (W) and around 090° (E). It seems that the usual south-westerly winds become southerly as they are deflected towards the north by the ridge of the Pennine Hills, the west edge of which is some 15 km to the east of the airport. Possibly the higher frequency of easterly and westerly winds at Manchester is caused by deflection of winds by the E–W ridge north of the city and by the more distant northerly edge of the Welsh mountains, or it may be a purely local valley effect.

It will be clear from this brief survey that wind direction at any place is influenced by local topography. The variations are such that they may be important for planning of towns and the siting of airports. There is a useful discussion of wind speed and direction in the London region in Chandler's (1965) *The Climate of London.*

*The usual wind-rose has lines which indicate by their length the frequency of winds from each of 8 directions, N, NE, E, etc. Fig 37 is based on an analysis by electronic computer in which it is more convenient to analyse in 12 thirty-degree groups centred on 0° (N), 030°, 060°, 090° (E), etc.

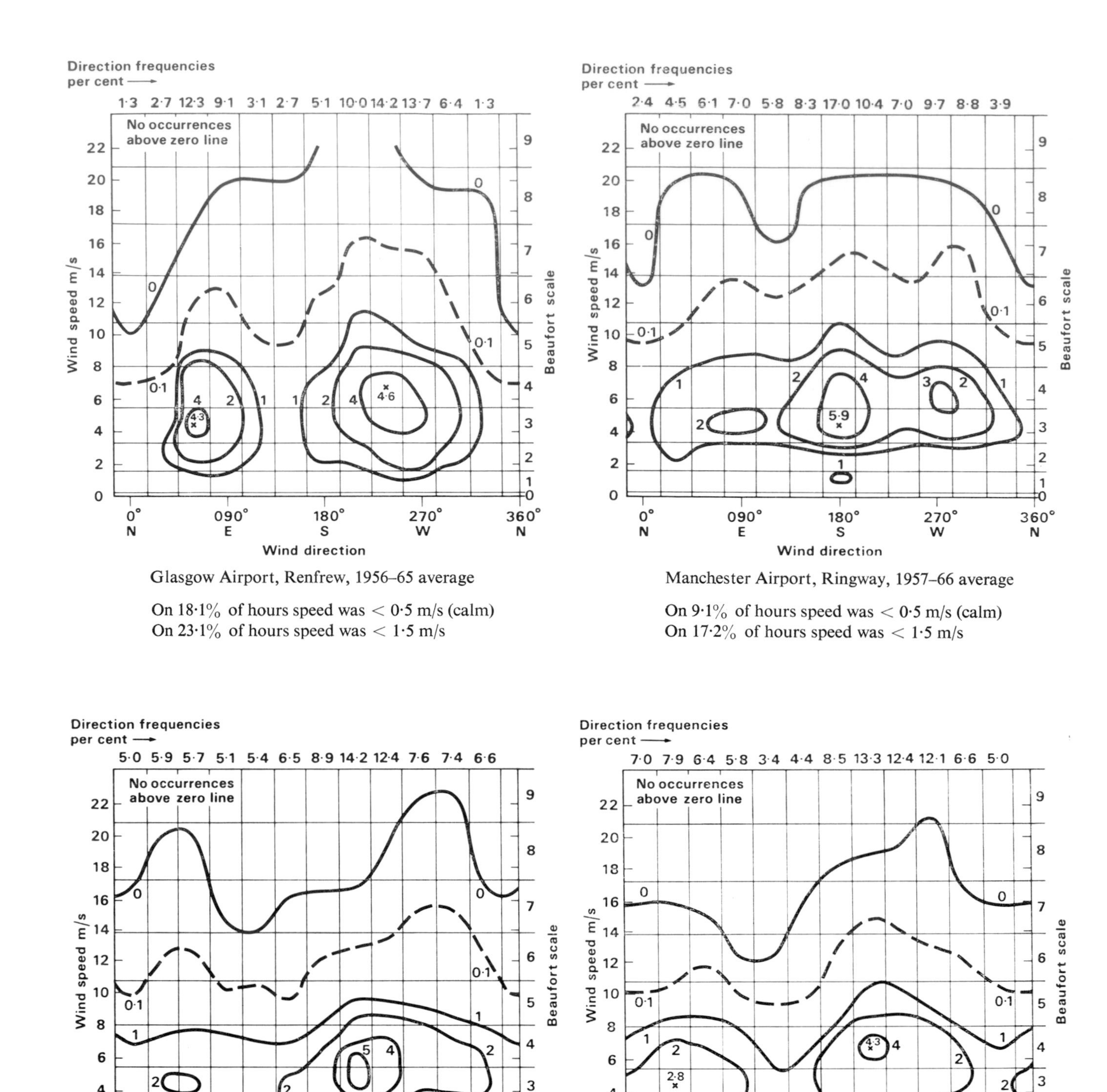

Birmingham Airport, Elmdon, 1957–66 average

On 9·3% of hours speed was < 0·5 m/s (calm)
On 16·8% of hours speed was < 1·5 m/s

London Airport, Heathrow, 1957–66 average

On 7·2% of hours speed was < 0·5 m/s (calm)
On 17·6% of hours speed was < 1·5 m/s

Figure 38 Wind speed and direction frequencies at four airports, all hours

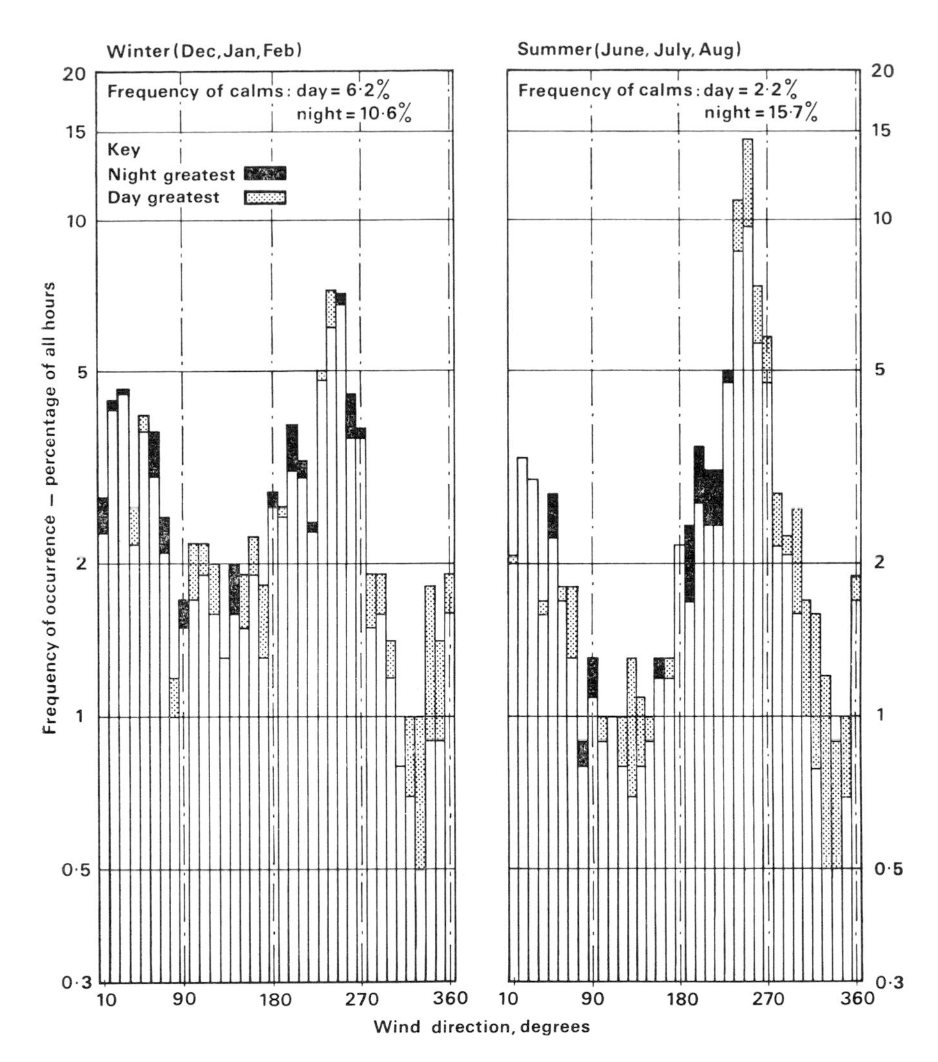

Figure 39 Frequency distribution of wind directions at Filton Aerodrome, Bristol. Means for period 1957–70. Data for day and night, summer and winter, shown separately. Day = 0700 to 1800, night = 1900 to 0600 GMT

There may be appreciable diurnal and seasonal variations in the direction of the wind. The effects of the daytime sea-breeze in warm weather and the corresponding land-breeze blowing towards the sea at night are familiar. Up-valley daytime winds caused by heating of the ground and the down-valley winds produced by night-time cooling may be less well known. These are generally relatively weak effects in Britain, only the sea-breeze being noticeable as a rule, and the effects of this may only be detectable more than a few kilometres inland with the aid of instruments (but see Chapter 2, page 71).

Figure 39 shows the results of an analysis of wind direction frequencies at Filton Aerodrome, Bristol. In this study each 10-degree direction estimate has been analysed separately and not grouped into classes of 30 or 45 degrees as is usually done. This finer analysis emphasises the sharpness of the peaks of wind-direction frequencies, and shows up the much more uniform distribution of direction in winter as compared with summer. Both in winter and in summer there is a greater frequency of winds from directions between west and north during daytime, than at night. There is a similar but weaker tendency for directions between east and south. It is not clear whether this is a local topographic effect or whether it is because there are more calms at night when the wind direction falls into one of these two classes. The frequency of night-time calms (speed less than 0·5 m/s) is much greater in summer than in winter at Filton, 15·7 per cent of hours against 10·6 per cent.

Similar analyses can be made for other places for which suitable records are available, although the great majority of anemographs are sited on airfields or other open places, well away from towns. Thus some skill is needed to interpret the results in terms of the conditions in inhabited areas. This is especially true when considering light winds. These will be more likely to follow tracks determined by topography than stronger winds. Thus, for example, a light wind from south-east at a particular place will not necessarily be carrying pollution originating from a source situated 10 or 20 km to the south-east of it. The air may be following a somewhat sinuous course so that its direction of movement at a given point may not be a good guide to the mean direction of its motion.

It should be noted that the pattern of wind-direction frequencies shown in diagrams such as Figures 37–39 may vary. In the years around 1970 it seems that the frequency of winds from approximately north and east increased at the expense of those from south and west. At the time of writing it is uncertain how long this anomalous spell will last.

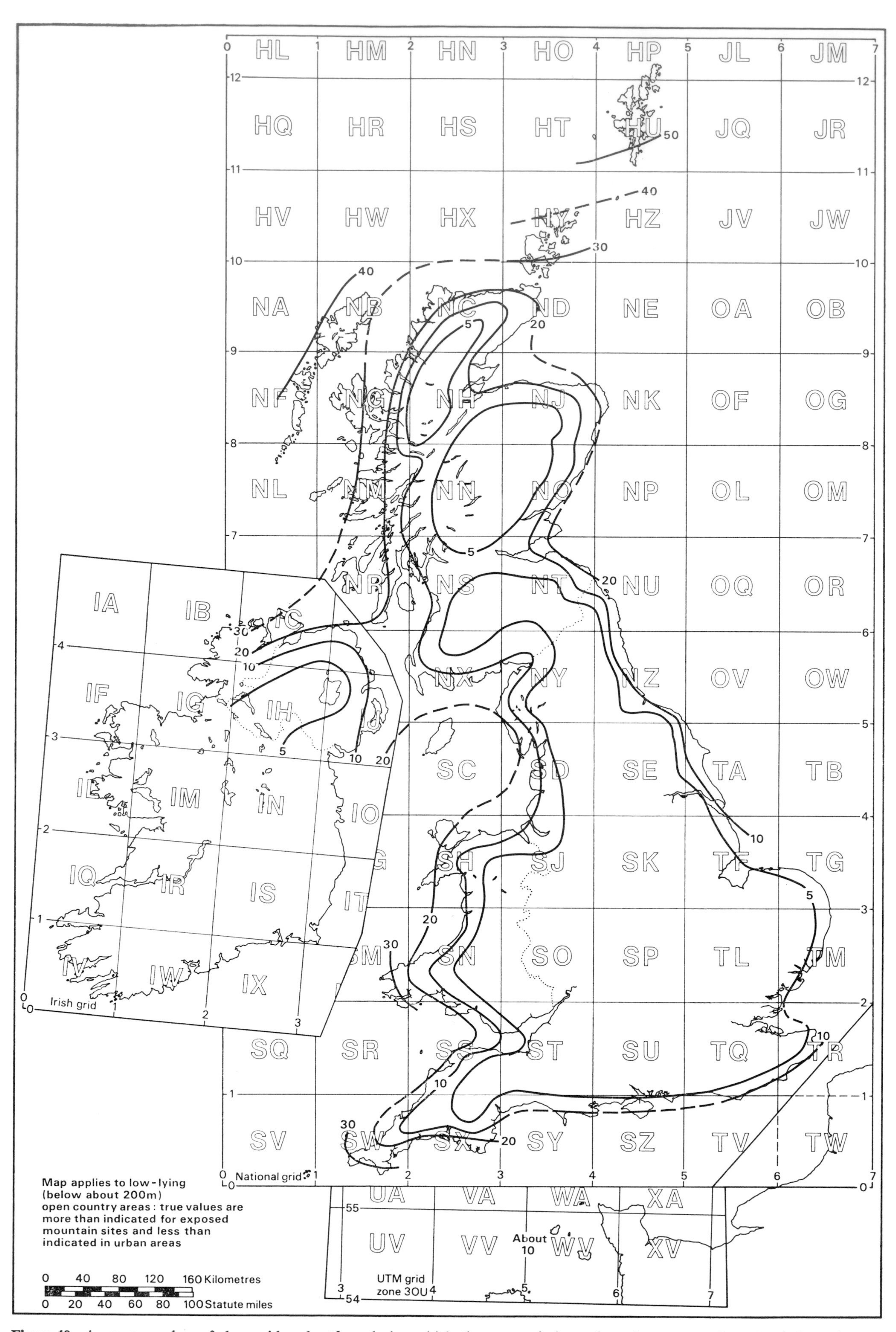

Figure 40 Average number of days with gale: (days during which the mean wind speed reaches or exceeds 17.2 m/s for at least 10 minutes), 1941-70.

Frequency of gales

Figure 40 shows the approximate average annual number of days with gale. If the mean wind speed exceeds 17·2 m/s for a period of at least 10 minutes, a day is recorded as a 'day with gale'. The map shows very clearly the effect of nearness to the coast and indicates that gales are most frequent on our western seaboard and least frequent well inland in England. Examination of data on the monthly frequency of gales shows that the maximum frequency at practically all places in Britain occurs either in December or January and that gales are least frequent in June or July. Gales may occur at any time of the year, however, and use of the forecasting services of the Meteorological Office is a wise precaution to take during construction work.

Sunshine and solar radiation

Measurement

The standard instrument for measuring sunshine in this country is the Campbell-Stokes sunshine recorder which consists of a glass sphere mounted in a frame made to hold specially printed cards. The sun's rays are focussed by the sphere on to the card and the total length of the burn (or burns) is compared with the time scale on the card to obtain the duration of bright sunshine. A new card is inserted each day. It is essential that the whole instrument should be securely fixed and properly adjusted and that it should be mounted in a position with an exposure which is as far as possible uninterrupted by obstacles; in the British Isles it is desirable to have a free horizon from north-east through east to south-east and from north-west through west to south-west, and to have no obstruction to the south with greater altitude than 6½° at latitude 60°N, or 16½° at latitude 50°N. Between south and south-west and between south and south-east the permissible altitude of an obstacle changes evenly between these limits and zero. Sunshine is rarely bright enough to record when the sun's altitude is below 3°, and obstacles below this altitude may, in general, be ignored.

The short-wave solar radiation falling on a horizontal surface is usually measured by means of a thermopile type of instrument such as the Moll-Gorczynski solarimeter. The receiving surface, which is shielded by two concentric hemispherical glass domes, is blackened and consists of alternate thin strips of manganin and constantin. These are connected to a suitable recording galvanometer or potentiometer and the electromotive force is proportional to the temperature difference between the central part of the surface and that of the rest of the instrument and its case; this temperature difference is a measure of the intensity of the radiation. The instrument is usually used to measure the *total* short-wave radiation falling on the horizontal, that is the diffuse sky radiation plus the direct solar radiation. By fitting a solarimeter with a shadow ring or disc it can be made to record the diffuse radiation alone. From the readings of the two instruments, the intensity of the radiation on surfaces of any orientation can be calculated.

A generally similar instrument is used to measure the solar radiation at normal incidence, but without the glass cover. It has shields so that only the direct beam of radiation reaches the element, and is fixed to an equatorial mounting and driven mechanically to follow the sun. Because of the extra mechanical complication of this instrument, it is only used at a few special observatories where it can be under constant supervision.

Table 6 Duration of possible sunshine (hours) on the fifteenth day of each month

	Latitude (North)					
	50°	52°	54°	56°	58°	60°
January	8.55	8.26	7.95	7.60	7.19	6.70
February	10·00	9·85	9·67	9·48	9·26	9·02
March	11·81	11·79	11·77	11·75	11·73	11·70
April	13·69	13·82	13·96	14·11	14·28	14·47
May	15·33	15·60	15·89	16·22	16·60	17·03
June	16·22	16·57	16·96	17·41	17·94	18·59
July	15·81	16·13	16·48	16·88	17·34	17·88
August	14·40	14·59	14·79	15·01	15·26	15·55
September	12·60	12·64	12·69	12·74	12·80	12·86
October	10·73	10·63	10·52	10·40	10·26	10·12
November	9·03	8·80	8·53	8·24	7·90	7·51
December	8·09	7·77	7·40	6·98	6·49	5·93
Year	12·20	12·22	12·23	12·25	12·27	12·30

A less accurate but simpler type of instrument, the bimetallic radiation recorder, may also be used to measure the total radiation on a horizontal surface. The receiving surface consists of two parallel rectangular white-coated strips of bimetal between which is placed a similar strip painted black. At one end the three strips are connected together while the other ends of the white strips are fixed to the frame of the instrument. The other end of the black strip is connected through a lever mechanism to a pen operating on a chart driven by a clock and so gives a continuous record.

Variation of length of day and times of sunrise and sunset with latitude and longitude

Table 6 shows the average duration of possible sunshine (length of day) on the fifteenth day of each month for every two degrees of latitude from 50°N to 60°N inclusive. It brings out clearly the important facts that in summer the day is much longer in the north than in the south – over two-and-a-half hours longer at Lerwick than at Scilly in June, and that in winter the day is longer in the south – over two hours longer at Scilly than at Lerwick in December.

Table 7 shows for the solstices and equinoxes the times of sunrise and sunset for every two degrees of longitude from 2° E to 8° W, the top part of the table referring to latitude 56°N (Edinburgh–Glasgow) and the bottom part to latitude 52°N (London–Birmingham). The table indicates, for example, how much later a start must be made in the west than in the east on work requiring daylight, and the necessarily very late start and short working day in the north in winter. The times given are in Greenwich Mean Time, one hour must be added to each one if British Summer Time is used. For example on 21 March at 2° E, sunrise is at 06 55 BST and sunset at 19 05 BST (7.05 pm BST).

Table 7 Times of sunrise and sunset at the solstices and equinoxes for latitudes 56°N and 52°N and for different longitudes
The times of sunrise and sunset in this Table are given in Greenwich Mean Time.

		Longitude				
		6°W	4°W	2°W	0°	2°E
a. Latitude 56°N						
21 March	sunrise	06 27	06 19	06 11		
	sunset	18 38	18 30	18 22		
22 June	sunrise	03 37	03 29	03 21		
	sunset	21 14	21 06	20 58		
23 Sept.	sunrise	06 09	06 01	05 53		
	sunset	18 23	18 15	18 07		
22 Dec.	sunrise	08 54	08 46	08 38		
	sunset	15 51	15 43	15 35		
b. Latitude 52°N						
21 March	sunrise	06 27	06 19	06 11	06 03	05 55
	sunset	18 37	18 29	18 21	18 13	18 05
22 June	sunrise	04 03	03 55	03 47	03 39	03 31
	sunset	20 48	20 40	20 32	20 24	20 16
23 Sept.	sunrise	06 10	06 02	05 54	05 46	05 38
	sunset	18 23	18 15	18 07	17 59	17 51
22 Dec.	sunrise	08 30	08 22	08 14	08 06	07 58
	sunset	16 14	16 06	15 58	15 50	15 42

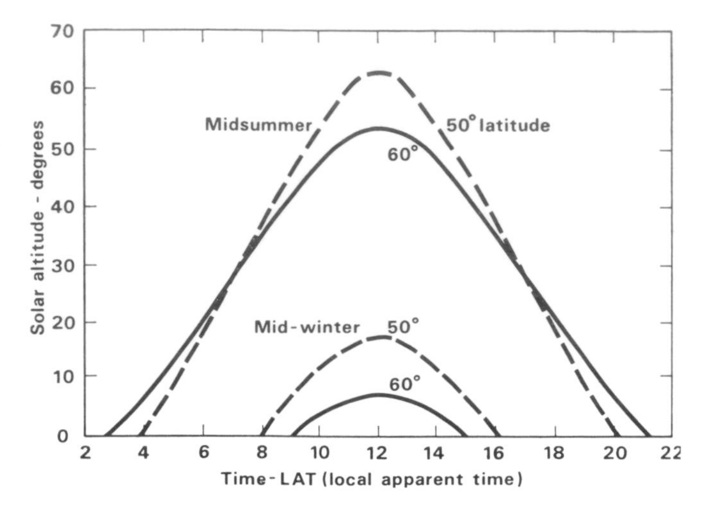

Figure 41 Midwinter and midsummer sunpaths at 50° and 60° latitude

Sunpaths

The path of the sun across the sky, as seen from any place, varies with the time of year, the length of the path being greatest at midsummer and least at midwinter. The position and length of the sunpath also varies with latitude, so that the maximum altitude of the midday sun is less at higher latitudes, even though the length of the path is greater in the summer (Figure 41). The position of the sun in the sky at any instant, and the angle made by its rays with any surface, may be calculated, but such calculations are tedious. For most practical purposes it is sufficient to use tables or diagrams.

Many different diagrams have been produced with these aims in view, and Figure 42 is an example of a diagram, drawn on a stereographic projection, showing sunpaths as seen from a place at latitude 55°N, and with an altitude and azimuth scale superimposed. Sunpath diagrams for 51, 53, 55, 57 and 59°N have been produced by Petherbridge (1969), together with transparent overlays for angles of incidence, radiation intensities, and transmittances of various arrangements of windows and sun-shading devices.

More recently, computer programs have been written, to produce tables of the requisite data. One program computes sunpaths, and causes them to be drawn on a stereographic projection by a curve-tracer. A second program can calculate, for any desired latitude, solar intensities on vertical, sloping or horizontal surfaces, making allowance for atmospheric clarity, cloudiness and ground reflection. Tables produced by this program have been printed in the *IHVE Guide 1970, Book A,* by the Institution of Heating and Ventilating Engineers.

Variations in sunshine across the British Isles

Figures 43–45 show average values of the daily duration of sunshine in hours over the British Isles in June, December and the year. At all seasons sunshine is greatest in southern coastal regions between Kent and Cornwall and least in the mountainous areas of Scotland. June is the sunniest month on the average, the mean daily duration of sunshine ranging from less than five to about eight hours, while December is the dullest, with mean daily duration everywhere below two hours and less than one hour in the mountainous areas.

In individual months sunshine totals may depart very considerably from the average values. In an exceptionally sunny month, such as July 1911 or June 1925, the daily average may exceed 12 hours at one or two places on the south coast. This represents about 75 per cent of the maximum possible duration. On the other hand in the Decembers of 1890 and 1912 the daily average amounts at Kew and Manchester, respectively, were only 0·01 hours, which is little more than one-tenth of one per cent of the maximum possible duration.

When using Figures 43–45 it should be remembered that they are based on measurements of sunshine made at generally unobstructed sites. Sunshine totals may be very much reduced at places which are obstructed by high ground or tall buildings, as shown, for example, in Figure 46, derived from records of sunshine at Garston in 1959–60. At some times of the year there was severe obstruction of the solar beam, especially near sunset. An extreme case would be the north-facing side of a valley running from east to west, which might receive no sunshine at all in winter (see Figure 58, page 60).

The variation in the average amount of sunshine during the day is also of interest. Over the year as a whole there seems to be a tendency at most places for the afternoon sunshine to exceed the morning sunshine by 5–10 per cent. At some stations, for example Kew, the difference is greatest in the winter months and this is probably due to the greater frequency of haze and fog near sunrise than near sunset, and also to the fact that there is a densely populated area to the eastward. At other places, of which Falmouth is typical, the difference is greatest in summer. At an inland station, such as Kew, the afternoon cloudiness is usually greater than the morning cloudiness in the spring and summer and this probably accounts for the fact that at Kew the afternoon sunshine does not appreciably exceed the morning sunshine in summer. At coastal stations on the other hand the afternoon cloudiness tends to be less than the morning cloudiness in the summer half-year and this could account for the appreciably greater amount of sunshine in the afternoons at such places (see also Chapter 3, page 132).

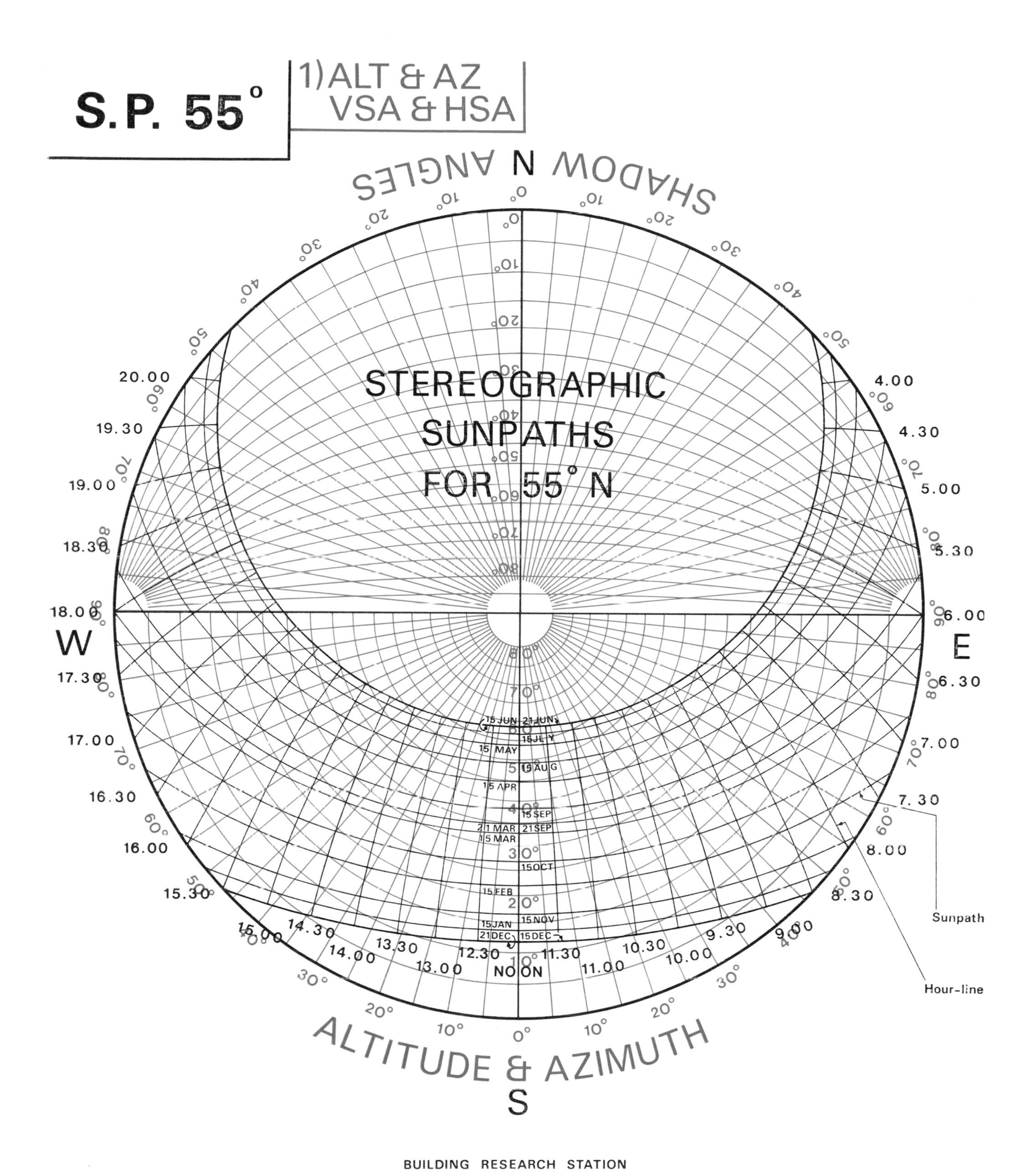

Figure 42 Stereographic sunpath diagram for latitude 55°, with altitude and azimuth scales superimposed. (Petherbridge 1969)

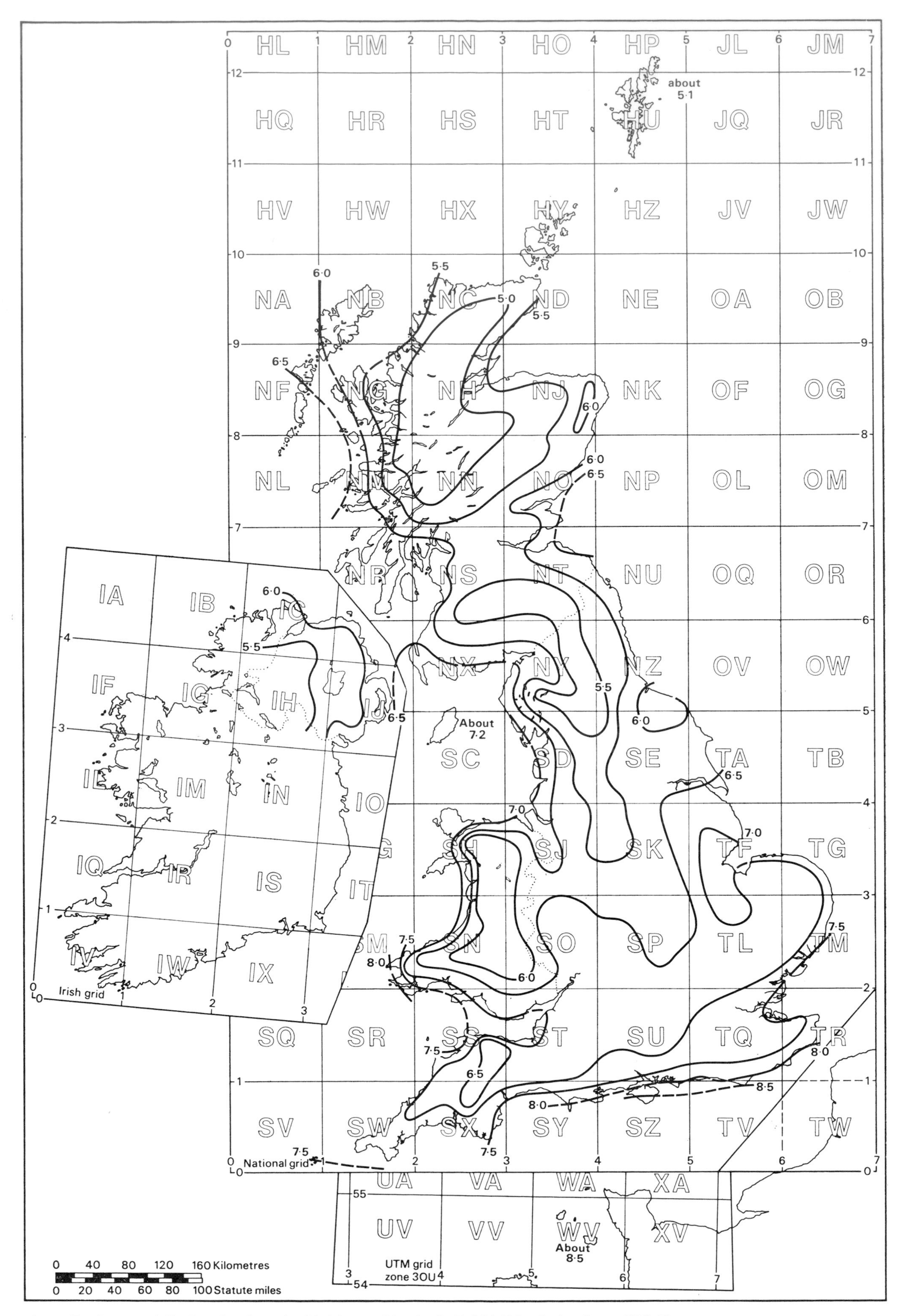

Figure 43 Average daily duration (hours) of bright sunshine in the British Isles during June, 1941–70

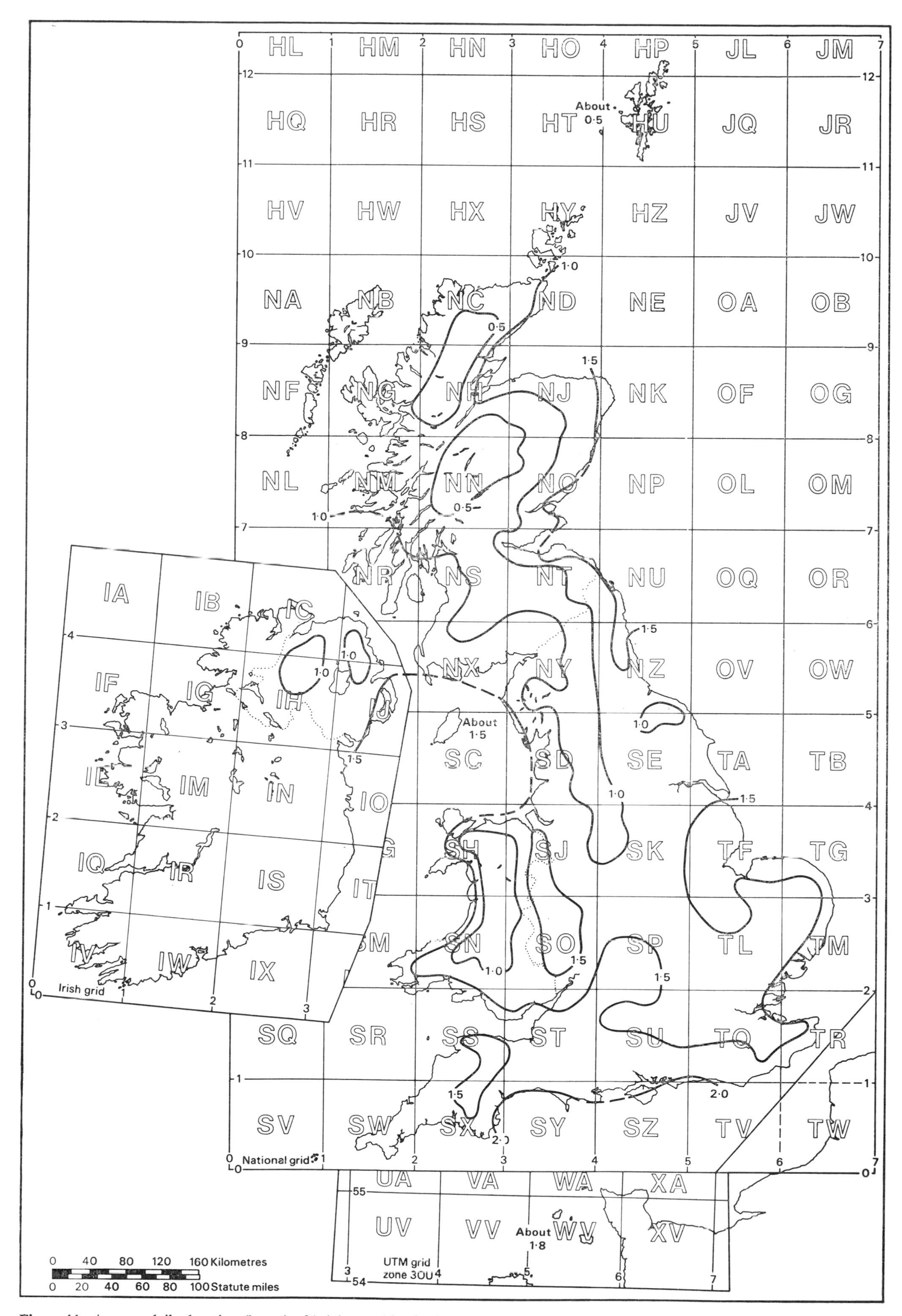

Figure 44 Average daily duration (hours) of bright sunshine in the British Isles during December, 1941–70

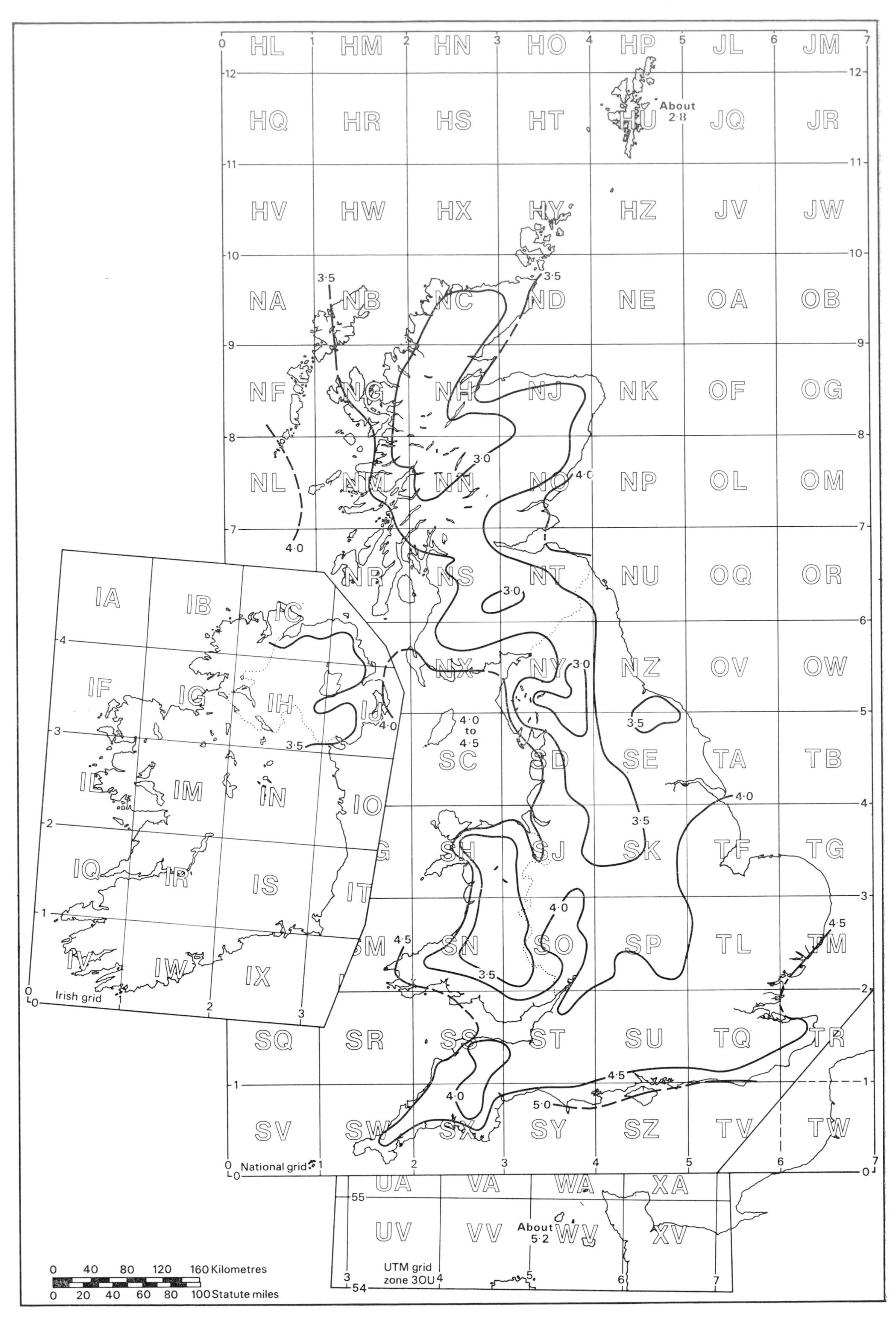

Figure 45 Average daily duration (hours) of bright sunshine in the British Isles for the whole year, 1941–70

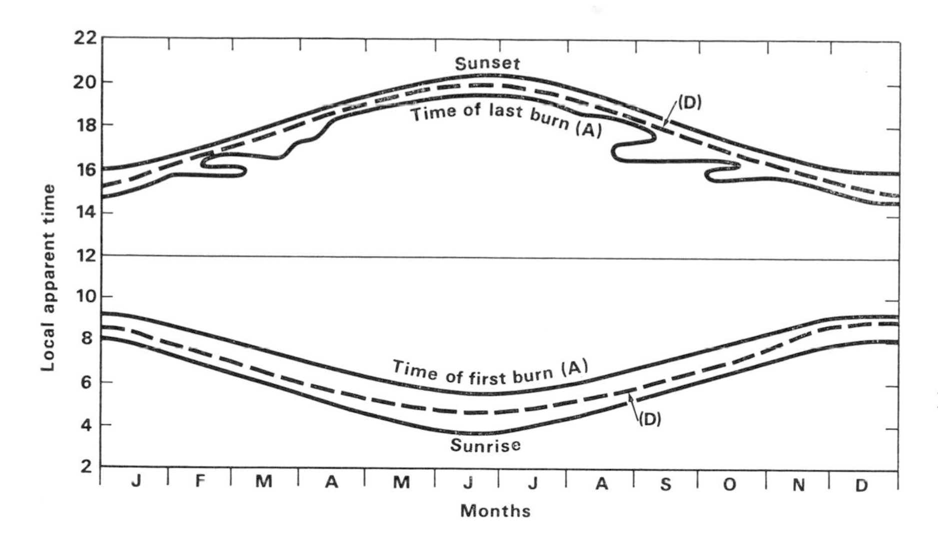

Figure 46 Diagram showing times of first and last burns on cards of the sunshine recorder at Garston, 1959–60, compared with true sunrise and sunset

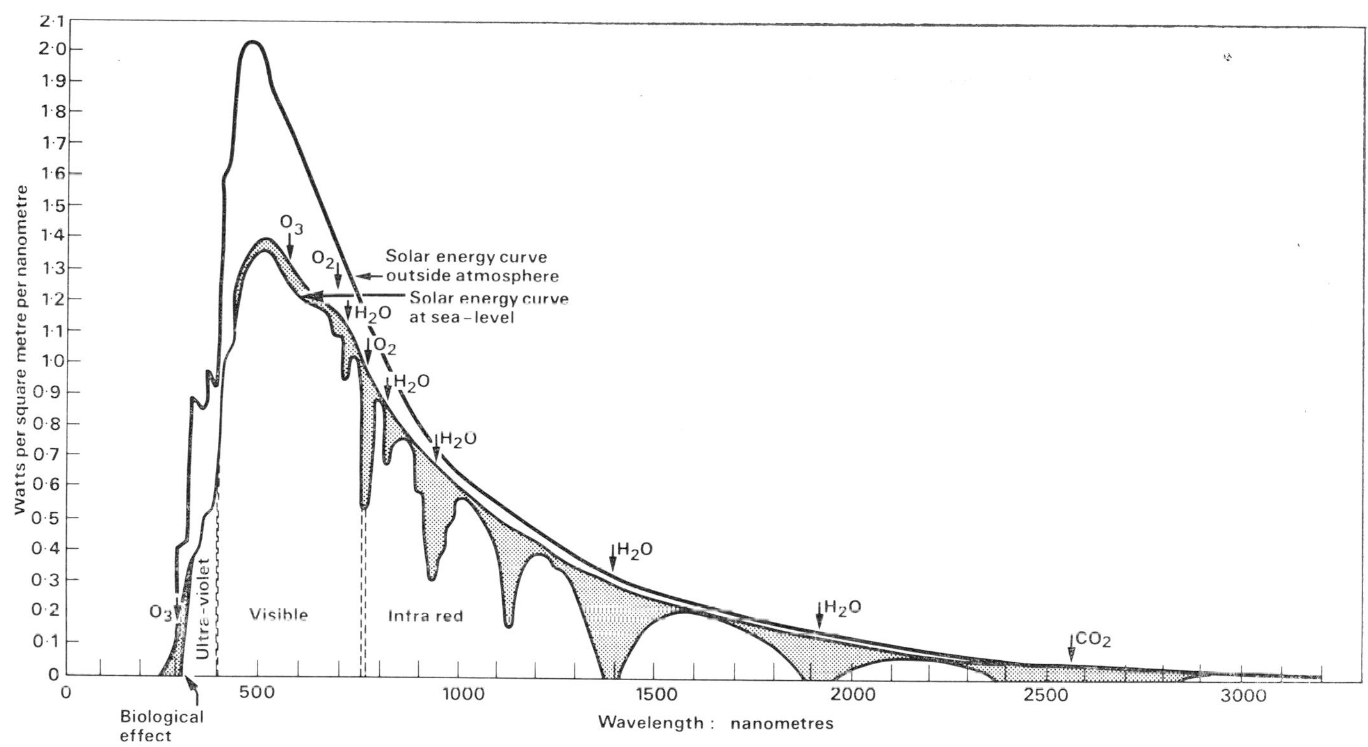

Figure 47 Curves showing the distribution of energy in the solar spectrum outside the atmosphere and at sea-level. The intensities in the latter curve are read from the lower edges of the darkened areas, which are due to selective absorption of the substances in our atmosphere which are indicated above the curve. (From Pettit, Edison: The sources of ultraviolet light. *Western Hospital and Nurses Review 12(5)*, January 1929. Corrected by Pettit on basis of later measurements, reported in Mount Wilson Contributions, No. 445 and 622, published in the *Astrophysical Journal,* vol. 75, 1932 and 91, 1940)

Solar radiation

Solar radiation is conventionally divided into 'short-wave' (in a band from about 0·29 to 4 μm wavelength) and 'long-wave' (in the band from 4 to 100 μm wavelength). The short-wave band contains about 99 per cent of the solar energy reaching the surface of the earth, about half of this being in the visible range, which has wavelengths from approximately 0·4 to 0·7 μm (Figure 47).

Radiation from the sun with a wavelength shorter than about 0·29 μm is cut off by the atmosphere, and most of that between this limit and about 0·4 μm (ultra-violet radiation) is excluded by ordinary window-glass. However, this ultra-violet radiation, although its intensity is relatively small, can have appreciable effects on some materials, particularly on paints and plastics (see Chapter 3), when they are exposed out-of-doors.

At the long-wavelength end, window-glass absorbs most radiation beyond about 3μm and is completely opaque to 'long-wave' radiation. As can be seen from Figure 47, the relatively smooth shape of the extra-terrestrial radiation curve is altered to a series of humps and hollows, especially at the long-wave end, because of absorption by molecules in the atmosphere. These molecules, mostly water vapour and carbon dioxide, absorb in definite bands of wavelength, and between these bands there is relatively little absorption of the radiation. It is because there is this absorption, and subsequent re-emission, of long-wave radiation, that the sky has an effective radiative temperature of some 20 deg C below air temperature at the zenith. (For example, if the mean air temperature is 10°C, the radiation received from the sky is the same as that which would be received from a black surface covering the sky with a temperature varying smoothly from

−10°C at the zenith to about +10°C near the horizon.) If these absorption bands did not exist, the effective temperature of the sky at night would be near to absolute zero.

The intensity of solar radiation

The intensity of the direct beam of solar radiation outside the earth's atmosphere is believed to be constant, or nearly so. The most recent measurements suggest that the intensity, on a surface normal to the rays, is 1·36 kW/m^2, at the mean distance of the sun (because the orbit of the earth is slightly elliptical, the intensity varies about the mean by about 3·5 per cent, being highest in the northern-hemisphere winter and least in the summer).

In the absence of an atmosphere, the intensity of the direct beam would be independent of the altitude of the sun above the horizon, but in fact the length of the path of the rays of the direct beam through the atmosphere increases as the altitude of the sun decreases. The greater the length of the path, the greater is the depletion of the solar beam by scattering and absorption in the atmosphere. That part of the scattering and absorption which is caused by dust and water vapour varies, according to the amounts of these materials in the atmosphere. Thus, in winter, when there is usually less dust and water vapour in the air than in summer, the intensity of the direct beam of solar radiation may be higher on a very clear day than it is on a similar day in summer, with the sun at the same altitude. However, in winter the maximum altitude of the sun is quite low, so that the highest intensity is actually experienced in summer, but at higher solar altitudes.

About half the radiation scattered by the atmosphere travels forward and forms the diffuse radiation from the sky. On an exceptionally clear day in Britain it forms about 15 per cent of the total radiation falling on a horizontal surface. Monteith (1962) has calculated the theoretical total radiation which can fall on a horizontal surface on clear days at latitude 51·5° and compared it with the radiation measured at Rothamsted, some 12 km north of Garston. Figure 48 is derived from his Figure 1, and the theoretical daily curves of radiation are compared with the measured curves on clear days in January, June and September. On the last two days, there is appreciable depletion of the total radiation in the middle of the day, which is presumed to be caused by the absorption and scattering of radiation by dust, pollen, spores and small insects carried up by convection in the air during the warmest part of the day. This effect does not seem to occur in winter when the soil is wet.

The exceptionally clear sky which can give rise to such a high intensity of solar radiation may only occur in Britain on about one day each year, and is perhaps most likely to occur in the months of May and June. For design purposes the intensity used will probably be a lower value than that experienced on the clearest days. In fact, the overlays produced by Petherbridge, which were referred to above, are based on measurements at Garston during those 5 per cent of days with the highest total radiation, and not on the extreme occasions.

For most purposes the radiation intensity on clear, cloudless days is the significant value, but the intensity of the radiation reaching the ground or a vertical surface may be much increased if there is a suitable reflecting surface. In particular, if a large cumulus cloud is suitably placed, the radiation reflected from it may increase the intensity of the total radiation by some 50 per cent. Such a high intensity will usually only persist for a few minutes, but it may cause a considerable rise in the temperature of a material which has a low heat capacity, such as a thin sheet of cladding. Normally, the total solar radiation received on a surface, averaged over an hour, is less when there is cloud in the sky than when the sky is free of cloud, or nearly so. It is possible to imagine a situation where a suitably placed cloud will give a large amount of reflected radiation for a long period, but such occasions will be so rare that they can be ignored for design purposes.

On average, the total solar radiation received on a horizontal surface on an overcast day is about one-third of the total received on the clearest cloudless day in the same month. On very dull days, the total received is much less than this. Figure 49 shows how the daily amounts of solar radiation received by a horizontal surface vary with the duration of bright sunshine in a typical month at Garston. The point marked by a cross (x) is the calculated clear-day total radiation, against the astronomically possible duration of bright sunshine on a cloudless day. The daily totals on partially cloudy days lie near the straight line joining this point and the mean overcast day value. There is always some variation about the line because of the effects of different types of clouds. For a given amount of cloud, the amount of scattered solar radiation varies with the type of cloud.

It is possible to classify the sky conditions into a number of categories, according to the type and amount of cloud present. This has been done by Lumb (1964) who developed equations to calculate the solar radiation under these conditions from

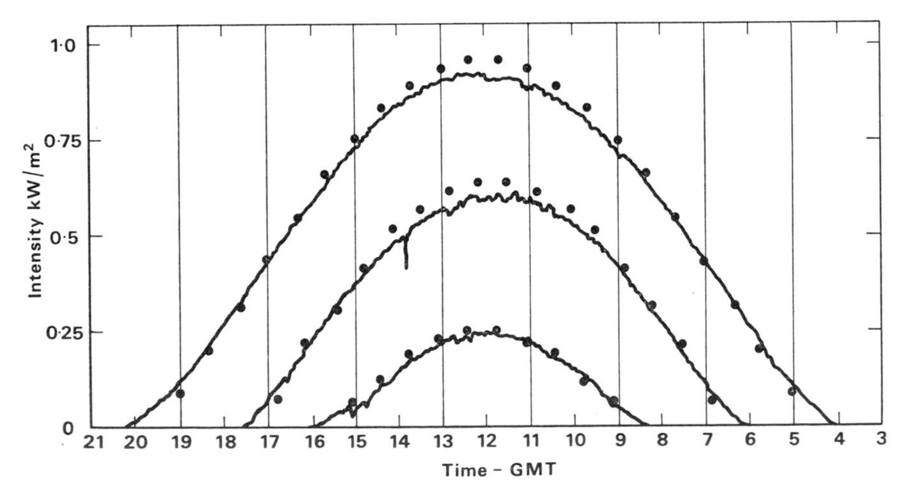

Figure 48 Calculated (dots) and measured diurnal variation of solar radiation on a horizontal surface at Rothamsted, on clear days in January, June and September. From a photograph of three superimposed potentiometer records. The time scale runs from 03 00 GMT (right-hand axis) to 21 00 GMT (left-hand side). (From Monteith 1962)

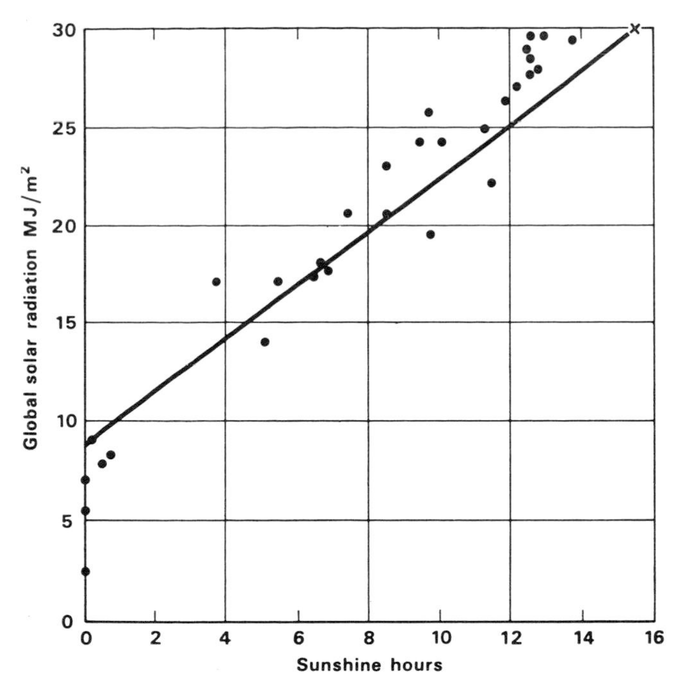

Figure 49 Relation between daily duration of bright sunshine and the total solar radiation on a horizontal surface at Garston in a summer month

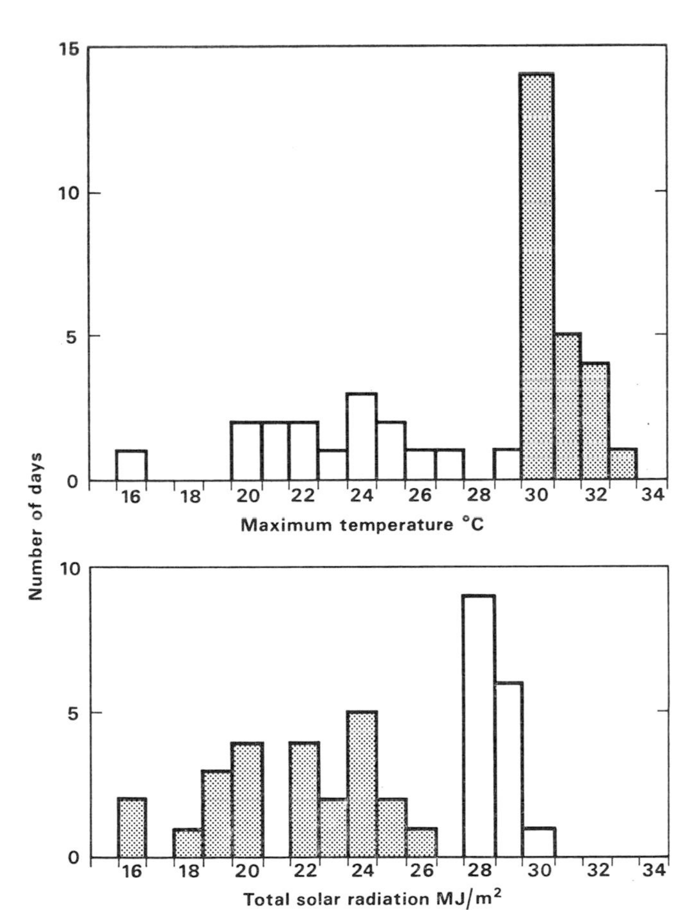

Figure 50 Frequency in 13 years, 1947–59, at Kew, of very hot days (maximum temperature greater than 29·6°C – shaded columns) and days of high global radiation (28·2 MJ/m² or more – open columns)

Table 8 Cloud categories and effective transmission factors

Category no	General description	Effective cloud transmission factor
1	Virtually clear sky (less than 2/8 cloud)	1·0
2	Well-broken low cloud (3/8 to 5/8), little or no medium or high cloud	0·8
3	6/8 to 8/8 cirrus-type cloud, but no cirrostratus	0·75
4	6/8 to 8/8 thin layers of medium cloud	0·7
5	Veil of cirrostratus covering the whole sky	0·6
6	7/8 to 8/8 stratocumulus, with or without some cumulus, little or no medium cloud	0·47
7	6/8 to 8/8 thick medium cloud, with or without layers of low cloud	0·35
8	Thick overcast of low cloud, perhaps with layered medium cloud, usually with drizzle	0·21
9	Thick overcast of low cloud, probably also thick layers of medium cloud, usually with rain	0·17

Columns 1 and 2 after Lumb (1964), column 3 after Loudon (1964).

the clear-sky values. The method was simplified by Loudon (1964), who showed that a simple multiplying factor can be used for each category of sky, without significant loss of accuracy.

Lumb's table of cloud categories is reproduced as Table 8, with Loudon's effective cloud transmission factors added. The accuracy of estimating 5-day totals of solar radiation on the horizontal by this method is usually better than 10 per cent, and of one-day totals better than 20 per cent.

The brightest days are not necessarily the hottest days. Indeed, a study of data from Kew Observatory showed that in the 13 years 1947–59 there were 24 days on which the air temperature exceeded 29·5°C, and 16 days on which the total global solar radiation (on a horizontal surface) exceeded 28·1 MJ/m². These 24 hottest days did not include any of the 16 days with high solar radiation. Figure 50 shows (above) the frequency distribution of the maximum temperatures on the 40 days. On the days with high radiation the maximum temperature ranged from 16° to 29°C, while the temperatures on the hot days were more closely grouped, in the range 30° to 33°C. The lower diagram in Figure 50 shows the distribution of daily global solar radiation amounts on the same days. On the hottest days there was a wide spread of radiation amount, from 16 to 26 MJ/m², while the radiation amounts on the bright days were closely grouped in the range 28 to 30 MJ/m². Clearly it would be quite wrong to design a cooling system to take account of the simultaneous occurrence of very high temperatures and solar radiation amounts.

The diurnal variation of global solar radiation intensity, air temperature, atmospheric dew-point and of wind speed on

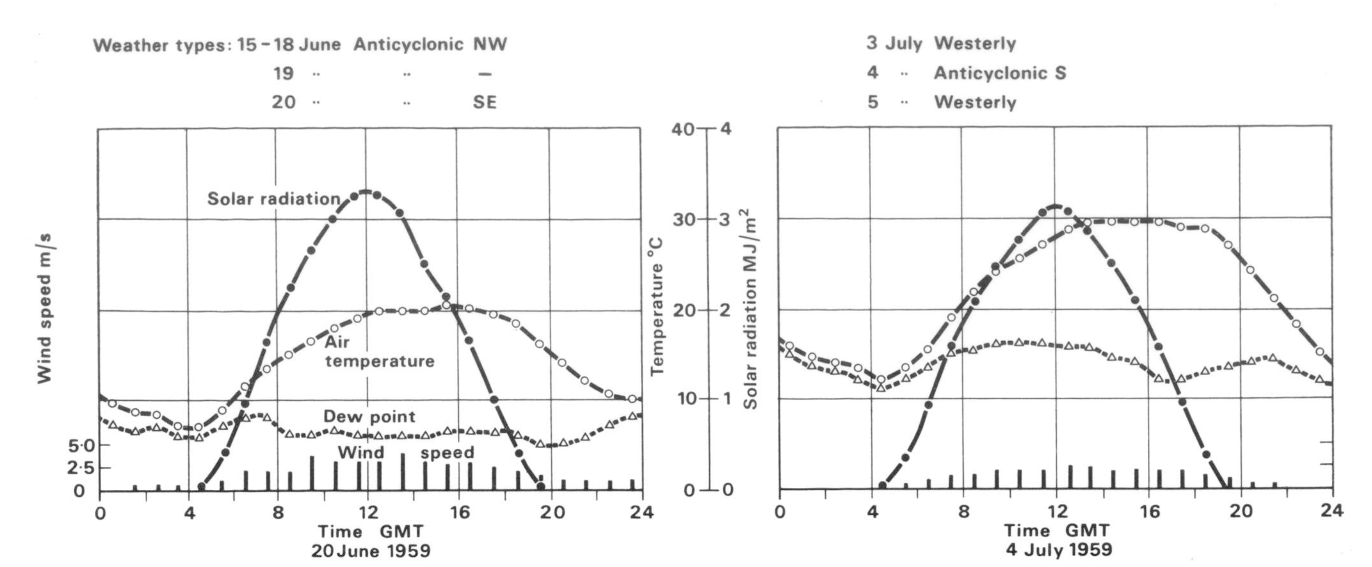

Figure 51 Diurnal variation of air temp, global solar radiation, dew-point and wind speed on two cloudless summer days at Garston in 1959. On 20 June the atmosphere was very clear with a high intensity of radiation. The dew-point averaged about 6°C and the air temperature rose from 8° at sunrise to 20° in the afternoon. On 4 July the air was not so clear and radiation intensities were about 6 per cent lower. The dew-point was between 11 and 16°, and the air temperature rose from 12° at sunrise to nearly 30° in the afternoon. Because of the higher moisture content of the air on 4 July, more long-wave radiation was re-radiated back to the earth, so that in spite of the reduced short-wave radiation intensity the air temperature rose far above that on 20 June

Table 9 Cumulative percentage frequencies of total amounts of solar radiation on horizontal, on sunless days in December

Total mWh/cm²	Lerwick 1956–64	Eskdalemuir 1956–64	Cambridge 1956–64	Aberporth 1957–64	Kingsway 1956–64	Kew 1956–64
5 or less	19	4	0	1	3	2
10 ,, ,,	47	20	4	6	9	16
15 ,, ,,	73	41	18	15	29	30
20 ,, ,,	93	65	36	24	42	51
25 ,, ,,	99	72	50	37	60	58
30 ,, ,,	100	86	62	41	71	74
35 ,, ,,		92	76	60	77	84
40 ,, ,,		96	83	72	82	87
45 ,, ,,		98	89	85	87	89
50 ,, ,,		100	93	93	95	93
55 ,, ,,			93	94	97	96
60 ,, ,,			94	95	98	97
65 ,, ,,			96	97	99	99
70 ,, ,,			98	97	99	99
75 ,, ,,			99	97	99	99
80 ,, ,,			99	97	100	99
85 ,, ,,			100	99		99
90 ,, ,,				99		99
95 ,, ,,				99		100
100 ,, ,,				99		
105 ,, ,,				100		
110 ,, ,,						
Average number of sunless days	15·3	14·2	13·9	10·3	15·4	15·4
Average radiation on these days mWh/cm²	11·2	19·3	28·2	32·3	25·6	24·4
Average ratio: diffuse/total (all days)	0·81	0·71	0·68	0·65	0·74	0·69
Lowest daily total in period mWh/cm²	0	1	7	4	3	4

Note: 1 mWh/cm² = 36 kJ/m².

two typical days at Garston, Hertfordshire, are shown in Figure 51. On both days the sky was cloudless, but on 20 June 1959 the air was relatively dry, while on 4 July 1959 it was more humid. The radiation intensity was rather higher on the former day, but air temperature did not exceed 20°C. On the second day the temperature rose to nearly 30°C, because the large amount of water vapour in the air emitted a great quantity of long-wave radiation back to the ground (the so-called greenhouse effect). On days with little water vapour in the air, more heat can escape by radiation to the sky, and prevent the air temperature from rising as high as it will on humid days. It will be noted that on 4 July the sultriness of the day was increased by the relatively low wind speed.

Solar radiation and illuminance on sunless days

A sunless day is one on which the ordinary sunshine-recorder registers no bright sunshine, although the sky may not be completely overcast during the whole of the day. In the winter, when the altitude of the sun is small, the direct rays of the sun may not penetrate holes in the cloud, because of the thickness of the layer of cloud. In summer, this is less likely to happen, except around sunrise and sunset.

There can be quite a large range of thickness in the clouds when there is a complete overcast and no bright sunshine is recorded. Because thick clouds reflect and absorb more solar radiation than thin ones, the total radiation received at a given place on any one day can vary over a wide range. Table 9 gives the frequencies of occurrence of daily amounts of total solar radiation (direct plus diffuse on the horizontal) at six places in Britain in the month of December. The data are grouped into classes of 5 mWh/cm^2 (equivalent to 180 kJ/m^2). The frequencies in each class are expressed as percentages of the total number of sunless days in eight or nine Decembers at each place, during the period 1956–64 (1957–64 at Aberporth).

From these figures, the daily total solar radiation which is exceeded on half the sunless days has been estimated for four of the places, Lerwick, Eskdalemuir, Aberporth and Kew (Table 10).

Table 10 Daily totals of solar radiation which are exceeded on half the sunless days in December at four stations. Averaged over 8 or 9 years

Station	Latitude	Period	Global solar radiation	
			mWh/cm^2	kJ/m^2
Lerwick	60·2°N	1956–64	11	400
Eskdalemuir	55.4°N	1956–64	17	610
Aberporth	52·2°N	1957–64	31	1120
Kew	51·5°N	1956–64	20	720

On the assumption that the hourly mean irradiance (intensity of radiation) on a horizontal surface on the sunless day in winter takes the form $I_h = (a + b\theta)$
where θ is the altitude of the sun above the horizon (degrees) and a and b are constants, the values of the constants have been found. The median total daily solar radiation data for Lerwick and Aberporth given in Table 10 were used in the computation, so that the resulting equation is

$$I_h = 0{\cdot}994 + 4{\cdot}133\theta \text{ W/m}^2 \quad (3)$$

If a luminous efficacy of 115 lm/W is assumed, the corresponding hourly mean illuminance (illumination) on a horizontal surface is

$$E_h = 114{\cdot}3 + 475{\cdot}3\theta \text{ lm/m}^2 \text{ (or lux)} \quad (4)$$

From this equation, the hourly mean illuminance at each hour of the day has been computed for a range of latitudes from 51° to 60°N, and is shown in Table 11. Because the basic data used to compute these values came from two stations which normally have exceptionally pure atmospheres, almost free from man-made pollution, the values in Table 11 are appreciably higher than those which are actually experienced at most inland areas in Britain. Measurements of radiation at Eskdalemuir, a high-altitude station in the Southern Uplands of Scotland, agree quite well with the computed values, but at Kew, the median sunless-day radiation total is only about two-thirds that at Aberporth, which is at almost the same latitude. Kew itself is in the western outer suburbs of London, and not in an exceptionally polluted neighbourhood. It is to be expected that in the centre of towns and highly industrialised areas the total solar radiation intensity, and hence the illuminance, will be half or less than the values given in Table 11.

In daylight calculations, an outdoor illuminance of 5000 lux is taken as the design value, which gives an indoor value of 100 lux when the daylight factor is 2 per cent. It can be seen

Table 11 Calculated illuminance in lux (lm/m^2) on a horizontal surface on a median sunless day in December, at a place with exceptionally clear atmosphere. Based on data from Lerwick and Aberporth

Latitude °N	Day length *n* hours	Time (local apparent time) 08 30/15 30 lux	09 30/14 30 lux	10 30/13 30 lux	11 30/12 30 lux
51	8·4	1400	3900	6400	7200
52	8·3	1100	3600	5900	6800
53	8·1	750	3200	5500	6300
55	7·7	100	2500	4600	5400
56	7·6	—	2100	4100	5000
57	7·4	—	1700	3700	4500
59	6·9	—	1000	2800	3600
60	6·7	—	600	2400	3100

from Table 11 that, even in an unpolluted atmosphere, the median outdoor intensity on sunless days in December barely reaches 5000 lux at 56°N, and does not attain this value further north (although it may do so on most of the 50 per cent of sunless days with higher radiation totals than the median days). Since many places in Scotland have about 15 or more sunless days in December (Figure 52) it can be expected that north of about 56° (which is the latitude of Edinburgh and Glasgow) there will be about 7 days in that month on average, on which the illuminance does not reach the design value. In the towns, more days still will not attain this value. At a given place, the illuminance will be approximately proportional to the total solar radiation for the day, assuming a uniform overcast all day. Thus on densely overcast days the illuminance may be less than 20 per cent of the values shown in Table 11.

In the more favoured areas of the coast of south and south-west England, south-west Wales and eastern Scotland (see (Figure 52) less than one-third of the days are sunless in December–January. It is likely that in these places the illuminance values of median sunless days are near those predicted by Table 11, except perhaps within the larger towns. In southern areas of Britain, daylighting design is based on the actual average illumination levels recorded at Kew over a period of 10 years (Illuminating Engineering Society 1962).

Atmospheric pollution

It is arguable whether atmospheric pollution should be considered as one aspect of climate. Certainly the effects of a given emission of pollution on the surroundings of the source are influenced markedly by the weather. However the subject is so big that the following notes are restricted to a brief account, and for full details the reader is referred to the comprehensive study by Meetham (1956).

Although there is always some natural pollution of the atmosphere, by various salts from sea spray, by dust blown up from soil, and by airborne spores, insects and so on, the term pollution is usually intended to refer to man-made materials. This artificial pollution is derived from a great variety of sources, and consists of a correspondingly large number of different materials.

The larger particles of grit and dust settle quickly, and therefore a large proportion of them fall to earth close to the chimneys from which they issue. According to Meetham (1956) the amount deposited each month may exceed 10 g/m^2 in towns, but is much less than 0·4 g/m^2 over country districts. In extremely polluted districts the fall of ash is hundreds of times as heavy as in the country.

On the other hand, smoke, gases and soluble matter are much more uniformly distributed, for unless they settle out in fogs or are washed out by rain they will travel mixed with the air. There may be only five to ten times as much dissolved sulphate or chloride deposited in an extremely industrialized area as in the country. Because the smoke and gases can travel for very long distances (hundreds of kilometres downwind of the source), their concentration in the air is comparatively uniform. Meetham shows maps of their concentration over the country. The summer map shows 10 to 20 $\times 10^{-6} g/m^3$ of sulphur dioxide in surface air over open country, and the winter map shows some two or three times this amount.

However, with the changing patterns of fuel consumption, and of the distribution of industry, it is to be expected that the picture will be changing continually and the most up-to-date figures should be obtained.

Notes on the variation of atmospheric pollution in different kinds of weather are given in Chapter 2, page 78. It is particularly important to note that the direction of drift of pollution, as well as its concentration, may depend very much on the type of weather.

Sea temperatures

The mean temperature of the surface waters of the seas around Britain generally is about 7 to 9°C in January. However, close inshore the mean temperature is 5°C or less in this month along much of the east coast, along the coast of Lancashire, and in the Severn estuary. It may fall to near −2°C in an exceptional winter near the south-east coasts, and small ice-floes may form, as happened in 1947.

In July–August the general sea-surface temperature averages about 12 to 14°C, but close inshore along parts of the southern and western coasts of England the average exceeds 16°C. However, the mean temperature of the surface waters in a warm August can reach about 19°C in more sheltered parts on the south and west coasts of England, depending on the depth of the water – higher water temperatures occurring in shallow bays. It is unlikely that even on very hot days the sea temperature ever exceeds 20°C. The data in these paragraphs are given by Lumb (1961), to whose paper reference should be made for more detailed information.

In addition to these natural variations, there may be local increases in temperature where large power stations discharge their cooling water.

Local climatological considerations

Local variations in climate due to topography

Superimposed on the regional climate of a district are local variations due to topography, type of soil and vegetation cover, proximity to coasts, lakes, forests and so on. In a region of diverse topography such local variations may be particularly marked. At night in quiet weather the surface air, cooled by contact with the ground, moves downhill and this may eventually produce large temperature differences, depending on elevation and on whether or not the downward flow of cold air is obstructed. During the day on the other hand slopes having different aspects may receive very different amounts of solar radiation especially when the sun's elevation is low. In disturbed weather, too, such regions may exhibit big variations in wind speed between sheltered narrow wooded valleys on the one hand and the tops of smooth hills and ridges on the other.

Hill and valley climates

Hill tops of moderate height are often sunny but their mean temperatures are a little lower than those of a neighbouring valley. (It is worth remarking that the mean winter temperature of a place on a hill about 200 metres above sea-level in south-central England will be about the same as that of a place near sea-level in southern or eastern Scotland – see Figure 4.) Temperatures are however less extreme on hill tops, maximum temperatures on warm sunny days being moderated by the breeze while minimum temperatures on clear nights are much higher because the colder air drains away downhill. When hills are more than about 150 metres above the surrounding country they are sometimes enveloped in cloud, and rainfall may be higher than in the valley. Also snow tends to lie longer and due to the stronger winds drifting may be serious especially on hill roads in cuttings. Wind speeds generally increase with height and hill tops tend to be more windy than lower-lying places. Buildings on hills are more liable to damage and require more protection against driving rain than similar buildings

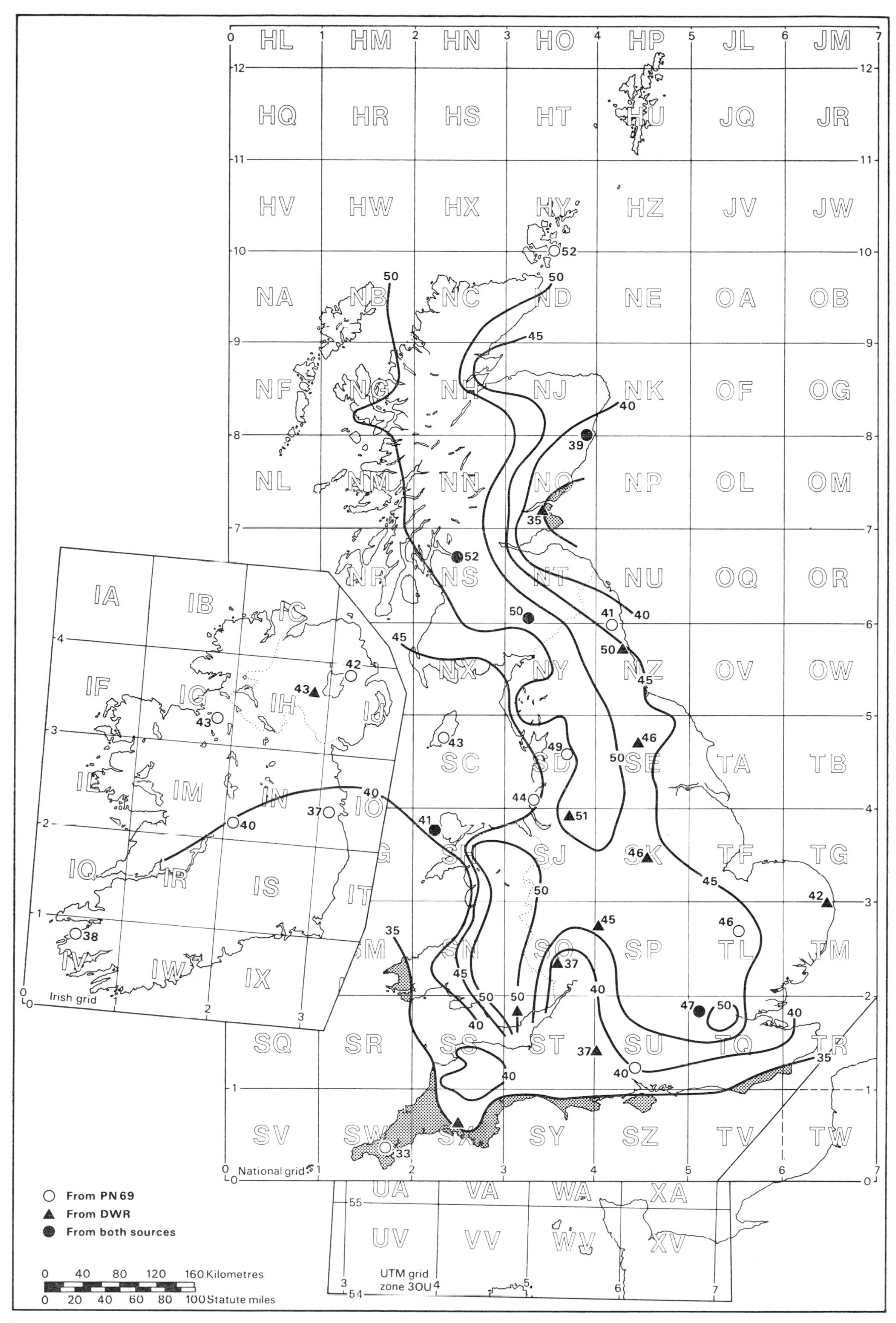

Figure 52 Percentage frequency of days in the British Isles during December and January with no bright sunshine.

on less exposed sites. The mean wind speed at a height of about 10 metres above a ridge which is 75 metres above the surrounding country and at right angles to the wind may be as much as 50 per cent higher than that at 10 metres over the plain. The wind speed over the summit of an isolated hill of the same height would probably be less than this. A position on the lee slope of a ridge or range of hills is not necessarily sheltered from the wind (see page 70 concerning lee-waves). Even on a small scale there may be a region on the lee slope of a small ridge where the wind flow is unusually turbulent and this probably occurs where the flow that has broken away from the top of the ridge rejoins the ground surface.

The climate of a valley is very dependent on the extent to which its topography helps or hinders the flow of air. When in clear quiet weather sloping ground is cooled by radiating to the sky at night the air nearest the ground is cooled rapidly and, being dense, flows down the slope. If at the bottom of the slope there is a hollow or valley bottom with no outlet to still lower ground, the cold air will accumulate and much lower temperatures will be reached than over flat ground in the same area. These katabatic (down-flow) winds attain speeds of little more than 1 m/s but when they flow into an open valley they may feed a down-valley wind whose speed may attain 2 m/s or more. They are likely to be strongest over snow-covered or grassy surfaces because these are the surfaces which cool most rapidly. The effects of a steady flow of cold air down a slope can sometimes be mitigated by placing a barrier such as a wall or thick hedge diagonally across it and by avoiding damming the flow at any point. For example, a terrace of houses along a contour might form an effective dam and thus encourage the formation of a deep pool of cold air on the upslope side during quiet frosty weather.

During the early morning on quiet clear days eastward-facing slopes will be the first to be warmed by the sun's rays and the air just above them will begin to flow up the slope. Eventually the whole of an open valley will become warmed in this way and a flow of air will develop up its whole length. The most favourable position in a valley is somewhere between the floor and the hill tops, provided that north-facing slopes are avoided. In windy weather the winds tend to blow more strongly along a valley if it runs in the general direction of the wind. Such 'funnel' winds probably occur in a number of east-west valleys in northern England and Scotland and their occurrence may be indicated by the presence of bent and wind-pruned trees.

Type of soil and vegetation cover

Different soil types affect the local climate because of their different thermal capacities and conductivities, both of these physical properties being increased by increased moisture and by compaction. Thus dry sandy soils have the greatest range of surface temperature being coldest at night and warming up most rapidly during the day. Wet compacted clay soils show the smallest variations. However the dry sandy soils are cooled to a lesser depth than the wet clay soils, during a cold spell. The effect of nature of soil may be greatly modified by a cover of vegetation. Grass, because of the layer of air trapped between its stems, favours lower minimum temperatures at the surface, immediately above the grass. A covering of snow also favours very low surface minimum temperatures at its upper surface, but permits only small changes of surface temperature by day because of its high reflectivity and its absorption of latent heat when the surface reaches melting point. Grass and snow cover both reduce the depth to which frost can penetrate into the ground, frost penetration being greatest in compacted moist heavy soils without snow or vegetative cover, and under pavements cleared of snow.

Coasts, lakes and rivers

Although the coast usually has cleaner air and more abundant sunshine the wind strengths are sometimes excessive. Exposure to on-shore winds, and in quieter weather the development of sea-breezes, are probably the main contributors to the marked climatic differences between the coastal strip and inland areas. In Britain sea-breezes are most vigorous along the east coast in spring and summer when the North Sea is relatively cool and their onset may bring a marked lowering of air temperature and sometimes an invasion of sea fog or very low cloud. The influence of lakes and rivers depends on their extent, depth and movement, but a good-sized area of deep water has a moderating effect on temperature and may provide some protection against frosts by encouraging breezes due to its temperature contrast with the surrounding land. Disadvantages of the proximity of water are the possibility of flooding and a tendency to encourage insects if the water is shallow.

Woods and trees

The influence of woods and trees is complex. Their main effect on temperatures is to decrease the diurnal variation. They also tend to increase the relative humidity and to decrease wind speeds. The latter effect may be put to good use by the judicious planting of shelter belts to act as wind breaks on otherwise exposed sites.

Effects of towns

Towns, large cities and continuous built-up areas influence the local climate to an important degree but scattered single buildings or small villages have little effect, except perhaps on wind in their immediate vicinity.

Atmospheric pollution is usually much more concentrated over towns: not only is the concentration affected by weather conditions, being increased when the wind is light and especially if a temperature-inversion forms, but the pollution in its turn affects the climate of the locality. The intensity of the solar radiation reaching the ground at the middle of a large city may be reduced by perhaps 15 per cent or more on average, and by double this amount on winter days. However, this may not have much effect on air temperature, because the reduction in input is at least partly compensated for by an increase in the amount of long-wave radiation which is scattered back from the sky. It is probable therefore that the modified radiation exchange has little effect on daytime temperature, but helps to make night-time temperatures higher than they otherwise would be.

Pollution has an important influence on fogs, which tend to be more frequent and more persistent on the outskirts of large towns than in the open country. They tend to be less frequent in city centres than in their suburbs, however, because of the higher temperatures. The fog frequency will also be affected by the surrounding topography. Towns in inland valleys will be more prone to suffer fog than those on exposed coasts. Even a wide shallow valley like that of the lower Thames helps to increase the frequency and severity of London's fogs.

Somewhat surprisingly, daytime maximum air temperatures in a town differ little from those in the surrounding country, if the whole area is at much the same altitude. At night, on the other hand, the town is often much warmer, sometimes to the extent of 6 deg C or more. In summer this is accounted for mainly by the solar heat stored by day in the buildings, which is released into the air at night. In winter the effect is normally less marked but night temperatures tend to be higher in towns, partly because of artificial heating of buildings and partly because of increased back radiation from the polluted atmosphere.

Mean wind speeds are appreciably lower in built-up areas, although there may be channelling and other effects in places causing relatively high wind speeds when the wind is blowing from a particular direction. Gust speeds may be little lower than over open country (see Chapter 2, page 63, for further discussions).

Human comfort conditions are often less pleasant in towns than in the open country during hot weather, even though the air temperature may be little different. This is because the comfort of the human body depends not only on air temperature, but also on the radiant temperature of the surroundings and on the rate of air movement. In the open country there is relatively free movement of air, while the body is exposed to almost the whole hemisphere of the sky, which in clear weather has an effective temperature much below that of the air. The effective temperature of the vegetation-covered ground may be no more than 6 to 9 deg C above air temperature. In consequence there is a net heat loss by radiation from the body to its surroundings, if it is shaded from direct sunlight, as well as a free exposure to the wind which promotes heat loss by evaporation from the skin. In towns on the other hand, much of the sky is obscured by buildings, and its effective temperature is in any case raised by pollution, so that the radiation loss from the body is much reduced. The surface temperature of those buildings exposed to the sun is much above that of vegetation, so that the effective temperature of the surroundings is fairly high. This, combined with the large amount of solar radiation reflected directly from light-coloured walls, and the much reduced air movement, leads to conditions which are much less comfortable than in the country. These less favourable conditions are likely to persist for a considerable time after sunset.

A note on available climatological information

Climatological information is available in a number of forms and from different sources, although naturally most of it, wherever it is published, is derived initially from Meteorological Office material.

Looking first at the regular publications of the Meteorological Office, there are the *Daily Weather Report* and the *Monthly Weather Report*, which are of the greatest value if information on current weather is required. The contents of these are summarised briefly below.

(i) Daily Weather Report
This is issued every afternoon by the Central Forecasting Office at Bracknell. Pages 1 and 4 of the *DWR* contain four tables giving the weather observations at 55 stations in Great Britain and Ireland, one each for the hours 00 00, 06 00, 12 00 and 18 00 GMT. The observations include wind speed and direction, cloud amounts, dry-bulb temperature and dew-point, weather at the time, rainfall and sunshine; maximum temperature in the day and air and grass minimum temperatures at night are also recorded.

Four weather charts are given on pages 2 and 3 of the *DWR;* the 12 00 GMT chart covers most of the northern hemisphere, those for 18 00, 00 00 and 06 00 GMT following cover north-west Europe.

Although referring to fairly widely scattered stations, the data in these reports are often adequate for giving the day-to-day weather in any part of the country.

(ii) Monthly summary
A brief summary of the weather for the month is issued a few days after the end of the month. It includes an analysis of the data from 20 stations.

(iii) Monthly Weather Report
Published about 4 months after the end of the month by the Climatological Services Branch of the Meteorological Office, this contains on the first page a general description of the weather of the month, recording the more notable happenings. The remainder of the Report consists of maps and tables, some of which are:

Maps showing the distribution of sunshine, temperature and rainfall for the month.

Table 1 – District values of air temperature, earth temperature, rainfall and sunshine, expressed as difference from the average (for this purpose the country is divided into ten districts).
Table 2 – Solar radiation and intensity of daylight, with means and extreme values for six stations.
Table 3 – Summary of observations of temperature, rainfall, sunshine and weather, for about 500 stations, including monthly mean maximum and minimum temperatures and the highest and lowest values for the month, number of frosts, and so on.
Table 4 – Summary of observations at fixed hours, for about 100 stations, with mean values of pressure, temperature, humidity and cloud amount at the observing hours, also frequencies of cloud amount, wind speed and direction.
Table 6 – Summary of autographic records of wind for about 120 stations, with highest gusts and hourly winds, and table of distribution of mean hourly wind speeds.

In addition to these, there are a number of other publications, which appear from time to time, including *Geophysical Memoirs* and *Scientific Papers,* which are reports of particular research projects. These and the *Monthly Weather Report* may be purchased from Her Majesty's Stationery Office.

A series of *Climatological Branch Memoranda* consists of reports of special investigations, some undertaken at the request of the Building Research Station. Subjects include combined frequencies of wet- and dry-bulb temperatures, frequencies of snow depth, extreme wind speeds, and averages of accumulated temperature. Copies of these Memoranda may be purchased from the Meteorological Office, Bracknell.

Tables of daily solar radiation and daylight illumination on a horizontal surface are available from the beginning of 1956 for the Meteorological Office network of stations.

Kew Observatory publishes weekly a brief summary of the main meteorological elements, pressure, temperature, rainfall, humidity, sunshine, wind and evaporation from the standard tank: a small charge is made.

Besides the published information enumerated above, the Meteorological Office holds a great many other records, extracts of which can be obtained. Since October 1958 many of the regular climatological data have been put on to punched cards on the Hollerith system, which gives the possibility of more elaborate analyses than was possible previously.

In addition, hourly data from about 20 stations have been put on to magnetic tape, which can be processed by high-speed computers to give even more detailed information, such as hourly frequencies of occurrence of specified temperatures. At the time of writing, the programs available have produced tables as listed below:

1. Fog frequency tables hour by hour (GMT). The ranges are in yards and the values given are the yearly average frequencies. The mean visibilities, standard deviations and absolute maxima are in miles. The absolute minima are in yards. Also printed is a table giving average frequencies over periods of hours: 19–00, 07–12, 01–06, 13–18, 19–06, 07–18, 06–18, 00–23, and daylight hours, which are derived from the main table by totalling over the requisite number of hours.

2. Similar tables of wind speed average frequencies in knots (spot hourly readings).

3. Similar tables of dry-bulb (air) temperature in degrees Centigrade.

4. Similar tables of dew-point temperature frequencies in degrees Centigrade. Readings only for the hours 03, 06, 09 and so on.

5. Similar tables of relative humidity frequencies.

6. Similar tables of vapour-pressure frequencies, in millibars.

7. Yearly average duration (in hours and tenths) of rainfall equal to or exceeding 20, 15, 10, 5, 4, 3, 2, 1, 0·5, and 0·1 mm, for every hourly observation. Values given refer to rainfall over the hour ending at the time of observation.

8. Average number of occasions when rainfall equal to or exceeding the above limits was observed.

9. For each observation hour, the percentage duration (the total duration of rainfall as a percentage of the number of hours observing time), for example $(D/31n) \times 100$ for January, where D is the duration totalled over n Januaries.

10. Three sets of tables derived from items 7, 8 and 9, for periods of hours 19–00, 01–06, 07–12, 13–18, 19–06, 07–18, 06–18 and 00–23, and for daylight hours.

Other programs are to be written to give analyses involving more than one variable, for example rainfall amount, rainfall duration and wind speed, to give information on driving-rain intensities.

Among useful collections of climatological data and statistics, the following may be mentioned:

The climate of the British Isles. E G Bilham. Macmillan, 1938.

London weather. J. H Brazell. HMSO, 1963.

The climate of London. T J Chandler. Hutchinson, 1965.

The English climate. H H Lamb. English Universities Press, 1964.

Average annual rainfall, 1916-50. Explanatory text No 2A and two maps to a scale of 1: 625 000. Ordnance Survey, 1967.

Also recommended in this field are the following publications of the Meteorological Office, obtainable from Her Majesty's Stationery Office:

Averages of humidity for the British Isles. MO 421. 1938.

Averages of rainfall for Great Britain and Northern Ireland, 1916-50. MO 635. 1958.

Averages of bright sunshine for Great Britain and Northern Ireland, 1931–60. MO 743. 1963.

Averages of temperature for Great Britain and Northern Ireland, 1931–60. MO 735. 1963.

Averages of earth temperature for the United Kingdom, 1931–60. Met.O.794. 1968.

British rainfall. A series of annual volumes beginning in 1860, listing and analysing the rainfall at all British stations. HMSO, from 1919.

Climatological atlas of the British Isles. 1952 (out of print).

Tables of surface wind speed and direction over the United Kingdom. Met.O.792. 1968.

Local climate surveys are given in **Climatological Memoranda** obtainable from the Meteorological Office at Bracknell, Berkshire:

The climate of Edinburgh. J A Plant. **Clim. Mem.** No 54A. 1968.

The climate of Glasgow. J A Plant. **Clim. Mem.** No 60. 1967.

The climate of East Lothian and North Berwickshire. F H Dight. **Clim. Mem.** No 49. 1966.

The climate of South-west Ayrshire and North Kirkcudbrightshire. F H Dight. **Clim. Mem.** No 52. 1966.

The climate of Kincardineshire and East Angus. F H Dight. **Clim. Mem.** No 53, 1966.

The climate of Central Ayrshire. F H Dight. **Clim. Mem.** No 58. 1966.

There are now a number of useful books in the field of applied meteorology and among these are recommended:

Technische Meteorologie. W. Böer. Teubner, Leipzig, 1964.

The climate near the ground. R Geiger. Translated from 1961 edn of Das Klima der bodennahen Luftschicht. Harvard University Press, 1966.

Microclimate of the USSR. I A Goltsberg (Ed.). Translated from 1967 edn pub. by Gidrometeorologicheskoe Izdatet'stvo, Leningrad. Ann Arbor-Humphrey Science Publishers, London, 1969.

Das Stadtklima. A Kratzer. Vieweg, Braunschweig, 1956.

Proceedings of the Symposium on Urban Climates and Building Climatology. Vol I – Urban climates; Vol II – Building climatology. **WMO Technical Notes** Nos 108 and 109. World Meteorological Organization, Geneva, 1970.

Chapter 2 Climate and town planning

Climatic effects on town planning

Climatic information may be required by planners to help them choose a site which will not be unduly affected by unfavourable weather conditions, such as strong winds, driving rain, severe frosts or prolonged fog. Alternatively, the site may be already determined and the need is to avoid a design and layout of buildings and roads which aggravates already severe weather conditions, or even produces bad conditions where they did not exist before. Preferably the design should ameliorate any climatic factors which would otherwise produce discomfort, for example provide shelter from strong winds, and should take advantage of any favourable factors – for example winter sunshine should not be excluded.

For town-planning purposes it is usually the local variations of climate within a region that are of interest. The approximate site of a new town, and especially of an extension to an existing one, will be mainly controlled by the availability of communications and of suitable land. In some parts of the country, however, there can be appreciable variations in the local climate over distances of a kilometre or two, or even shorter distances in hilly country, see for example page 59.

Differences in temperature are the ones which spring readily to mind. In Britain the mean temperature of the air at a metre above ground decreases by about 0·6 deg with every 100 m increase in height of the land. Thus the mean temperature at East Kilbride, which lies at heights of 150 to 180 m above sea-level, will be about a degree below that of similar towns in the Clyde Valley some 7 km to the north, which are only about 20 m above sea-level. This in its turn means that there will be a greater frequency of snowfall at the higher place, and snow will lie for a longer time. As a recompense, however, the higher place will probably be less subject to dense fogs (if not so high that it becomes subject to hill fogs, or low cloud).

These, and other aspects of local climate are dealt with in more detail in the following paragraphs.

The final decisions on town planning (and indeed on most aspects of building) will depend on the relative costs of different schemes. It follows that the costs attributable to climatic effects must be assessed. Generally this will only be possible if we know the precise relationship between the climate and a given aspect of the behaviour of buildings, or of the inhabitants.

Fairly accurate estimates can probably be made of heating costs and of the variance in these caused by climate, and perhaps of some other items, such as main drainage and the costs of the clearance of snow from streets. But when the problem involves human sensations, which may be physiological or aesthetic, it may be much more difficult to define the relationships in physical terms and hence assess costs. For example, if one site is more exposed to wind than another, it may be more uncomfortable for shoppers. It is possible (see below) to define a wind speed beyond which people will not willingly venture into a given street, and the cost of providing shelter will have to be balanced against the loss of trade if the shelter is not provided. It is not certain if such a speed would be a constant one. It is possible that people used to a more windy climate may accept a higher wind speed in a given situation than those accustomed to calmer conditions. Again, the wind speed acceptable to people walking about is higher than that which would be tolerable to a bus queue. However, it must be admitted that it is not often possible to specify exact climatic criteria which must be satisfied in order to produce acceptable conditions out of doors in a particular case. For example, on page 62 the effects of wind in towns are discussed. It is suggested that pedestrians (or people waiting in bus queues) are distinctly uncomfortable in wind speeds greater than about 5 m/s. For a given layout of buildings which has been tested in a wind tunnel, it will be possible to say what speed of the wind in the free air (outside, or well above, the town) will give rise to this local wind at various points among the buildings. Hence the frequency with which the discomfort level will be exceeded can be estimated. But this cannot be done, except in a rather general way, for other layouts which have not been tried out in the tunnel.

What do we mean by town climate?

Before looking at the details of local climatic variations, it is worth while trying to decide what we mean by town climate, and how it differs from other climates.

It is useful to look at a list prepared by Landsberg (1962) showing the principal elements of town climate, and how they differ from the rural climate (Table 12). The differences are derived from a variety of sources and can only be regarded as general guides. Moreover, with changes in human habits and in technology, there will be corresponding changes in the effects on the town climate. For example, in recent years the introduction of smokeless fuels on a large scale has much reduced the fogginess and smokiness of some towns in Britain, with a corresponding increase in sunshine and solar radiation reaching the ground. Furthermore, although it is true that there is in general a reduction of wind speed in towns, certain configurations of building layouts can cause marked local increases in the speed of the wind, creating quite unpleasant and even intolerably strong winds.

It must be pointed out that, although certain obvious features of town climate can be noted, it is not always easy to say in a particular case whether all such features are due to the presence of the town, or whether they would be there anyway. It should be remembered that towns are usually sited where they are because of the presence of certain geographical factors—for example, a strategic valley, a river crossing, or a sheltered bay on the coast. Thus the local climate of the town site in its original, undisturbed condition may differ appreciably from that of the surrounding country, and setting up a reference meteorological station in the country outside an existing town may not give a true comparison.

Certain real features of town climate can be distinguished, such as those listed in Table 12. These are dealt with in more detail below. It should not be overlooked that non-meteorological factors may be significant, for example if the noise-level from traffic or industrial activities is high, it may not be possible to open windows, so that buildings may become uncomfortably hot in quite a moderate outside air temperature, while if the windows could be opened conditions would have been tolerable. Likewise, changing fashions in architectural design alter the significance of the individual climatic elements.

Table 12 Average changes in climatic elements caused by urbanisation (after Landsberg 1962)

Element	Town environment as compared with rural
Contaminants	
condensation nuclei and particulates	10 times more
gaseous admixtures	5 to 25 times more
Cloudiness	
average cloud cover	5 to 10 per cent more
fog in summer	30 per cent more
fog in winter	100 per cent more
Precipitation	
total amount	5 to 10 per cent more
days with less than 5 mm	10 per cent more
snowfall	5 per cent less
Relative humidity	
in winter	2 per cent less
in summer	8 per cent less
Radiation	
global solar radiation	15 to 20 per cent less
ultra-violet in winter	30 per cent less
ultra-violet in summer	5 per cent less
duration of bright sunshine	5 to 15 per cent less
Temperature	
annual mean	0·5 to 1·0 deg C more
average winter minima	1 to 2 deg C more
heating degree days	10 per cent less
Wind speed	
annual mean	20 to 30 per cent less
extreme gusts	10 to 20 per cent less
number of calms	5 to 20 per cent more

Temperatures in towns by day

A number of records of the air temperatures in towns during the day have now been obtained, using instrumented motor vehicles to get a cross-section of the temperature distribution during a limited period of time. This is the cheapest and most convenient way of getting a sufficient number of nearly contemporaneous readings.

All the measurements agree that at the hottest time of the day, in the early afternoon, there is little difference between the temperatures of the air at about 1 to 2 m above ground in the different parts of the town and in the surrounding country, in the absence of marked topographical differences (Figure 53). Usually the middle parts of the town are a little warmer than outside, but the difference never seems to exceed 2 deg, and is usually less. However, recent measurements in two American cities have shown that on hot days the core of the town was not the hottest part. Rather, there was a ring, perhaps 1 to 2 km wide, which was between 0·5 and 1·0 deg warmer than the central core, which was itself some 2 km in diameter. The author of the report on this survey (Ludwig 1968) suggests that the reason for this is that the central core of these towns (Dallas and Fort Worth, Texas) consists mainly of high buildings along narrow streets. He suggests that, except for short periods, the sun does not reach the streets, but that the direct solar radiation is intercepted by the upper parts of the buildings, so that little is available for warming the lowest levels of the air. If this is so, one would expect to find this 'warm ring' effect only in towns with this particular configuration of buildings. It does not appear to have been noticed in any European survey.

But temperatures in towns measured with a shielded thermometer do not give a realistic impression of people's sensations. It always feels much warmer in the town than in the nearby country on a hot sunny day. This is because a human being is a heated, sizeable object fully exposed to the exchange of heat and moisture with his surroundings. He absorbs short-wave solar radiation from the sun, or reflected from surfaces around him and exchanges long-wave, infra-red radiation with his surroundings. On a sunny day, buildings and the surfaces of streets are usually appreciably hotter than the surface of his skin and clothing, so that he absorbs more long-wave radiation than he loses. The sky is a cool surface, but usually he sees little of this in a town. Finally, because air movement is often much reduced in the town on quiet, warm days, he can get little relief from the evaporation of moisture from his skin.

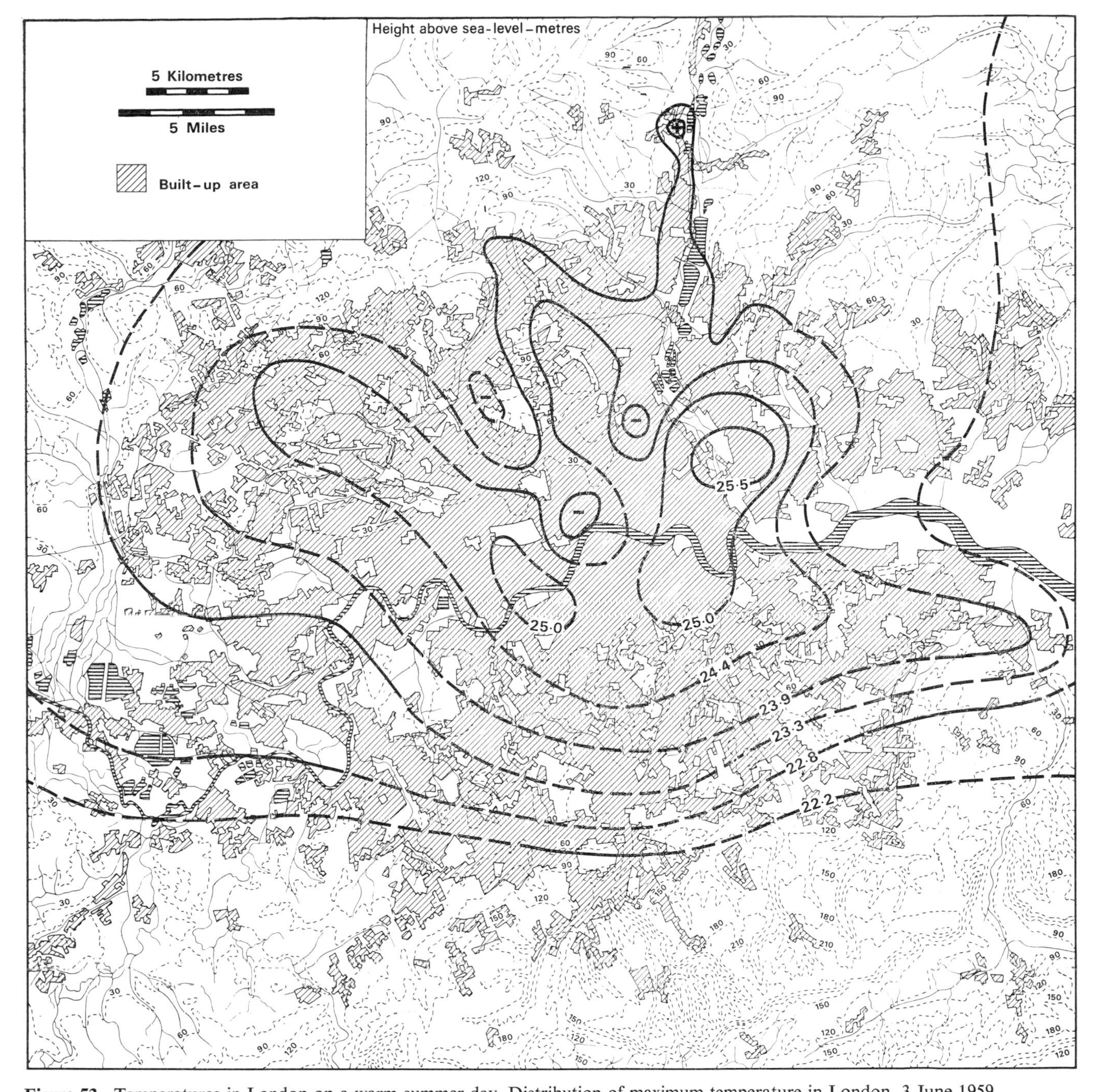

Figure 53 Temperatures in London on a warm summer day. Distribution of maximum temperature in London, 3 June 1959. Broken lines indicate some uncertainty of position. Isotherms are numbered in °C.
(From Chandler 1965 – based on readings at official stations and supplementary stations of the London Climatological Survey)

In the country, on the other hand, the objects surrounding him usually have a lower temperature than those in the town, he can probably see more of the cool sky and lose radiant heat to this, while he is probably more freely exposed to any breezes which help to cool him by evaporation of moisture from the skin. Thus even if the air temperature is the same, he will probably feel much cooler in the open country.

It seems likely that trees planted along streets in towns will help to ameliorate the high effective temperatures on hot days resulting from the large amount of radiation emitted and reflected from buildings and pavements. Their effectiveness arises in part from the direct local shading of passers-by from the rays of the sun and from the radiation emanating from surrounding surfaces, and in part from the shading of parts of the buildings and streets from solar radiation, so that these do not get so hot as if they were directly exposed to the sun. Measurements of the effectiveness of such shading are scarce. Gajzágó (1973) has reported measurements of the surface temperature of a lawn on a sunny September day in Budapest, when the air temperature was 17°C. The temperature at the surface of the grass was 24° in the open, away from trees, 27° on the sunny side of the tree, just outside the boundary of the leafy crown, and 14° under the tree on the shaded side. Dirmhirn and Sauberer (reported in Steinhauser, Eckel and Sauberer 1959) measured temperatures in Vienna on a hot August day at around 14 00 to 15 00 local time, and recorded air temperatures in narrow lanes some 8 to 9 deg lower than those in wide streets. The temperature of the asphalt street covering ranged from about 50°C in the main streets (where it was fully exposed to the sun) to 25° in the shady narrow lanes.

However, the beneficial shading effect of trees is to a small extent offset by radiation downwards from the leaves. They also tend to reduce the speed of the wind, so perhaps slightly increasing 'stuffiness'. Clearly more study of this problem is needed in order to give a more exact estimate of the beneficial effects to be expected from planting trees in streets. It is probably desirable to use deciduous trees as a

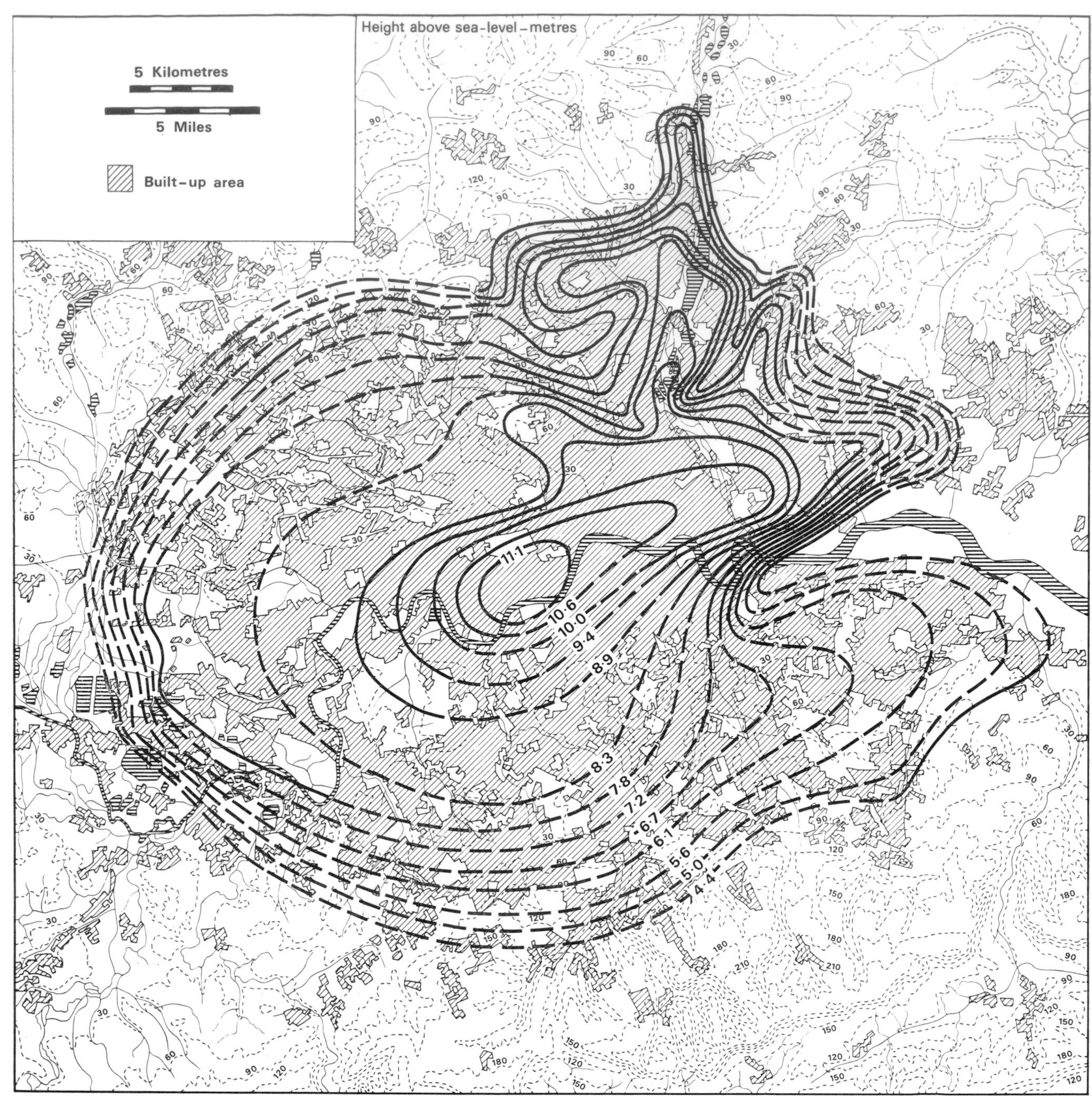

Figure 54 Temperatures in London on a clear summer night. Distribution of minimum temperature in London, 14 May 1959. Broken lines indicate some uncertainty of position. Isotherms are numbered in °C. (From Chandler 1965)

rule, so that when the leaves fall in autumn the sunshine can reach the buildings and streets.

During winter in Britain, certainly in the more northerly parts, strolling or sitting out of doors may be possible only if the sun is shining. The solar radiation may make little difference to the air temperature, but will increase the effective temperature as sensed by people, and may be especially important for older people with their lower level of activity.

Uncomfortable conditions for pedestrians on hot days are sometimes locally increased by the discharge at street level of hot air from air-conditioning plants.

Temperatures in towns at night

Some 80 to 85 per cent of the heat from the rays of the sun which reaches the surface is absorbed during the day by the buildings and streets of a town. Some of it is used to evaporate rainwater absorbed in the fabric of the buildings, but it seems that most must go to heating the air, some by direct conduction to the air in contact with the surfaces, the rest being emitted as long-wave radiation which is largely absorbed by water vapour in the air. In the open country, on the other hand, most of the solar radiation goes to evaporating water from plants (transpiration) during the day and little of this heat is available at night for warming the atmosphere. The vegetation-covered countryside cools rapidly at night, when skies are clear, because the grass and other vegetation hinder the conduction of heat from the soil while the town cools less rapidly, so that as the night progresses a considerable difference of temperature may develop between town and country. Figure 54 shows the distribution of air temperature in and around London in the early morning of 14 May 1959, after a night with clear skies and very light north to north-easterly winds. There is a difference of nearly 7 deg C between the warmest part of the town and open country on the outskirts. Chandler, from whose book *The climate of London* this Figure is taken, reproduces in his book a number of maps showing the distribution of temperatures in and around London, measured during the course of the London Climatological Survey, on days and nights under a variety of weather situations. These large differences of temperature are only developed when the skies are clear and winds light. When the sky is covered by thick clouds, and especially when the wind is strong, the town is little warmer at night then the country.

The details of the exchanges of heat and humidity in a town are far from being elucidated completely, and the above explanation is much simplified (see also page 80). However, the storage of solar heat within the buildings is a fact which is obvious to anyone who has tried to sleep in a town after a series of really warm days. The difference in temperature is, in fact, immediately obvious as one drives into the town from the country, particularly in the late evening.

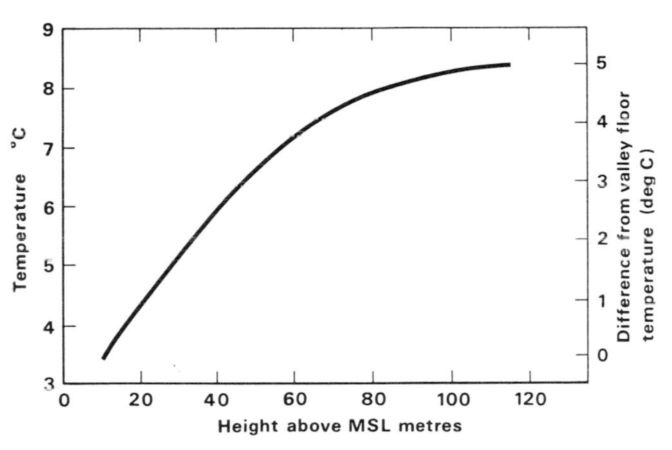

Figure 55 Relation between air temperature and altitude in a valley on radiation nights. (After Harrison 1967)

Valley temperatures on clear nights

At night when the sky is clear, the earth's surface loses heat rapidly by radiation, the cooled ground extracting heat from the air in its turn. Thus the layer of air nearest the ground cools, contracts and becomes denser. Where the ground slopes, this denser, cooler air tends to flow downhill, eventually filling the lowest parts of hollows and valleys with a pool of cold air, which becomes deeper and colder as the night goes on.

This 'katabatic' flow of air is most pronounced when there is little or no wind, for in moderate or strong winds the air near the ground is vigorously mixed with that above it. Then no very cold layer can form, and the air in the valley bottom remains relatively warm. Indeed, on clear windy nights there will usually be a 'lapse' of temperature upwards, with temperature decreasing with height at a rate of about 0·6 deg C for each 100 m, as it does on cloudy nights.

The intensity of the katabatic flow on a clear, calm night and of the resultant cold pool also depends on the ease with which the cold air can flow down the slopes. Thus the angle of the slopes and the smoothness of the surface are both important. Katabatic flows are most pronounced over short grass and over snow. They are markedly hindered by hedges, woodland, walls and buildings. Indeed, dense obstructions act like dams, ponding the cold air and holding it back until it becomes deep enough to overflow the obstruction, or diverting it on to a new course.

Given stable conditions, the cold pool of air in a valley will be most intense at the end of the night, just before sunrise. Measurements in a number of places have shown that temperature may increase by as much as 0·17 deg C per metre change in height over a range of 20 m or more (see Figure 55). In a valley 100 m deep the bottom may be

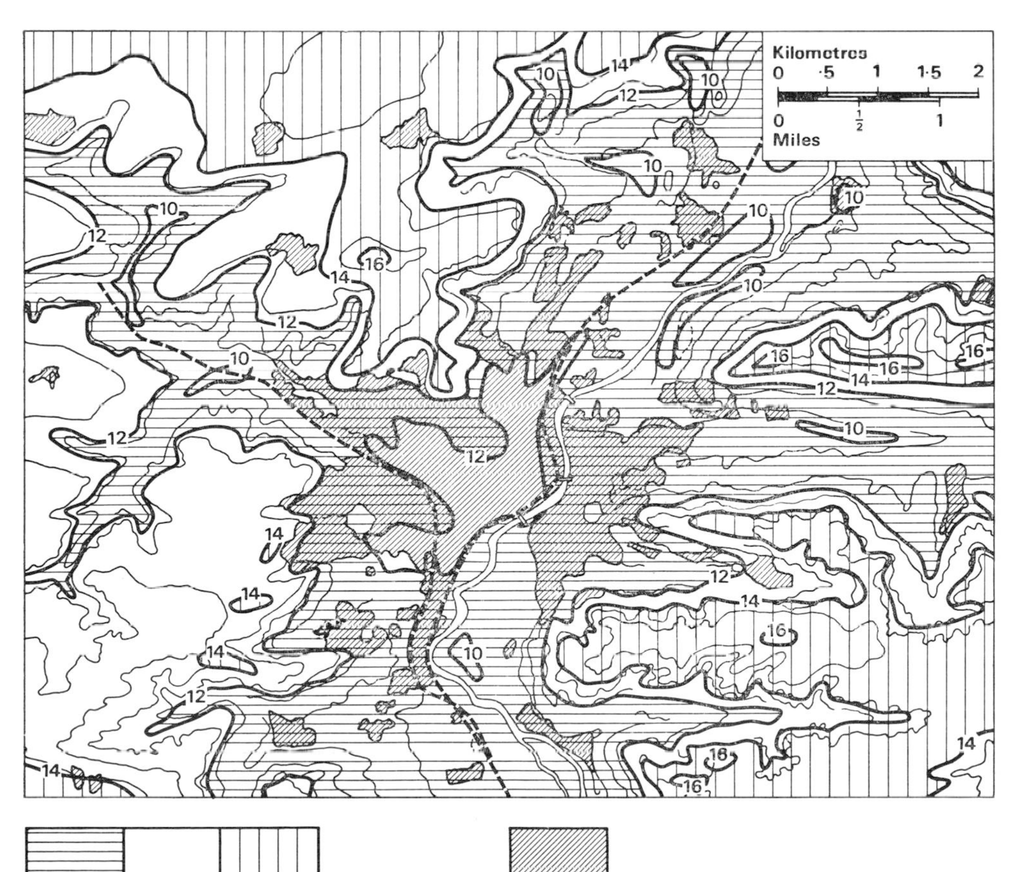

Figure 56 Night-time temperatures in a valley and the effect of a town. The temperature is lower in valleys than on hills, and is higher in the town than at places outside the town at the same altitude. (Kratzer 1956)

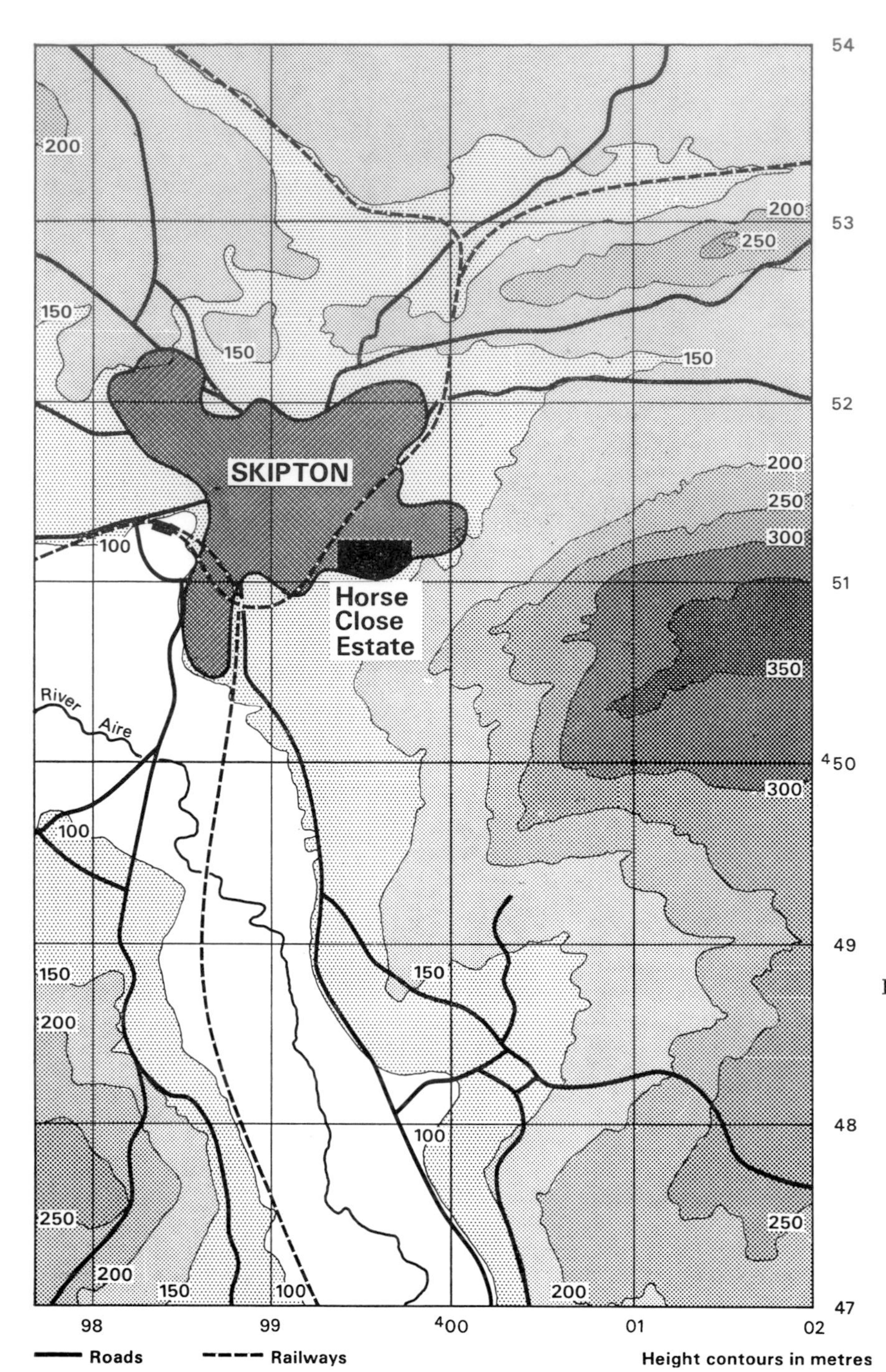

Figure 57 Plan of Skipton and its environs, with hill slopes subject to katabatic airflows. (Based on Ordnance Survey maps)

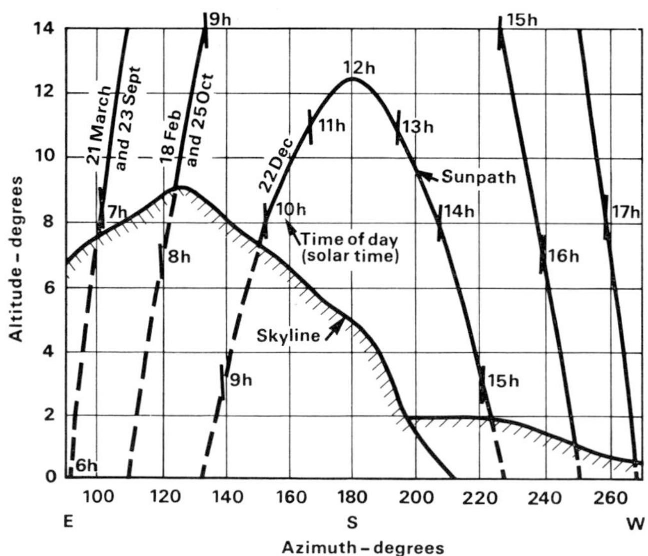

Figure 58 Diagram showing obstruction of the sun in winter by the hills behind Horse Close Estate, Skipton, Yorkshire, 54°N

about 5 deg C colder than the top on a radiation night (Harrison, 1967, 1971).

If there is a town in a valley bottom, the 'heat-island' effect will counteract the katabatic influences to an extent depending on the size and density of building. Kratzer (1956) shows how a town the size of Jena produced air temperatures 3 deg C above those at other places in the valley bottom which were well away from the town (Figure 56). Even so, the tops of the ridges above the valley were some 2 to 3 deg C warmer than the town centre.

A large city will completely overcome katabatic influences, to judge from measurements quoted by Chandler (Figure 54). He shows that on one still clear night, when temperatures in the Thames valley below London were 4 °C or less, in central London the air temperature was as high as 11 °C. One might expect that in the absence of the town the temperature all along the valley bottom would have been about 4 °C, so that the town has increased the night minimum temperature on this occasion by some 7 deg C.

In hilly country, buildings at the foot of long steep slopes of open moorland may be very exposed to cold katabatic airflows. It was observed a few years ago that severe damage had been caused by frost to most of the concrete footpaths on a housing estate at Skipton in Yorkshire. The houses lay on sloping ground, at an altitude of 130 to 150 m above sea-level (Figure 57). Above them the open hillside, facing approximately north-west, rose by some 200 m to a height of about 330 m in a distance of 1300 m. This slope receives very little sunshine in winter, and there can be no doubt that there is vigorous katabatic flow of cold air down it and over the houses on many nights of the year. Figure 58 shows the path of the sun across the sky on certain winter days and how it is obstructed by hills behind the houses.

In such a situation, which is typical of many towns in hill country, a carefully sited plantation of frost-resistant trees could perhaps be arranged to divert the flow of cold air. However, it might be better to build the houses on a south-facing slope if this were possible. Such a slope receives more sunshine than a north-facing one, and the extra warming of the ground delays the onset of the cold downslope winds. This would somewhat reduce the number of very cold nights. However, whatever the orientation, any site at the foot of such a slope is bound to be very subject to katabatic winds and thus increased liability to frost.

Mean temperatures and accumulated temperatures in towns

Because the daily minimum air temperature in a town is on average higher than that in the country around, while daytime maxima are usually similar to those in the country, it follows that the mean air temperature in a town is on average higher than in the country. The amount varies with time of year in a rather irregular fashion – Kratzer (1956) quotes some figures showing an almost constant difference through the year, and others showing a bigger difference in summer. Annual mean differences for about 25 towns range from 0·4 to 1·8 deg C, with no obvious correlation between temperature difference and size of town or climatic

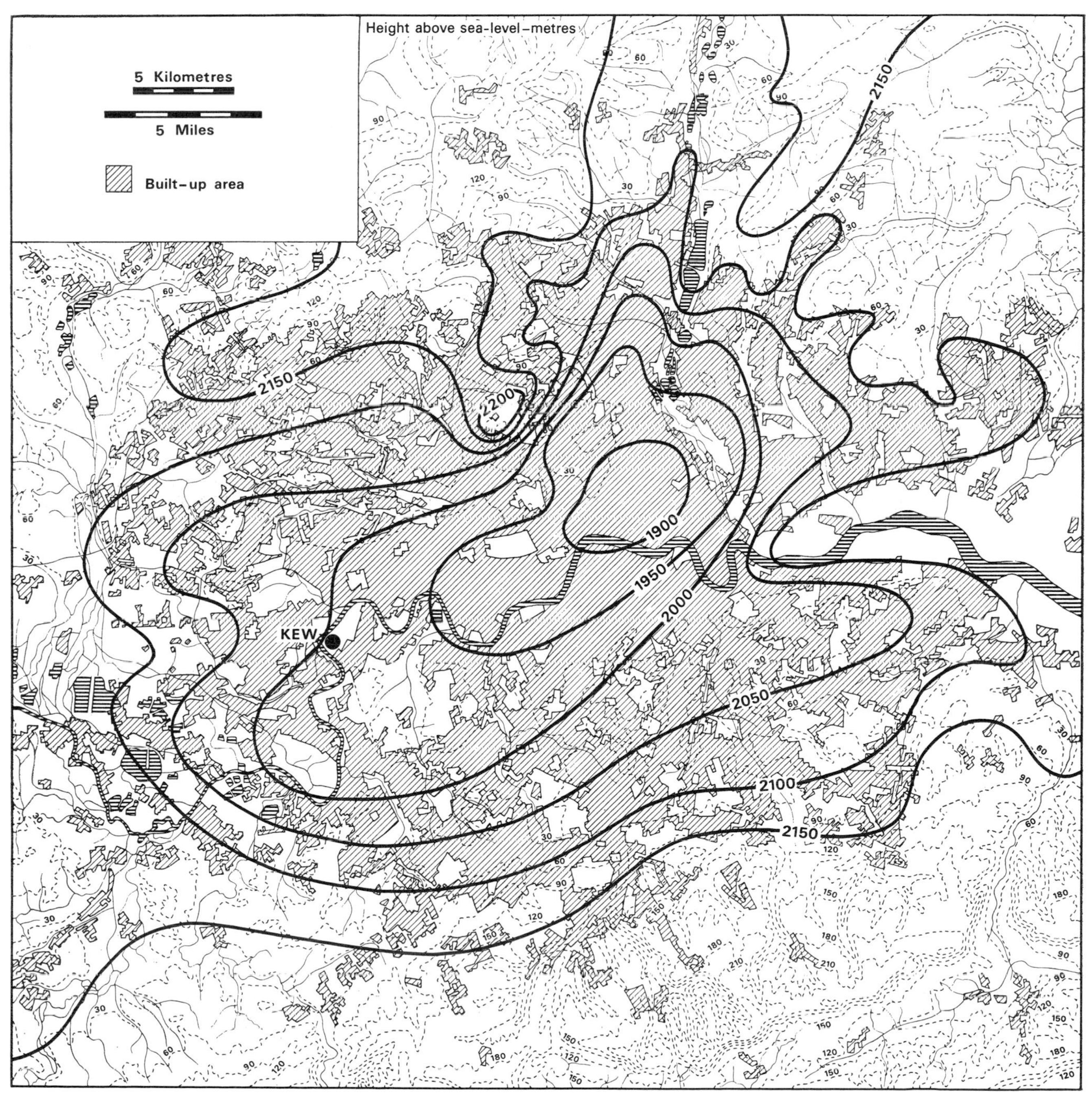

Figure 59 Degree-days in London. Distribution of annual accumulated temperature below 15·6°C in London, 1951–60 (Celsius degree-days). (From Chandler 1965)

zone. Data quoted by Chandler (1965) suggest a difference between country and central London of about 1·0 deg C, after correcting for difference in altitude. It is quite uncertain how much the differing values quoted by Kratzer are caused by slight differences in the exposure of the sites used in the comparisons.

However, Oke (1972), using only temperature data for calm, clear nights, when the difference in temperature between town and country is a maximum, has been able to demonstrate a logarithmic relationship between the population of a city and the maximum temperature difference. For European cities, the maximum difference (the so-called 'heat-island' effect) is approximately proportional to the one-fifth power of the population. He found that for places in the USA and Canada the difference in temperature was proportional to the one-fourth power – he suggested that the different values arise from differences in the density of population in the two regions and possibly to different rates of usage of artificial heating. It is clear that on average the town is warmer than the country, and the extensive survey by Chandler (1965) enabled him to draw average isotherms for the London region and, hence, average isolines for degree-days (Figure 59).

Degree-day values (see Chapter 3, page 145) may be used to estimate fuel consumptions during a period of 'average' weather, and the fluctuations from average in extreme seasons. The data are normally calculated from temperature-averages obtained from climatological stations in well-exposed sites, usually in open country. As shown by Figure 59, there is an appreciable variation in the average annual degree-days over the London region, from about 2100* in the valley east and west of the built-up area to 1900 in and around the City. Thus there is a reduction of some 10 per cent in annual heating requirements in the middle of the town – as estimated from air temperature (reduction in wind speed might lower the heat requirement still further). The map shows, however, that variations in altitude can complicate the picture. North-west of the central area lies Hampstead, at an altitude of some 130 m above sea-level, and 125 m above the land close to the river. Here the annual average degree-days amount to about 2200, a difference of 15 per cent in a distance of 5 km.

It is worth noting that published values of degree-days for 'Thames Valley' are normally based on figures from Kew Observatory. Figure 59 shows that the annual average degree-days at Kew amount to 2000, about 5 per cent less than in open country at the same altitude, and 5 per cent more than in central London.

Discomfort caused by wind in towns

Even over the most uniform terrain there are fluctuations in the speed of the wind, with alternating lulls and gusts. The rougher the ground, the more intense are the fluctuations. Buildings form an aerodynamically rough surface, and give rise to an extremely turbulent wind-stream in the lower atmosphere. In addition, because of the bulk of individual buildings, there are marked local deflections of the airflow.

Thus in a town there will be considerable variations in the speed and direction of the wind from place to place, and from time to time. Tall buildings may bring down to ground-level fast-moving air streams which frequently cause discomfort and inconvenience to pedestrians, and damage doors and windows. There have even been occasions when the local winds produced by high buildings have blown people off their feet, and at least two deaths of old people have been reported in such incidents. Local intensification of air pollution may also occur when such air streams bring down smoke and fumes from higher levels.

In a previous section (page 56) it was noted that on hot days in towns comfort might often be reduced because wind speeds were too low – the relatively low wind speeds outside the town on such days being further reduced by the obstruction to free airflow by the buildings of the town. Conversely, on windy days when the conditions indicated by the thermometer might suggest reasonably comfortable conditions, there may be local winds, resulting from particular configurations of buildings, which may create unpleasant blusteriness.

During the last few years there have been reports from several cities in Britain of excessively strong winds in streets and shopping precincts. These caused considerable inconvenience to people walking about – indeed, it was suggested that at one place it was difficult to let some new shops because shoppers were deterred from visiting the precinct by the winds. In all cases there was a tall new building which was thought to be the cause of the strong winds, although when these troubles were first noticed, in about 1964, there was no guidance available to designers on airflow in these circumstances. Table 13 shows what effects may be observed at different wind speeds. The effects with winds of classes 0 to 3 are derived from observation of the effects of airflow on people in a wind tunnel. The effects for Beaufort classes 4 to 9, from 6 m/s upwards, are largely based on the standard meteorological tables for assessing the speed of the wind by non-instrumental means (World Meteorological Organisation 1971).

The observations suggest that 'mechanical' discomfort sets in at a wind speed of about 5 m/s, and that above 8 m/s conditions could reasonably be described as very uncomfortable while above about 20 m/s they can be dangerous. However, the degree of discomfort or annoyance

Table 13 The degree of mechanical discomfort caused to pedestrians by wind

Beaufort Number	Wind speed m/s	Effect
0, 1	0–1·5	No noticeable wind
2	1·6–3·3	Wind felt on face
3	3·4–5·4	Hair is disturbed, clothing flaps
4	5·5–7·9	Raises dust, dry soil and loose paper Hair disarranged
5	8·0–10·7	Force of wind felt on body Limit of agreeable wind on land
6	10·8–13·8	Umbrellas used with difficulty Difficult to walk steadily
7	13·9–17·1	Inconvenience felt when walking
8	17·2–20·7	Generally impedes progress
9	20·8–24·4	People blown over by gusts

* In Figure 59 the degree-days are given in Celsius units, having been converted from Chandler's Figure 62 by multiplying his data by 0.555 and rounding off results to the nearest 50 units.

felt will depend on the activity of a particular person. If waiting in a bus queue he may well be affected more strongly by a given wind speed than if he is walking more or less briskly, while if he is walking by the sea or in the mountains on holiday he may even derive pleasure from the buffeting and regard the experience as 'exhilarating'. But for the purposes of every-day life, the figures in the Table are a reasonable guide.

At lower temperatures, wind will increase discomfort by carrying away more heat from the body, and so chilling the skin or, in extreme cases, the body itself.

The degree of discomfort can be assessed approximately by estimating the so-called 'wind-chill index', which is described in Chapter 4, page 166. Figure 158 shows, for example, that at an air temperature of 10°C, with no wind, the heat loss from bare skin is about 250 W/m^2, giving a 'pleasant' sensation, but if there is a wind of 5 m/s at the same temperature, the heat loss is about 700 W/m^2 and the sensation is 'very cool'. It is true that the wind-chill index does not take into account the effect of clothing, which would normally be varied to suit the prevailing or expected weather, but it is a simple and widely used index. Since the hands, which are particularly sensitive to cold, are often less protected than the rest of the body, the index may well give a reasonable estimate of how they will be affected (see also Penwarden 1973). Tall buildings not only create these unpleasant winds; they also shield the ground near them from sunshine. This may be an advantage in summer, but on bright winter days with a chilly wind there can be the double disadvantage of being exposed to artificially strong cold winds while being sheltered from the warming rays of the sun. Careful planning of building layouts could allow the maximum use of open spaces on the few fine days that we have in winter.

The method of assessing the modifications to the speed of the wind by various configurations of buildings is discussed in the next Section.

Wind speed around buildings

Both the layout and the configuration of buildings are important in determining the details of the pattern of wind flow among the buildings. It is found that the highest speeds occur (a) in the vortex which forms on the windward side of the high building, (b) close to the windward corners of a high building, and (c) in any space beneath (or through) the building. In these three regions the speeds can be two or even three times the speed in the undisturbed airstream with no buildings present. Information now available enables the wind patterns in the neighbourhood of a tall building among lower buildings to be predicted, although in more complicated cases involving several high buildings it may still be necessary to make tests on models in a wind tunnel. Several examples of layouts which have been tested are illustrated in Building Research Station Digest 141 *Wind environment around tall buildings* (1972) and in a text book *Wind environment around buildings* (Penwarden and Wise, 1975).

Aerodynamic tests on models of aircraft are done in wind tunnels in which great care has been taken to make the air-flow as uniform as possible, simulating conditions well away from the surface of the Earth. Buildings, however, are within the atmospheric boundary layer in which the speed of the wind increases with height above the ground. The rate of increase of mean speed may be represented, when the wind is strong, by the relation

$$V_1/V_2 = (h_1/h_2)^\alpha$$

where V_1 and V_2 are the speeds of the wind at heights h_1 and h_2 respectively and the exponent α has a value which depends on the roughness of the ground. In moderate to high winds (above about 5 m/s) the value of α may be taken to be 0·16 or 0·17 in open smooth country (for example grassland with little other vegetation). In suburban areas of towns the value is appreciably higher and it has been suggested by Davenport (1968) that 0·28 is appropriate. This is the value that has been used in tests at the Building Research Station. Buildings of heights up to 100 metres, modelled to a scale of 1/120, have been tested.

It is essential that in such tests the speed of the wind in the tunnel should vary with height, otherwise the distribution of pressure on the models, and hence the airflow produced by the models, will be quite unlike those experienced in practice. On the real buildings the pressure on the windward face increases with height, up to a region at about two-thirds or three-quarters of the way up. Because the pressure on the face is higher near the top than at the foot, air is forced downwards. It is this current of air flowing down the windward face that produces the strong winds at the foot of the building. Figure 60 shows a typical airflow pattern on the windward face of a slab-shaped building which has a low building at a distance in front of it about equal to the height of the tall building. There is a region A in the space in front of the tall building, with a strong vortex with horizontal axis, a region B at each corner where strong corner streams of wind form, and if there is an opening through the building at ground-level at point C, even stronger winds will flow through the opening.

The strength of these local winds is dependent on various factors, but in particular on the height and breadth of the tall building. The values quoted below are for a building four or more times the height of the surrounding low buildings and with a height to breadth ratio of 1 to 2, so that it is slab-shaped.

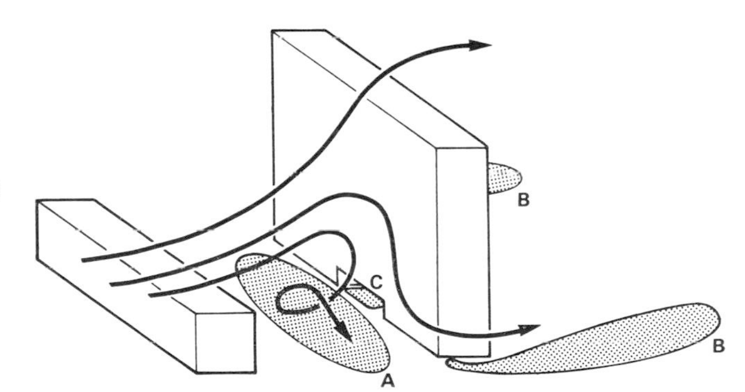

Figure 60 Typical pattern of airflow round a tall slab block with a low building on its windward side. The regions A, with a vortex in the courtyard between the buildings, B with a stream of air deflected down the face and round the corner, and C with the through flow in the passage under the building, are particularly subject to strong winds. Region A will be subject to the strong winds when the wind blows from the low building towards the high one. Region B will be subject to strong winds with almost all wind directions, the exact position of the severely affected regions depending on the wind direction. Region C will be affected with winds blowing towards either of the two main faces of the building. Typical values of the wind-speed ratio R_H are: region A, 0·5; B, 0·95; C, 1·2.

In order to compare the local winds produced under different conditions it is desirable to refer them to some standard value of the wind. One possibility is to compare them with the local wind that would exist if the buildings were all removed. This was the procedure used in Building Research Station Digest 141 *Wind environment around tall buildings*. An alternative is to relate the local wind to the wind speed at the height of the tall building, which is the procedure used by Penwarden and Wise (1975) and is the one adopted here. The reader is referred to their work for a full explanation of the background, and only a brief outline will be given here.

The local wind speed near the building, in the regions A, B and C referred to above, is expressed as a ratio R_H with respect to speed at the same time in the free air at the height of and in the neighbourhood of the top of the building under study. The appropriate value of R_H is derived in each case from wind-tunnel tests or can be estimated from experience based on case studies of similar situations. In its turn, the wind speed at the height of the building is derived from the open country wind speed at a height of 10 m above ground by the application of a correction factor S. This factor S takes account of the fact that the variation with height over the country differs from that over the town. The procedure is to estimate the ratio of the gradient wind (at a height of 275 m) to the open country value at a height of 10 m, and the corresponding ratio of the gradient wind to the wind over the town at a height of H metres. The ratio of these two ratios is the factor S, and the relation between S and the height H of the building is given in Figure 61. The wind speed at a meteorological site which will produce a local wind of 5 m/s (assumed to be the highest wind for comfort as described on page 62) near a tall building of height H in a suburban area, can now be calculated as

$$V_{10} = 5/R_H S \ldots\ldots\ldots\ldots\ldots\ldots (6)$$

where the appropriate value of the speed ratio R_H is inserted. In the absence of exact measurements, the following values of R_H are likely to be about right for tall, wide buildings: for high-speed region A a value of about 0·5 could be adopted; for region B, 0·95; and for region C, about 1·2. It has been found desirable to express the open country or airfield wind speed at a height of 10 metres in terms of an average wind speed denoted by $\bar{V}_{10}$.

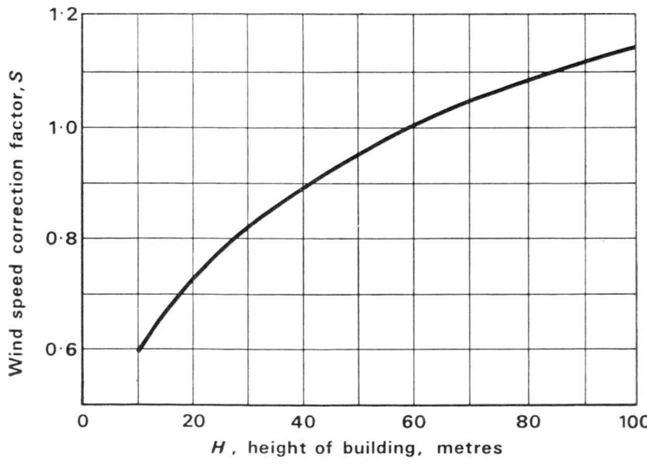

Figure 61 Relation between the height, H metres, of a building and the wind-speed correction factor S, which relates the wind speed at a height of 10 metres in open country to the wind speed at a height of H metres in the suburban environment

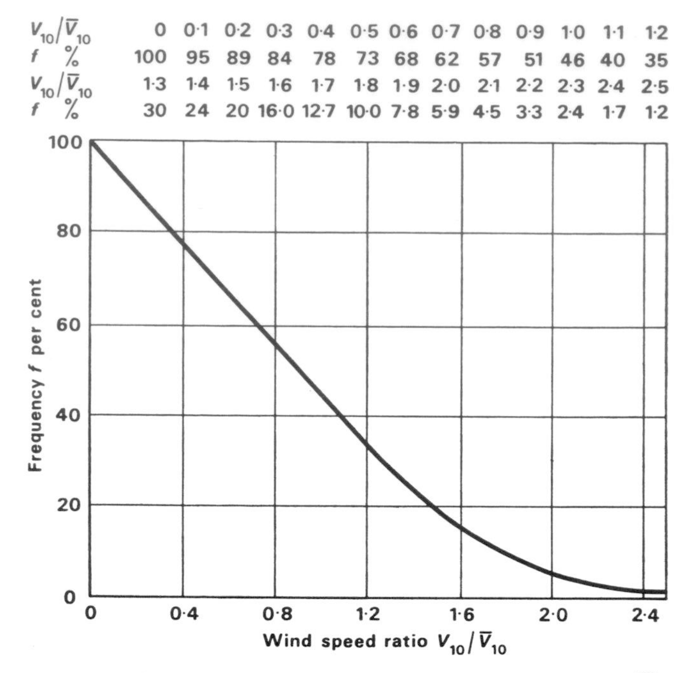

$V_{10}/\bar{V}_{10}$	0	0·1	0·2	0·3	0·4	0·5	0·6	0·7	0·8	0·9	1·0	1·1	1·2
f %	100	95	89	84	78	73	68	62	57	51	46	40	35
$V_{10}/\bar{V}_{10}$	1·3	1·4	1·5	1·6	1·7	1·8	1·9	2·0	2·1	2·2	2·3	2·4	2·5
f %	30	24	20	16·0	12·7	10·0	7·8	5·9	4·5	3·3	2·4	1·7	1·2

Figure 62 Cumulative percentage frequency f of the ratio $V_{10}/\bar{V}_{10}$ where V_{10} is the hourly mean wind speed and $\bar{V}_{10}$ is the average wind speed. Based on data from 35 anemograph stations. (Penwarden and Wise 1975)

The average wind speed $\bar{V}_{10}$ for each station is obtained from the cumulative percentage frequency distribution of hourly mean speeds at the station (Meteorological Office 1968). When the cumulative percentage frequency f is plotted against $V_{10}/\bar{V}_{10}$ for all stations on one graph it is found that a common line can be drawn which represents all the stations with sufficient accuracy. This line is shown in Figure 62 with, for convenience, a corresponding tabulation.

A convenient way to find the average wind speed $\bar{V}_{10}$ for any place is to read the value from a map. Figure 63 is derived from the data for the 35 stations used in Figure 62 (Meteorological Office 1968), the contours being drawn in with guidance from a map of extreme hourly speeds (Shellard 1965) and one of mean speed for an earlier period (*Climatological atlas of the British Isles*, Meteorological Office 1952). The map is the best that can be drawn from existing information, although no doubt a better one can be produced when more data are available and when account can fully be taken of the effects of topography, surface roughness and non-typical exposure*. At present these effects have to be estimated in a somewhat subjective fashion. Speeds read from the map apply to open country and, as explained below, can also be used to estimate the speeds in suburban areas and in smaller towns. The validity of the speeds read off over the middle of large cities is more doubtful, though they probably represent speeds at a height of about 10 m above the general level of the roof-tops. It is better to use wind speeds measured in and over the city if they are available. However, the values read from the map have been found to give consistent results when applied to practical cases.

It may be thought desirable sometimes to make a further adjustment to take account of the fact that on average the hourly mean wind speeds during the daytime are some 10 per cent higher than the 24-hour means.

*Another map of annual mean wind speed is reproduced as Figure 31 in Chapter 1. In this an attempt has been made, not only to reduce all values to a common height of 10 metres above ground, but also to adjust each value so as to be representative of open unobstructed country. Thus the values in Figure 31 are generally somewhat higher than those in Figure 63. The calculation procedure in this Section is based on the values of the latter map.

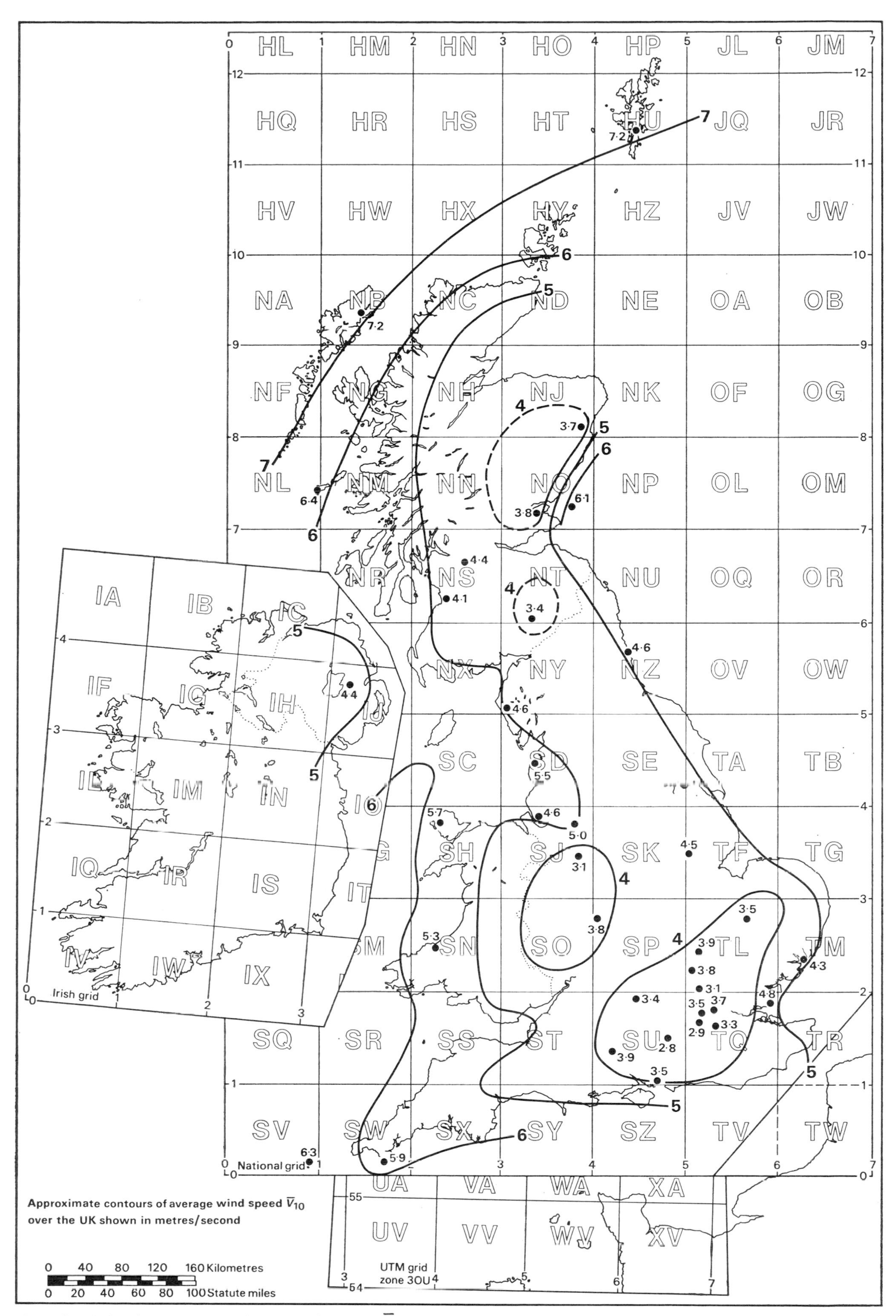

Figure 63 Approximate contours of average wind speed $\overline{V}_{10}$ over the UK, in metres/second. All speeds are reduced to an 'effective' height of 10 m using a 0·17 power law. No adjustment has been made for the effects of topography, surface roughness or non-typical exposure

Table 14 Example of wind-speed frequency table. Percentage frequencies of winds at Manchester (Ringway)

Mean wind speed m/s	Percentage number of hours with winds from												
	350° to 010°	020° to 040°	050° to 070°	080° to 100°	110° to 130°	140° to 160°	170° to 190°	200° to 220°	230° to 250°	260° to 280°	290° to 310°	320° to 340°	All directions Total
							December						
Calm													2·1
0·5–1·7													8·4
1·8–3·3	0·9	1·5	1·4	0·6	0·7	1·5	2·9	1·6	0·7	1·0	0·9	0·8	14·5
3·4–4·8	0·2	1·2	1·3	1·2	1·5	2·9	6·5	3·5	1·8	1·1	1·4	0·8	23·4
4·9–8·5	0+	0·4	0·6	1·6	2·1	3·4	8·7	5·4	2·9	3·3	2·2	0·6	31·2
8·6–11·0		0·1	0·1	0·7	1·2	1·4	2·7	2·2	1·5	2·8	1·0	0·1	13·8
11·1–14·1				0·3	0·9	0·8	1·0	0·7	0·5	0·9	0·3	0·1	5·5
14·2–17·2				0·1	0·1	0+		0·1		0·4	0·2	0+	0·9
17·3–20·8										0+	0·1		0·1
20·9–24·4													
24·5–28·5													
28·6–32·7													
>32·8													
Total	1·1	3·2	3·4	4·5	6·5	10·0	21·8	13·5	7·4	9·5	6·1	2·4	99·9
								Percentage number of hours missed					0
							Year						
Calm													3·3
0·5–1·7													9·3
1·8–3·3	0·8	1·5	2·0	1·4	1·2	1·4	2·2	1·7	1·2	1·3	1·8	1·3	17·8
3·4–4·8	0·7	1·6	2·2	1·9	1·8	2·7	4·9	3·3	2·2	2·1	3·2	1·8	28·4
4·9–8·5	0·5	0·9	1·1	1·7	1·8	3·2	5·8	4·0	2·5	3·2	3·6	1·5	29·8
8·6–11·0	0·1	0·1	0·1	0·6	0·5	0·9	1·8	1·3	0·7	1·2	0·9	0·3	8·5
11·1–14·1	0+		0+	0·1	0·2	0·3	0·5	0·4	0·2	0·5	0·3	0+	2·5
14·2–17·2	0+			0+	0+	0+	0·1	0·1	0+	0·1	0+	0+	0·3
17·3–20·8				0+		0+	0+	0+	0+	0+	0+		0+
20·9–24·4													
24·5–28·5													
28·6–32·7													
>32·8													
Total	2·1	4·1	5·4	5·7	5·5	8·5	15·3	10·8	6·8	8·4	9·8	4·9	99·9
								Percentage number of hours missed					0·1

From Meteorological Office 1968a.

Thus the whole procedure for calculating the frequency with which an hourly mean wind speed of 5 m/s will be exceeded near a building in a suburb or in a moderately large town is this: first decide an appropriate value of R_H to give the local speed ratio and the value of the correction factor S and calculate the corresponding open country hourly mean wind speed V_{10} from

$$V_{10} = 5/R_H S.$$

Next, select the appropriate value of the average hourly mean wind speed $\overline{V}_{10}$ for the area from the map in Figure 63. If thought necessary, increase this by 10 per cent to take account of higher daytime wind speeds. Calculate the ratio $V_{10}/\overline{V}_{10}$ and determine from Figure 62 the frequency with which, on average, this value will be exceeded.

This procedure is based on the assumption that the wind-speed ratio R_H does not vary with the direction of the wind. In the case of the corner streams in region B this is broadly true, but the high-speed regions A and C will usually have values of the ratio which differ markedly with direction of the wind, so that the simple procedure will overestimate the frequency of local winds of over 5 m/s (unless it should happen that all winds which give rise to such a high local speed blow from directions within the limited sector which is critical). It will therefore be necessary to use combined statistics of wind speed and direction, of the kind shown in Table 14, so that the frequency of wind speeds above the calculated value for the directions which are critical can be estimated.

The frequencies obtained in the above calculations apply to hourly mean wind speeds. It will be expected that during an hour with a mean speed of a given value, there will be a five-minute period with a speed some 10 per cent higher than the hourly value, while higher speeds still will occur for even shorter periods.

The statistics used to derive the frequencies can be used in other ways. For example, Figure 64 is a map showing what height of building can be set among low-rise housing in different parts of the country, without causing corner streams of wind of 5 m/s or more to occur for more than 20 per cent of the time. Similar maps could be prepared based on other frequencies of occurrence, and for other local wind situations (in the high-speed regions A and C shown in Figure 60 for example) or for specific configurations of buildings.

Wind speeds above the town

In the air above the general mass of buildings, over a town which is built up to a fairly uniform height and density of buildings (apart perhaps for some isolated towers), it seems likely that wind speeds are little different from those outside the town, although the wind will be much more turbulent.

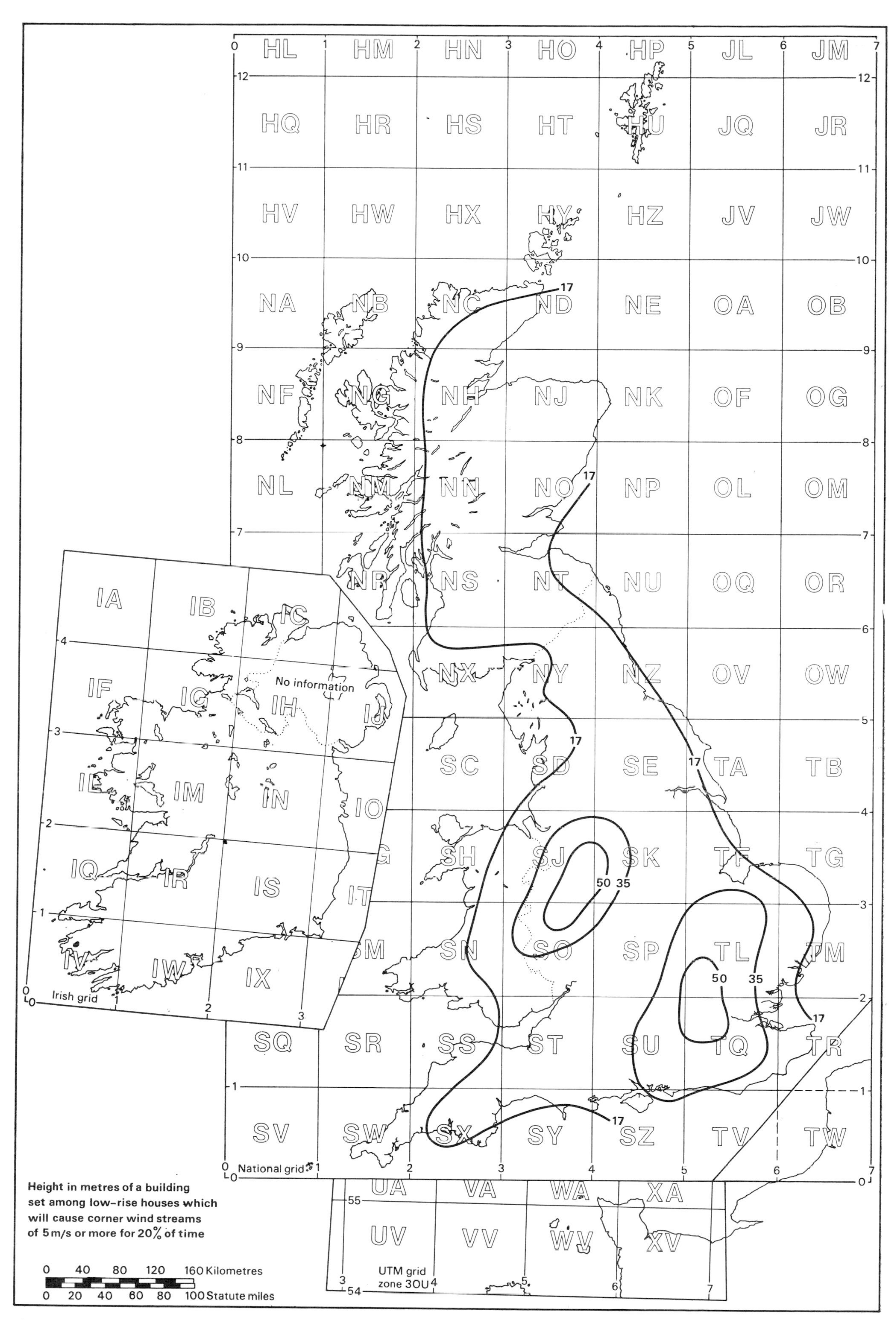

Figure 64 Height of a building which, when set among low-rise houses, will cause corner wind streams of 5 m/s or more for 20 per cent of the time. Height shown in metres

Table 15 Wind speeds at the Post Office Tower, London, on 23 February 1967, during 10-minute period beginning 04 56 GMT (from Shellard 1967)

Height above ground	Height above 'zero plane'	10-minute mean speed	Estimated hourly speed	Maximum 2-sec gust
m	m	m/s	m/s	m/s
189	151	19·7	18·6	27·8
61	23	12·9	12·2	21·1
43	4·5	11·5	10·8	20·6

Table 16 Maximum wind speeds in the London area on 23 February 1967

Place	Height above ground	'Effective' height	Time	Direction	Speed
	m	m	GMT	deg	m/s
Maximum hourly speeds					
London Weather Centre	70	38	04 00–05 00	230	13·4
Kew Observatory	23	15	03 00–04 00	190	13·9
Hampton	31	30	04 00–05 00	220	16·5
Maximum gusts					
Heathrow Airport	10	10	03 35	220	26
Hampton	31	30	03 45	200	27
Gatwick Airport	12	10	04 25	250	27

From *Monthly Weather Reports* (Meteorological Office).

These remarks are based on some measurements of wind speed over London, quoted by Shellard (1967).

The measurements were made on the top of the Post Office Tower, at a height of 189 m above ground, and on a second, nearby tower at heights of 43 and 61 m. The general height of the roof-tops around was some 30 m above ground, and the measurements were made with electrical cup-anemographs which could respond fully to gusts of about 2 to 3 seconds' duration. The records were obtained on a windy day, 23 February 1967, when gusts of up to 27 m/s were observed at various places around London, on anemographs at heights of between 10 and 30 metres above ground. Observations from the Post Office Tower are shown in Table 15 and those from 'ground-level' anemographs in Table 16. It will be observed that on this day the peak gust at a height of 189 m above the town was little above that observed at 10 m above ground on airfields within 50 km of the town centre. At lower elevations the gusts were less strong than those over open country. The hourly mean speeds at heights of 43 and 61 m in the town were little different from the maximum hourly speeds observed outside the town. (The hourly mean speeds in the region of the Tower were estimated from the 10-minute mean speeds given by Shellard by dividing them by a factor of 1·06 as recommended by Shellard in his 1965 paper, Table 5).

It is interesting to note that the hourly mean speeds at the three heights above the town fit a power law with respect to height, if it is assumed that the wind acts as though there is a 'zero plane' at a height of about 38 metres, instead of at ground-level. We have then the relationship

$$V_1/V_2 = (h_1/h_2)^{0.15} \quad \dots\dots\dots\dots\dots (7)$$

approximately, where V_1 and V_2 are the hourly mean wind speeds at heights of h_1 and h_2 above this assumed zero plane.

Alternatively, as proposed by Helliwell (1971), a logarithmic relationship can be fitted to the data, of the form

$$\bar{u}(z) = (u_*/\mathrm{k}) \log_e (z/z_0)$$

Helliwell found that for sites in central London the friction velocity u_* was 1·13 m/s and the roughness length z_0 was 0·78 m. Thus the mean hourly speed $\bar{u}(z)$ at height z can be found, using a value of 0·4 for von Kármán's constant k.

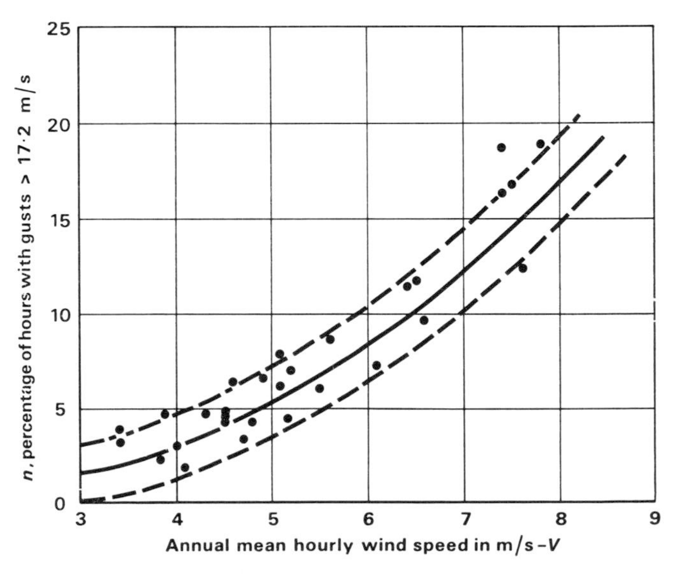

Figure 65 Relation between annual mean wind speed and percentage number of hours with gusts exceeding 17·2 m/s, for 29 stations in the British Isles. Data for the 10 years 1954–63, the relationship being $n = 0{\cdot}11\ V^{2.44}$. The dotted lines are the 95 per cent confidence limits

Local variations of wind speed

The general wind speed varies appreciably from one part of the country to another. Figure 31 (page 32) shows the mean annual wind speed over Britain. In itself a knowledge of mean wind-speed may not be of great use, but the map shows how sheltered most inland areas are compared with the coasts. The frequency of high winds and strong gusts is much higher where the mean annual speed is high than in more sheltered regions (Figures 40 and 65). For example, analysis of data from 29 stations in Britain during the 10 years 1954–63 indicates that the percentage frequency of hours each year during which gusts of 17·2 m/s or more occur (n) is related to the mean annual wind speed (V) by the relation

$$n = 0{\cdot}11\ V^{2.44} \qquad (8)$$

The line represented by this relation is drawn in Figure 65, together with the 95 per cent confidence limits.

Maps can only show regional variations in wind speed. Local topography and smaller-scale roughness (buildings, vegetation and so on) make the actual pattern of wind speed much more varied than appears from these maps. In particular, a place which is locally sheltered so that its mean wind speed is lower than elsewhere in the region is not necessarily sheltered from extreme gusts. In the free air, wind speed increases with height by an amount which depends on the roughness of the underlying ground. The average speed over an isolated hill 120 m high may be some 50 per cent higher than over the surrounding country, so that a town on a hill top will be much more exposed to strong winds than one on lower ground in the same region. We may expect therefore that such a town will be more expensive to build because more care will have to be taken to make the buildings strong enough to resist increased wind pressures, they will be more difficult to make weather-tight, and moreover the running costs will be greater because heat losses will be greater (not only will the wind be stronger, but air temperatures will generally be lower on the hill when the wind is strong).

On the other hand, if the town is more subject to strong winds, there may usually be a smaller risk of bad fogs. However, the limited climatological data so far obtained from Cumbernauld suggest that though it is appreciably more exposed to wind than Stirling (which is about 17 km to the north), it has as many foggy mornings as Stirling. Thus there is no simple correlation between frequencies of fogs and of strong winds.

The extremely local variation in wind speed that may be experienced in hill country was well demonstrated in a recent investigation of driving rain at Tredegar in Gwent. This is described in detail in Chapter 3 (page 98), but it may be remarked here that in a distance of 700 metres the amounts of driving rain intercepted differed on average by over six to one, and on some days there was 15 times as much driving rain at the exposed site as at the more sheltered one. The mean wind speeds at the two places may have differed in about the same proportion, since there can have been much less difference in the rainfall over this distance.

These local differences may be of vital importance when the winds are strong. In the lee-wave storms of February 1962 (see page 70) it was observed that there was a great deal of damage to brand-new houses, and to many others built in the preceding 10 years or so. Many of the damaged houses were built on ridges and other high places. No doubt in some cases the sites were chosen because they gave good views of the surrounding country, but often the higher land must have been used because the sheltered valley sites were already built upon. A brief description of some of the damage and the presumed meteorological reasons for it is given in *Gales in Yorkshire in February 1962* (Aanensen 1964). This account shows that regions in the lee of hills may be more subject to strong winds than corresponding places on the windward side of the hills. In Britain the east or north-east side will usually be the lee side in strong winds, but the well-known Helm Wind of Cross Fell in Cumbria is caused by easterly winds.

It was noted in an investigation of atmospheric pollution at a site in Scotland that two lead-peroxide gauges, exposed in a field only a few metres apart at the same level, gave consistently different records of the amount of sulphur dioxide absorbed each month. Careful comparison checks confirmed these results, and it was concluded that one gauge was in a place which on average had a slightly higher wind speed than the other. Consequently, as the sulphur dioxide content of the air passing the two gauges must have been the same, it was concluded that the one exposed to the higher wind speed sampled more air and therefore extracted more sulphur dioxide than the other. Similarly one may expect the attack on buildings by gaseous or aerosol pollutants to vary with local wind speed, other factors being equal, and even that different parts of the same building will be attacked in a way which will reflect, in part, the microclimate of the wind close to the surface.

Direction of light and strong winds

When the atmospheric pressure-gradient is slight, for example under anticyclonic or 'col' conditions, the resulting light winds are usually quite variable in direction (unless a sea-breeze develops). In such circumstances the wind direction at a place may well be determined by local topography rather than by the regional pressure-distribution.

Thus the local winds will probably need to be studied on the site, although large-scale maps may be used to give a general idea of the local winds. The map study can be especially useful for assessing winds arising from daytime heating (up-slope winds) and nocturnal cooling (down-slope or katabatic winds), or from sea-breezes.

The proportion of time during which light winds or calms occur varies considerably from one place to another. See for example Figure 38: at Birmingham the wind speed was 1·5 m/s or less for about 9 per cent of the time, while at Glasgow these calm or low-speed conditions were experienced for 23 per cent of the time on average over the year. Not only are fogs more likely to form and persist when winds are light, but difficulties may arise with natural ventilation systems in buildings.

The general direction of strong winds is determined primarily by the vigorous weather situations which give rise to them, with modifications caused by local topography. The predominant direction from which the very strongest winds blow varies from one place to another (Figure 38). Generally speaking, this direction is the same as that of the most frequently occurring wind. Thus at Manchester high wind speeds are almost equally likely from any direction except north-east. At London they are most likely to occur when the wind direction lies between south-south-west through west to north. In fact at any given place the distribution of wind-direction frequencies at high speeds is much the same as it is at lower speeds, down to about 4 m/s.

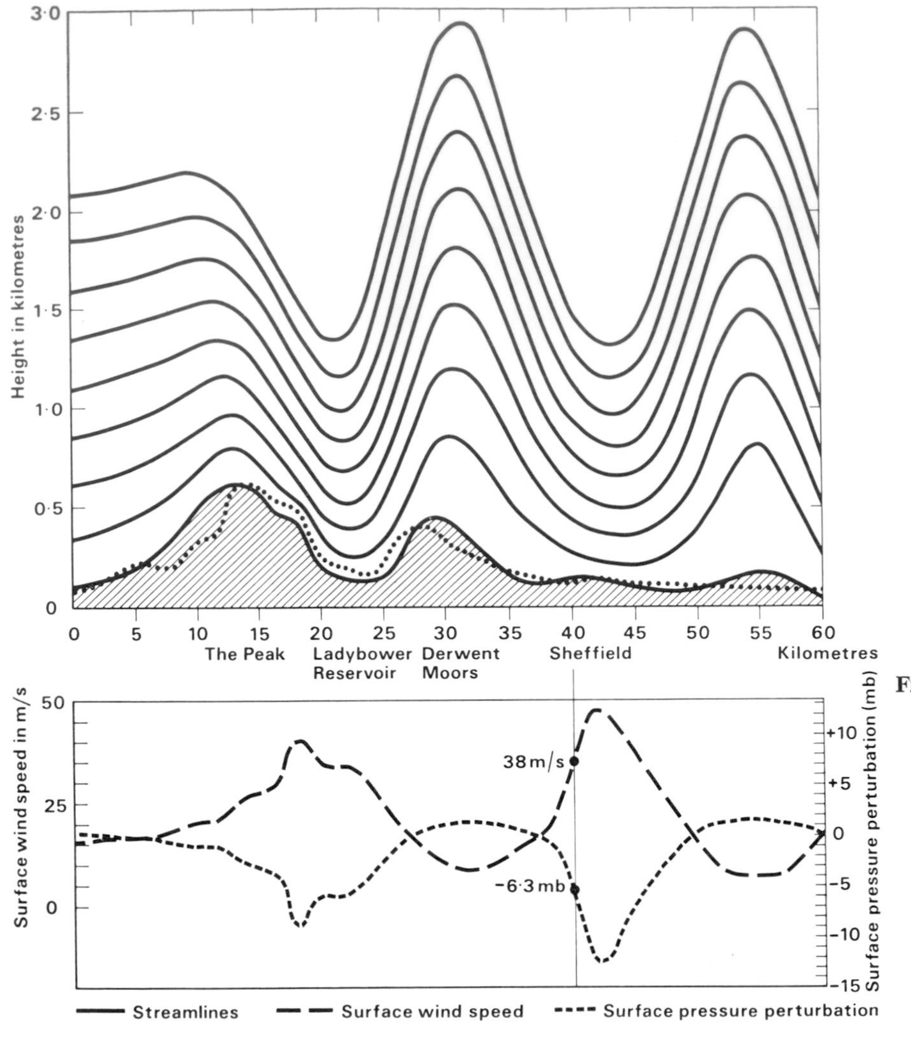

Figure 66 Lee-wave winds in the vicinity of Sheffield, early morning, 16 February 1962: (*a*) calculated streamlines and (*b*) surface wind speed and perturbation of surface pressure for flow across the Pennines in the vicinity of Sheffield. In (*a*) the actual terrain is indicated by the dotted line while the terrain effective in the computation is represented by the shaded area. (From Aanensen 1964)

Lee-wave winds

Places on the lee side of hills with respect to strong winds are often thought to be sheltered. While this is generally true, in some weather conditions standing waves may be set up in the atmosphere as the air flows over the hills. Often those are quite innocuous, their only manifestation to a person on the ground being the 'wave-clouds' which they produce, these often having a smooth, fish-bodied outline. These clouds appear most often around sunrise and sunset, and differ from other types of cloud in that they are stationary while the wind blows through them. The cloud is continually forming at its up-stream edge, where the air is rising, and decaying at its down-stream edge where the air is sinking towards the trough of the wave.

However, when the wind is strong, and the variations of wind speed and temperature with height have a suitable relationship, vigorous standing waves of great wavelength and amplitude may be produced (Figure 66). Below the troughs of these waves, strong winds may be brought down to ground-level (possibly with severe turbulence in 'rotors'), while only a few kilometres down-wind there is a zone of light winds under the peaks of the waves (Figure 67). It was estimated that on the occasion of the Yorkshire gales of February 1962, the strongest winds in the Sheffield area had speeds twice those experienced on the windward side of the hills (Aanensen 1964).

Because of this effect, the wind speeds to be expected in a band of country some 20 km wide, on the eastern flanks of the Pennine Hills, are comparable with those on the coasts (Figure 86). The topography of this region is particularly favourable for the formation of lee-waves, although storms as severe as those of 1956 and 1962 are unlikely to recur often, possibly only once in every century or so on average. However, it is clear that the likelihood of such strong winds should be considered when choosing the sites of buildings in the region.

As far as is known, no other parts of Britain are as suitable for the formation of lee-waves in storm conditions, although the so-called Helm Wind, which occurs in the Eden Valley with an easterly wind blowing over the Cross Fell range, is locally well-known. However, perhaps because the area is relatively little populated, and because it is customary to build well in this rather windy region, no failures from wind are reported.

Tornadoes

Tornadoes are rapidly rotating columns of air, made visible by the condensation of water vapour at the core of the column. Within a metre or two of the centre, the wind speed is extremely high, perhaps as much as 100 m/s, but this falls off rapidly with distance from the centre. Tornadoes experienced in the central parts of the USA are much larger and very destructive, the damaging winds extending over a band which may be 2–3 km wide. The diameter of the region of damaging wind speeds is rarely more than about 10 metres in Britain. However, damage to buildings is usually caused, not by any high wind speed, but by the much-reduced air pressure in the middle of the tornado. As this region of low pressure passes

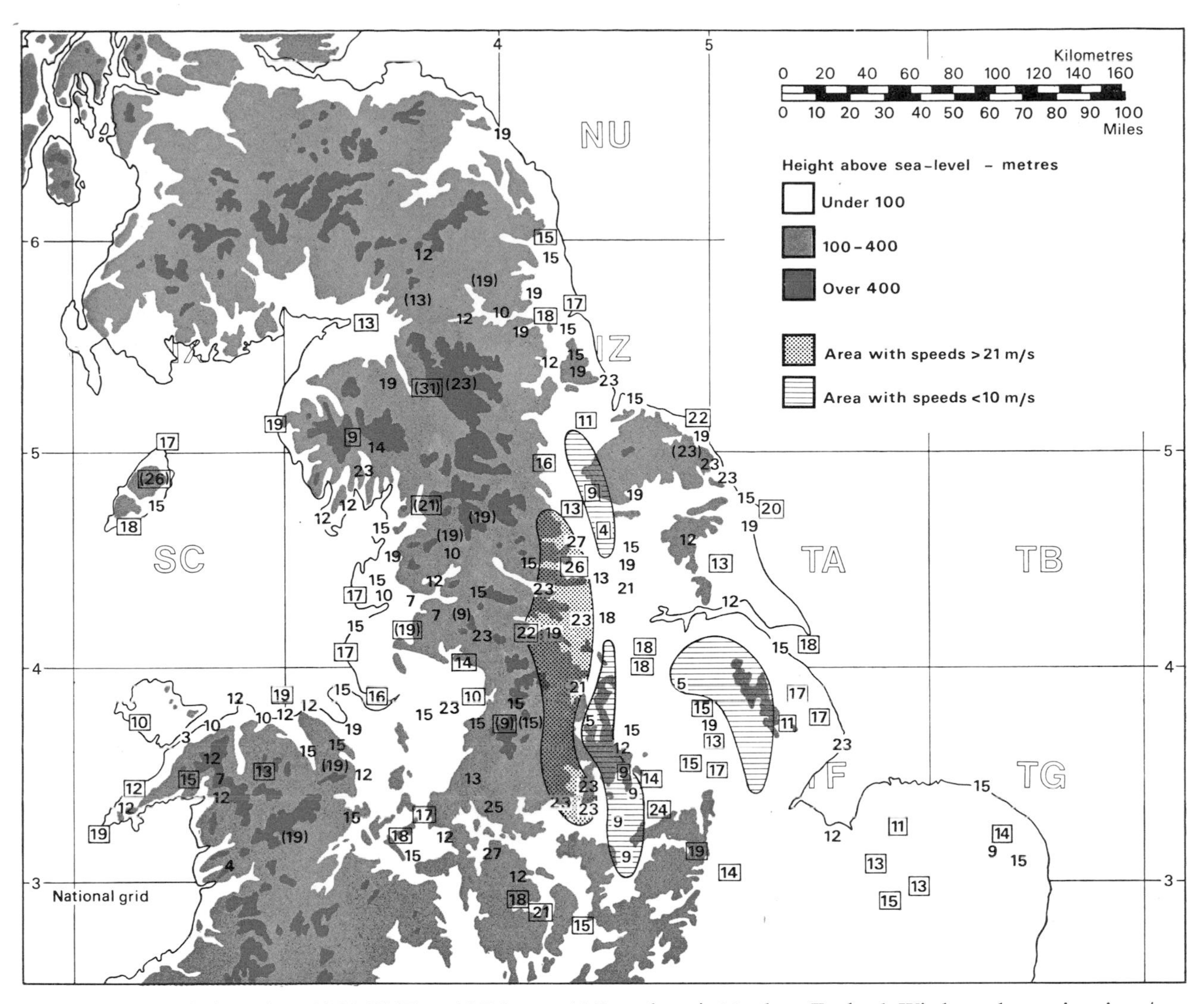

Figure 67 Mean wind speeds at 09 00 GMT on 16 February 1962 at places in Northern England. Wind speeds are given in m/s. Bracketed values are from stations over 245 m above MSL. Values in boxes are from Meteorological Office or auxiliary stations, others are from climatological stations. (From Aanensen 1964)

over a house, the difference between this and the normal atmospheric pressure in the house gives rise to an outward force on the structure. Windows and doors are likely to be burst open, and buildings of lightweight panels may even explode.

Tornadoes seem to occur on about 10 days in each year somewhere in Britain, but most of these are too weak to cause serious damage; usually all that happens is that tiles are removed from roofs or a chimney is blown off. Occasionally one will cause more damage, for example a tower of a school was partly demolished on 15 November 1966 at Leicester. However, although the results of such phenomena can be spectacularly destructive, the path of the tornado is so narrow and the frequency of occurrence so low that the chance of an individual building being affected is quite negligible.

The available evidence (Lacy 1968) suggests that tornadoes can be experienced anywhere in Britain, but that they are rather more frequent in England than elsewhere in the British Isles.

Sea-breezes

The sea-breeze, a cool, moist wind blowing from sea to land, is essentially a summer daytime phenomenon, although a weak sea-breeze can occasionally be observed near the coast at almost any time of the year except midwinter. The sea-breeze arises from the temperature-differential between land and sea and takes the form of a shallow airstream, only a few hundred metres deep. It is most pronounced on warm days when the general wind is light. When the sea-breeze forms, usually some time before midday (Figure 68), it keeps the air temperature at the coast lower than it would otherwise become. As the day progresses, the sea-breeze penetrates further and further inland, the weak sea-breeze 'front' often being marked by a line of haze and cloud. On occasion the sea-breeze front penetrates 100 or even 150 km from the coast, but by this time it is late evening and the resulting slight changes in temperature, humidity and wind are of little practical significance.

Presumably the sea-breeze does give appreciable cooling on hot days in places some distance from coasts, but there seem to be no statistics available.

However, the effects of the sea-breeze are not all beneficial. Sea-fog may come in with the breeze and produce unpleasantly cool conditions. At other places the results may be much worse. Currie (1955) has described how many a pleasant summer day in the northern part of Yorkshire, some 40 km from the coast, is quickly spoilt at about 13 00 GMT by the arrival of the sea-breeze, bearing with it the accumulated smoke and chemical fumes from the industrial plants of Middlesbrough and Billingham. The temperature falls by some 4 to 7 deg C in less than 10 minutes, while the visibility falls from perhaps 20 km to about 800 metres. The remainder of the day passes hazy, smoky and unpleasantly cool.

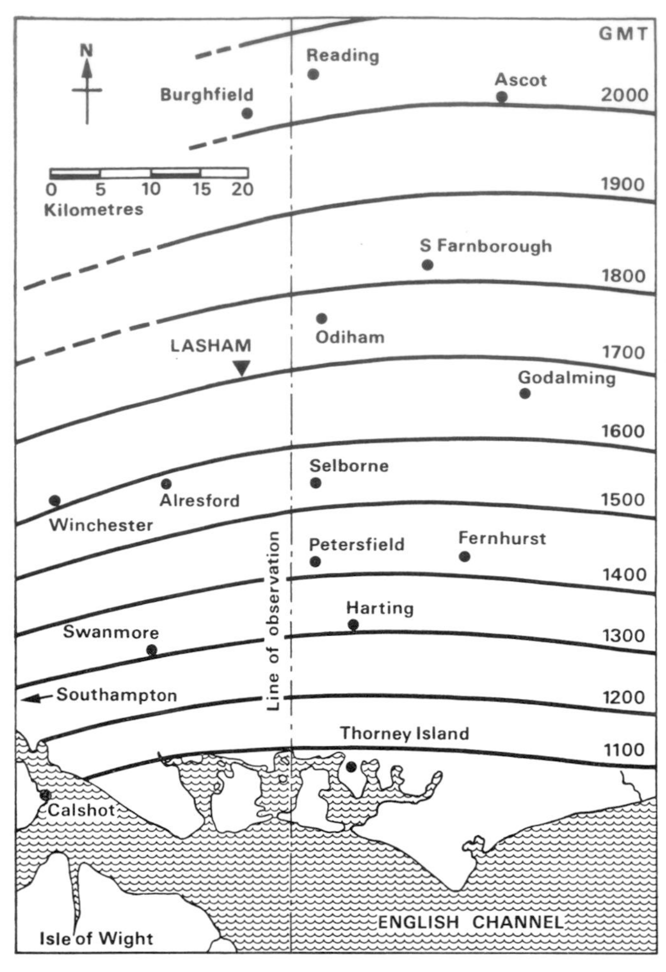

Figure 68 Average GMT times of passage of 41 sea-breeze fronts in southern England, 1962–66. (From Simpson 1967)

The variation of air temperature with wind direction

The temperature of the air on any day depends broadly on the type of weather, and the direction from which the wind blows on that day in its turn depends on the weather type. Particular weather types tend to be associated with a limited range of wind directions. Thus one might expect that each wind direction will, at a given time of the year, be associated with a certain temperature. It is popularly thought that east winds are associated with low temperatures in winter, and that south-westerlies occur in mild winter weather. These beliefs are to some extent supported by the data shown in Table 17 and Figure 69.

These figures are based on measurements made four times a day at 10 places in Britain, during Januaries in the 10 years 1930–39. Each value in Table 17 is the average of all the temperatures measured when the wind direction was that stated at the head of the column – for example, at Wick during these 10 months of January, the mean temperature while a north wind was blowing was 3·7°C. In Figure 69 a wind-rose is drawn centred on each of the 10 stations, the relative lengths of the eight arms being proportional to the frequency of winds blowing from each of the eight directions. The mildest and coldest winds at each station are marked with the relevant temperature, rounded off to the nearest degree. In Table 17 the highest value for each station is shown in bold type, the lowest in italic.

At the only inland station for which data were analysed (Croydon), and for all west-coast stations, south-west winds were indeed the mildest. At most east-coast places south winds were the mildest, or as mild as south-west ones. The coldest winds were consistently those from east or north-east, except at Wick and Tynemouth, where west and north-west winds respectively were the coldest, because winds from these directions were cooled by their chilly hinterlands, and winds from easterly quarters were somewhat warmed by their long sea-passages. The cold north-west winds at Tynemouth may be partly katabatic in origin. However it is worth noting that at Edinburgh, north-west winds blowing from the snow-covered Scottish Highlands were almost as cold as those from north-east and east. Even at the south-coast stations Lympne and Portland winds blowing off the land were appreciably colder than those off the sea – at Lympne north-west winds were 2 deg C colder than west winds. It appears that there are local differences in the direction of the coldest winds in winter which may be significant in town planning, and perhaps more especially in the planning of open spaces intended for recreation. As a further stage in this study, it would probably be worth while considering the frequency with which bright sunshine is associated with various combinations of wind speed and direction and with air temperature, for in northern Britain sitting or strolling out of doors in winter is probably only tolerable in sunny weather.

Table 17 Mean air temperatures (°C) associated with different wind directions in January Averages for 10 years 1930–39

Place	Wind direction								Percentage
	N	NE	E	SE	S	SW	W	NW	of calms
Wick	3·7	4·3	4·7	5·3	**6·0**	4·9	*3·1*	3·3	2·3
Tiree	5·3	4·9	*4·3*	5·5	6·7	**7·3**	6·8	6·3	9·9
Edinburgh	3·5	*3·2*	*3·2*	5·1	**6·1**	5·6	4·4	3·3	29·8
Tynemouth	4·7	4·9	3·9	4·6	5·6	**5·9**	4·8	*3·5*	5·3
Gorleston	4·6	*2·9*	3·2	5·4	**6·1**	5·9	4·7	4·7	7·4
Croydon	3·9	*2·1*	2·7	5·2	6·4	**7·1**	5·9	5·2	10·8
Lympne	3·1	*1·2*	2·2	5·2	**6·4**	**6·4**	4·9	2·9	5·5
Portland	5·0	6·1	*4·5*	7·5	8·7	**9·3**	8·5	7·1	8·7
Scilly	6·7	*6·1*	*6·1*	7·5	9·1	**9·4**	8·9	7·9	2·6
Pembroke	5·7	4·0	*3·7*	5·9	8·3	**9·0**	8·3	7·1	9·7
Holyhead	6·1	5·3	*3·5*	5·2	7·6	**8·1**	7·1	6·9	6·9

The highest value for each station is shown in bold type, the lowest in italic.

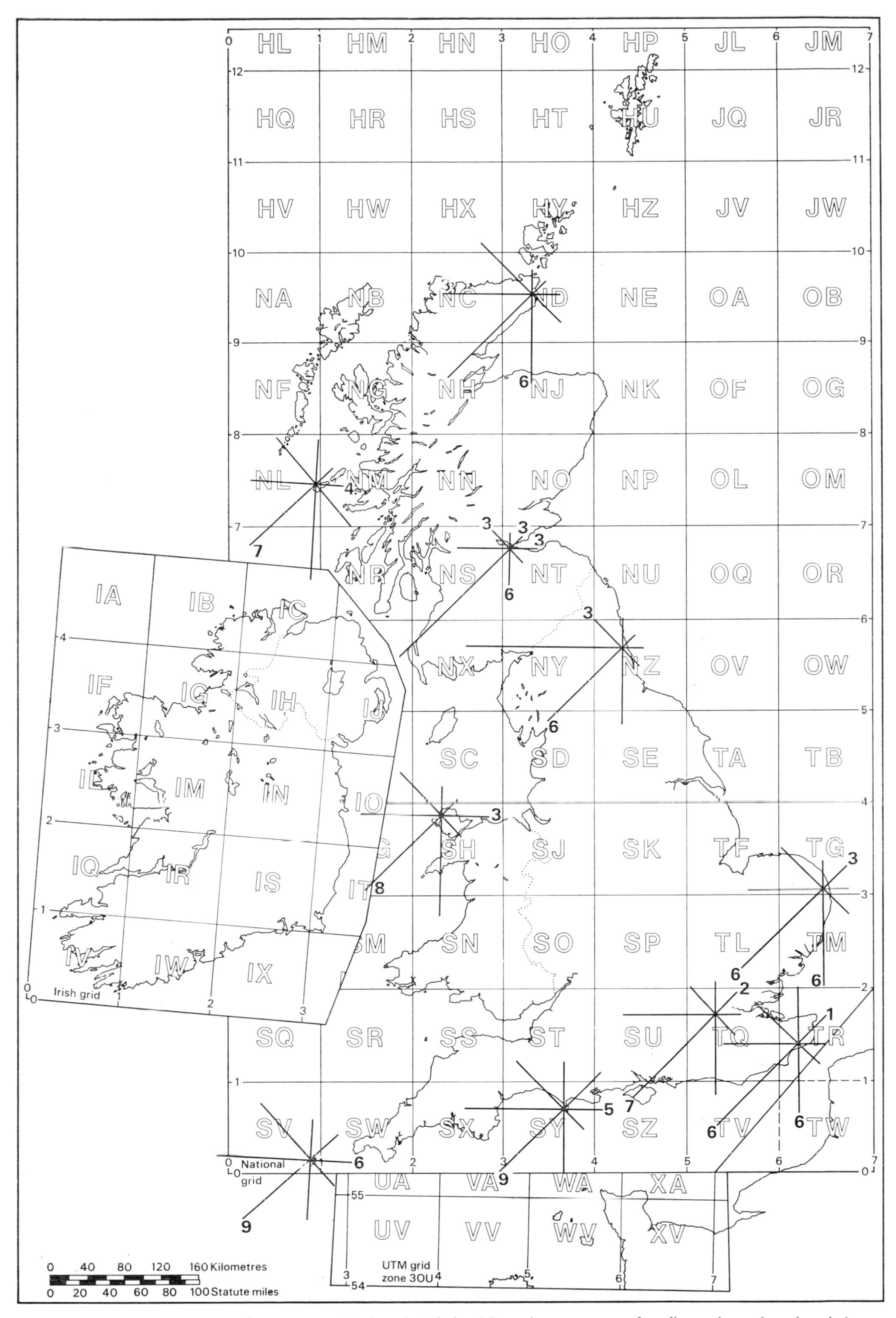

Figure 69 Wind-temperature roses for January at 10 places in Britain. Mean air temperatures for all occasions when the wind was blowing from each of eight directions were computed for each place, but in the figure only the values for the warmest and coldest directions are shown (see also Table 17)

Strong winds with rain, snow and hail

In Britain as a whole wind speeds are somewhat higher (by about 20 per cent on average, see Lacy and Shellard 1962) when it is raining than when it is dry. Conversely, when the wind is strong there is a greater chance that it will be raining than when the wind is light (see Chapter 3, page 116). As on average both wind speed and rainfall amount increase with altitude in hilly country, elevated places experience much more driving rain (see Chapter 3, page 106).

In lowland Britain, drifting of snow is relatively uncommon, so that it seems that in this region appreciable quantities of snow fall only with rather light winds. In upland regions, however, especially near eastern coasts, there may be severe drifting in strong winds. Occasionally, too, the Channel coast may suffer when a wintry depression tracks up-Channel.

Small hail, up to about 10 mm in diameter, does little or no damage to buildings, and in any case usually falls with rather light winds. The rare 'severe local storm', which is an exceptionally violent type of thunderstorm, can produce hailstones 75 mm across, or even larger. These stones may weigh upwards of 100 g. In a storm of this kind on 1 July 1968 in South Wales, violent squally winds drove the giant hailstones against windows, many of which were shattered. Corrugated plastic rooflights were perforated and it is believed that there was some damage to tiled roofs.

On average, one of these severe storms may occur at some place in southern Britain about once in 5 years. Thus, although the 5 km or so wide track of the storm may extend over 50 km or more, the likelihood of one occurring at any given place is very small indeed.

Interference with traffic by snow

Recent experiences show that a shallow snow cover, 20 mm or less in thickness, may be sufficient to hinder or even halt road traffic. This seems to be in part due to the inherent unsuitability of current designs of motor vehicles for use on slippery roads, but in part also to the fact that the number of vehicles on the roads is so great that a slight disturbance of the traffic can quickly grow to block the flow completely

When snow is cleared from roads it may not always be easy to find space to store it: compressed by the clearing process, a given volume melts more slowly than uncompacted snow. It is obviously not acceptable to pile it on to footpaths, but to leave it in gutters restricts the flow of traffic and may interfere with drainage of melt-water.

Falling snow as such is not usually significant, except inasmuch as heavy snow reduces visibility (especially at night) and may cause drivers to go more slowly. Serious interference with traffic does not occur unless there is a snow layer on the roads. Snow falling overnight is likely to cause most trouble, because a layer can build up without disturbance by traffic, and because it may be more difficult at this time to mobilise snow-clearing teams. Thus the statistics of the number of times that snow is lying at the morning observation hour, 09 00 GMT, are likely to be the most significant.

Such statistics can be presented in the form of a map showing the average number of days in a year when snow is lying (Figure 24). In this context a day with snow lying is defined as one on which, at 09 00 GMT, half or more of the country representative of the station is covered by snow.

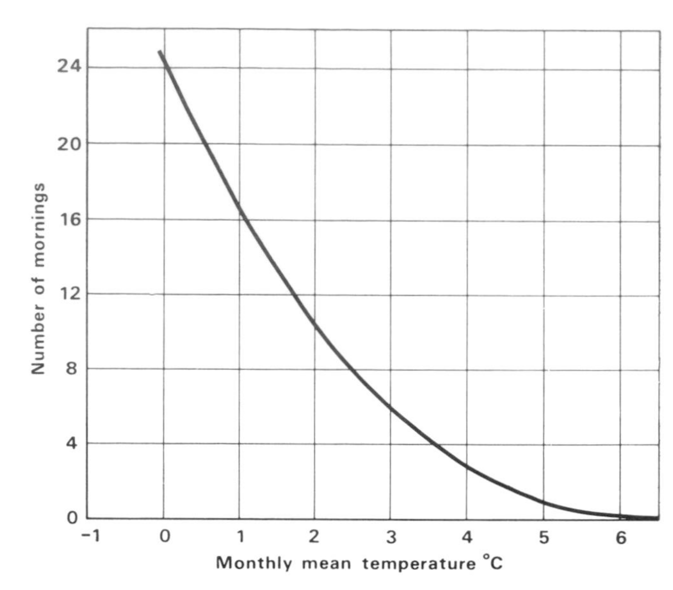

Figure 70 Relation between monthly mean air temperature and number of mornings (09 00 GMT) with snow lying (ie half or more of country covered). Data from period 1912–38. (After Manley 1939)

The observation would be confined to country at the same altitude as the observing station.

Manley (1939) produced figures relating the monthly mean air temperature at a place with the number of mornings when snow covered half or more of the country at 09 00 GMT. These have been used to give the curve in Figure 70. It must be recognised that this relationship is only a rough guide, for duration of snow cover also depends on the frequency of snowfall, on the amount of snowfall, and on the character of the station and its surroundings (such as height, aspect and proximity to hills and to eastern coasts). However the curve does demonstrate the rapid increase in duration of snow cover which may be expected with decrease in monthly mean temperature, and hence with increase in altitude (monthly mean temperature decreases on average by about 0·6 deg C with every 100 m increase in altitude). Figure 71 shows how the number of mornings on which snow was lying varied with

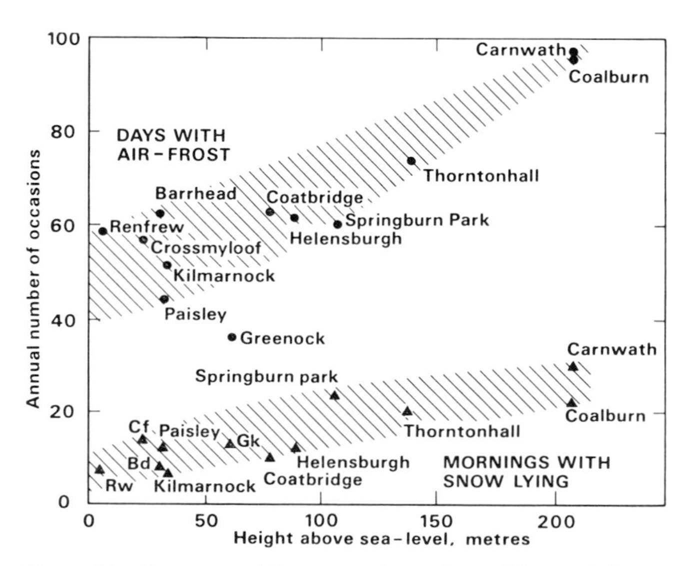

Figure 71 Mean annual frequency of mornings with snow lying (at 09 00 GMT) and number of days with air frost (temperature 0°C or below), 1954–67, related to altitude, for places in Glasgow and the Clyde valley

Table 18 Number of days with snow lying at 09 00 GMT in 10 winters

	Height m	1946 –47	1947 –48	1948 –49	1949 –50	1950 –51	1951 –52	1952 –53	1953 –54	1954 –55	1955 –56
Fort William	8	7	19	10	7	31	19	16	5	29	6
Craibstone, nr Aberdeen	91	60	31	16	18	43	42	24	24	58	59
Balmoral	283	71	53	22	26	102	52	61	40	82	58
West Linton	244	66	25	10	20	64	38	34	25	58	46
Edinburgh, Royal Botanic Gardens	26	—	—	—	2	21	27	5	6	33	23
Renfrew	8	—	—	—	—	—	—	5	12	17	8
Acklington, Northumberland	42	50	3	4	4	16	16	10	11	33	28
Finningley, Lincs	6	53	4	4	5	12	5	4	19	24	30
Manchester Airport	75	17	4	2	1	17	6	7	11	22	18
Birmingham, Edgbaston	163	57	10	4	2	16	12	14	12	36	29
West Raynham, Norfolk	78	55	1	0	1	18	7	13	17	36	25
Ross-on-Wye	68	43	5	1	0	12	4	15	13	17	16
Plymouth (Mt Batten)	27	7	4	0	0	3	4	1	1	5	12
Croydon, Surrey	60	40	6	3	0	7	7	5	7	16	10
West Malling, Kent	92	44	9	2	3	6	8	9	13	20	11
Mean, 4 Scottish Stations		51	33	15	18	60	38	34	23	57	42
Mean, 9 English Stations		41	5	2	2	12	8	9	12	23	20

Data from *Monthly Weather Reports.*

height at stations in the Clyde valley. The data used were given in the *Monthly Weather Report* and cover the years 1954–67. It will be observed that the mean annual frequencies lie, not on a single line, but in a band which slopes upwards as the altitude increases. There is a range of values at any given altitude because of local differences in the sites. Similar diagrams can be drawn for other districts, data for which can be extracted from *Monthly Weather Reports.*

The duration of snow cover varies markedly from one winter to another (see Table 18). The variability is greatest at low-lying stations.

Fog in towns

Years ago, large towns were notorious for their severe, dense fogs, known in London as 'pea-soupers'. Now however, and apparently as a result of the spread of smokeless zones, fogs are not as dirty as they once were. It is not certain whether the frequency of fogs has changed.

London fogs have been more studied than those of other towns and, as was remarked in Chapter 1, page 52, the frequency of fogs in the densely built-up central area is now less than in the suburbs of London, probably because the former is warmer at night. The occurrence of fogs is also affected by topography. This is noticeable even in the shallow

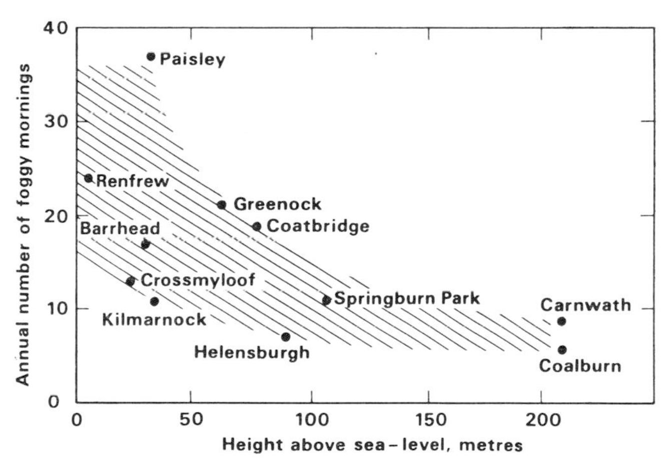

Figure 72 Variation with altitude of mean annual number of mornings with fog in the Clyde Valley, 1954–67. Visibility less than 1 km at 09 00 GMT.

Thames Valley, but it is extremely marked in a valley such as that of the Clyde wherein Glasgow lies (Figure 72). Here there is a fourfold range of frequency of foggy mornings from one place to another, which seems in the main to be correlated with altitude. Places on the upper parts of the sides of the valley are relatively free of fog. Balchin and Pye (1947) in their classic study of the local climate of the Bath region also showed this clearly.

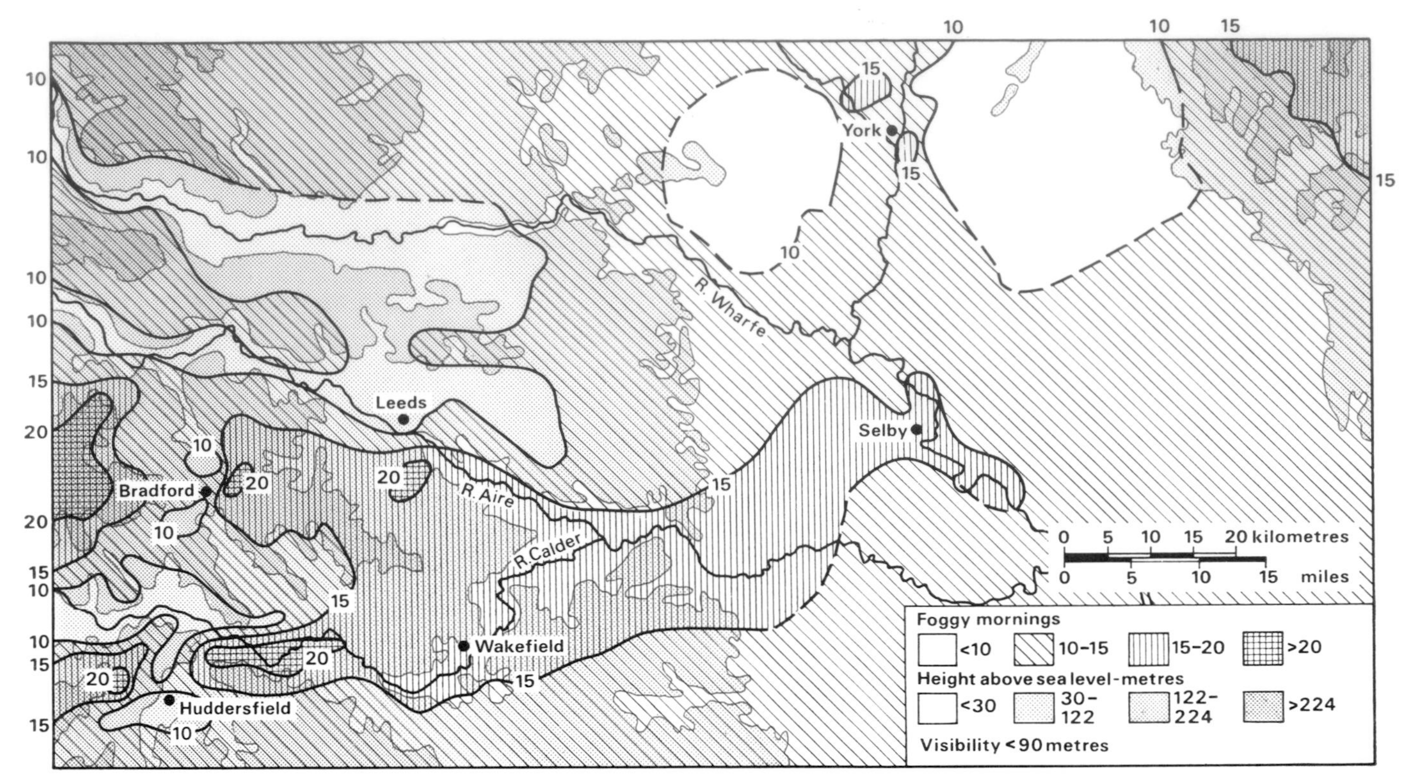

Figure 73 Geographical distribution of the number of foggy mornings in Yorkshire, 1960–61, with visibility less than 90 m (Eyre 1962)

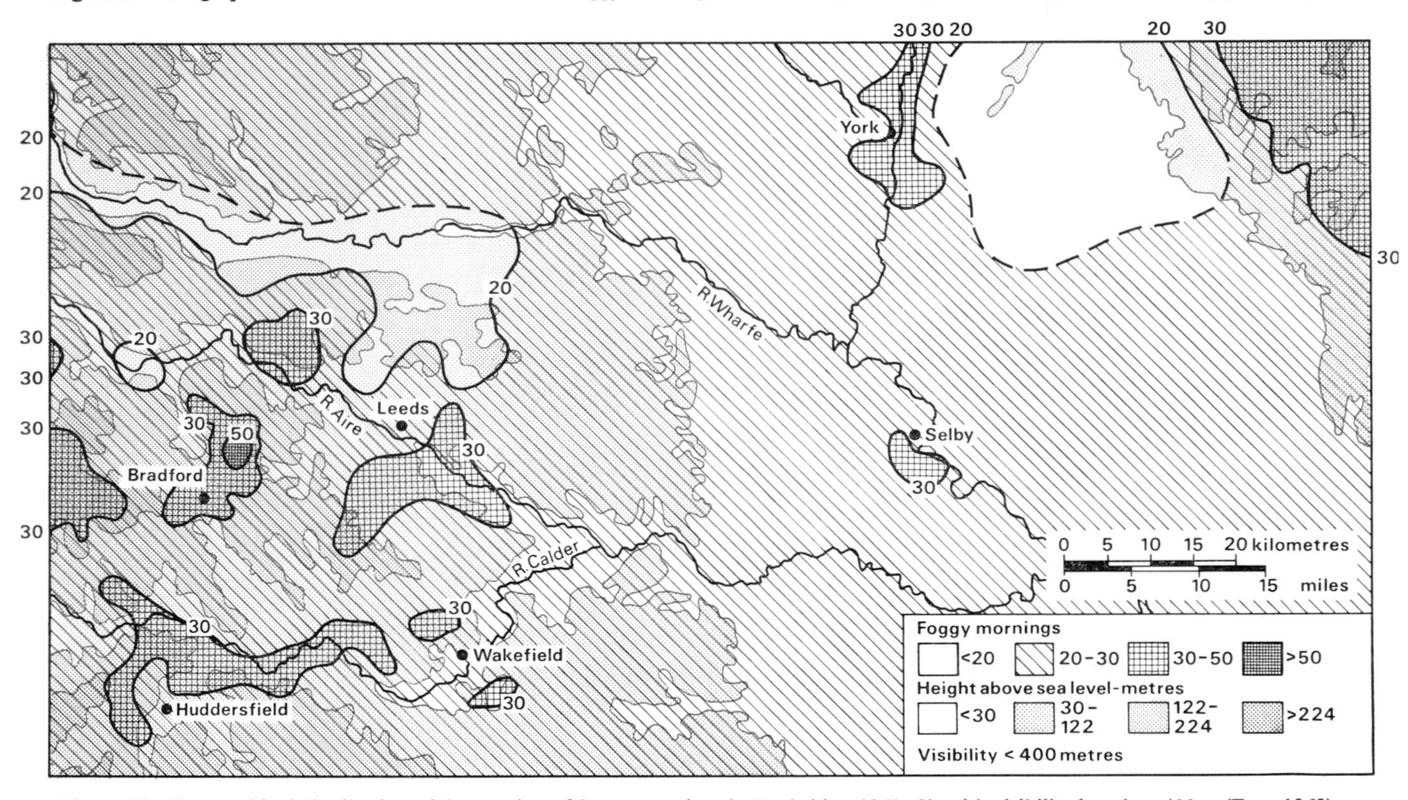

Figure 74 Geographical distribution of the number of foggy mornings in Yorkshire, 1960–61, with visibility less than 400 m (Eyre 1962)

There is not always a simple relation between topography and fog. Eyre (1962) reported a study of the occurrence of fog in parts of Yorkshire (see Figures 73 and 74) during one winter. This showed clearly that, although topography and the presence of towns play an important part, other factors are of significance. These probably include aspect and wind, for there seems to be a suggestion that south-facing slopes have less fog than others, while the tongue of higher frequency of dense fog stretching to Selby suggests a drift of foggy air eastwards from a source in the industrial belt. Note too the increased frequency of fog on the highest ground, this being in fact due to low cloud. A study such as the one reported by Eyre requires a dense network of observers; he had over 800 in an area about 120 km long by 70 km wide, but local and perhaps important variations in the frequency of fog can only be determined by a such a detailed study.

A recent analysis by Atkins (1968) suggests that at Ringway Airport, about 13 km south of the centre of Manchester, poor visibility during light winds had become less common since about 1960 (see Table 19). This analysis is on the basis of wind speed only, a more elaborate one including wind direction might throw more light on the reasons for this change. On the other hand, these improvements have not occurred in smaller towns to the same extent (Corfield and Newton 1968). But this is more a matter of atmospheric pollution, which is

Table 19 Visibility at Ringway Airport, Manchester. Cumulative percentage frequencies of hourly observations of poor visibility, combined with wind speeds

Period	Maximum visibility in metres	Frequency of observation % Wind speed 0–1·5 m/s	Frequency of observation % Wind speed 2–5 m/s
1949–59	200	14	1
	400	19	2
	1000	37	6
	2000	61	20
1960–64	200	11	0·5
	400	15	1
	1000	31	5
	2000	53	15

Percentage frequency of wind speeds

	Wind speed m/s	Percentage of occasions
1949–59	⩽ 1·5	10
	⩽ 5·0	15
1960–64	⩽ 1·5	15
	⩽ 5·0	64

(From Atkins 1968)

Table 20 Frequency of fogs at Heathrow Airport, during daylight hours. Percentage number of hourly observations during the 10 years 1957–66

Month	Ranges of visibility – metres 0–22	23–40	41–94	95–199	200–399	400–1000
January	0	0·5	3·9	2·2	5·5	12·4
February	0	0	1·7	2·6	1·7	9·3
March	0	0·2	2·0	1·4	1·6	7·8
April	0	0	0·6	0·3	0·9	4·4
May	0	0	0·5	0·7	0·3	1·5
June	0	0	0	0·9	0·4	0·5
July	0	0	0	0	0·3	0·3
August	0	0	0	0·4	0·5	1·1
September	0	0·1	2·6	1·9	2·1	5·2
October	0	0·9	4·0	4·0	3·8	6·4
November	0·2	1·8	2·7	2·8	2·6	9·2
December	1·1	1·9	5·0	4·1	4·3	14·2
November, December and January means	0·43	1·40	3·87	3·03	4·13	11·93
Cumulative means for November, December and January	0·43	1·83	5·70	8·73	12·86	24·79

discussed in the next Section (page 78).

There is an apprcciable variation in the frequency of fog from year to year. Lawrence (1966a) has suggested that there is an approximately 11-year cycle in the frequency of bad fogs in the London area. He states that persistent atmospheric stability, light winds and low temperatures are necessary for the development of bad fogs, and that these conditions seem to reach a peak about two years before sunspot minima. Thus there were peaks of smoke pollution at Kew around the winters of 1932, 1942 and 1952. In 1962 there was a high concentration of sulphur dioxide, but after some years' operation of the Clean Air Act, concentration of smoke was no higher than normal. No such analysis has been made for other places and it is too early to say whether there is any real relationship or whether it is purely fortuitous.

Evans (1957) has analysed hourly reports of the occurrence of fog at London Airport (Heathrow). A striking feature is the double maximum in the time of bad fog, one just after sunrise and another at about midnight, although smoke fogs often thin after about 21 00 GMT. On Sunday mornings the poorest visibilities occur an hour or more later than during the week. These facts strongly suggest that domestic smoke takes a significant part in the thickening of fogs.

Table 20 shows that November and December have the greatest frequency of fogs at London, Heathrow, whatever the density, although the first four winter months are all bad in this respect. It is likely that this pattern is broadly true for the whole of outer London, except perhaps for the higher parts such as Hampstead and the North Downs.

Fog, though so frequent, usually lasts only a few hours. However it does occasionally, in calm conditions, last for many hours on end. Kelly (1963) studied the occurrence of spells of fog lasting for at least 12 hours in the London area (at London Airport (Heathrow), Croydon Airport, and at Kingsway in central London). He considered separately the records of thick fog (visibility less than 200 m) and of dense fog (visibility less than 50 m). He used the normal practice of calling fog lasting for an unbroken period of 24 hours 'persistent', and in addition called a spell of at least 12 hours 'semi-persistent'. Table 21 lists the number of all spells of thick fog lasting at least 12 hours, and therefore includes both categories. The number of thick fogs lasting 24 hours or more at the three places were respectively 10, 3 and 5. The longest durations recorded were respectively 69, 44 and 57 hours, the first and last being during the disastrous London fog of December 1952. The longest duration at Croydon was in November–December 1948.

Table 21 Frequencies of persistent thick fogs in the London area, 1947–56. Number of occasions and (in brackets) total number of hours the fogs lasted

	October	November	December	January	February	March	Total
Heathrow	2 (30)	14 (290)	8 (216)	8 (191)		2 (28)	34 (755)
Croydon	3 (36)	8 (139)	6 (119)	4 (77)		3 (43)	24 (414)
Kingsway		5 (111)	5 (120)	1 (30)	1 (18)	1 (12)	13 (291)

Thick fog — visibility less than 200 m.

Persistent fog — fog lasting continuously for 24 hours or more.

(From Kelly 1963)

Table 22 Cumulative frequencies of prolonged continuous spells of fog in the London area in 10 years 1947–56. Number of fogs with visibility less than 200 m and (in brackets) less than 50 m

Duration of spell (hours)	Heathrow	Croydon	Kingsway
12 or more	34 (7)	24 (4)	13 (4)
18 or more	18 (2)	7 (2)	9 (3)
24 or more	10 (1)	4 (2)	5 (2)
30 or more	7 (1)	3 (1)	2 (1)
36 or more	4 (1)	1 (0)	1 (1)
42 or more	3 (0)	1	1 (1)
48 or more	3	0	1 (0)
54 or more	1	0	0
66 or more	1	0	0

From data quoted by Kelly (1963).

The frequencies of continuous spells of fog during the 10 years at the three London stations are given in Table 22. From this it would appear that a continuous spell of 12 hours or more fog with a visibility of less than 200 metres can be expected about 3 times a year at Heathrow, twice at Croydon and once at Kingsway. A 24-hour spell (persistent fog) can be expected once a year at Heathrow and once in 2 years at Croydon and Kingsway. It is not known whether there have been any changes in these frequencies in more recent years.

Corresponding data for visibilities of less than 50 metres (a density which would be expected to interfere seriously with most wheeled traffic) are given in brackets in Table 22. The frequencies are much lower, with a persistent spell (24 hours) to be expected only once a year at Heathrow. However, the frequency of such a spell at Croydon or Kingsway in this period was twice as high as at Heathrow, whereas with thinner fogs the opposite was the case. The longest duration recorded for a fog with visibility less than 50 metres was 45 hours, at Kingsway, whereas the longest duration at Heathrow was 37 and at Croydon 32 hours.

It appears on the basis of this evidence that prolonged 'thick' fogs with a visibility of less than 200 metres are commoner on the outskirts of London than in the middle of the town, but that prolonged 'dense' fogs, with visibility less than 50 metres, are rather more common in the middle of the town than on the outskirts. There are no corresponding statistics available for other towns.

Atmospheric pollution

In the preceding Section, fog has been discussed. While fog is often associated with artificial, ie man-made, pollution, this is not always so. Conversely, atmospheric pollution is not always associated with fog. This is the justification for having separate Sections on atmospheric pollution and fog, although obviously there are links between the two.

The town planner is concerned to place a source of pollution where it will not cause excessive concentrations of pollution over built-up areas, or alternatively to site new developments where they will not be subject to pollution from existing sources. At one time it was customary to place industrial premises and other sources of pollution downwind of residential and commercial areas. In this context, downwind refers to the most frequent or 'prevailing' wind.

The prevailing wind in Britain is usually taken to be from SW or W, irrespective of place or season. As described in Chapter 1, this can be misleading, for the prevailing wind is not the same at all places, while it may not blow from the same quarter at all times of the year. There may also be a well-marked diurnal variation on certain days, in particular at places subject to the sea-breeze.

The annual mean frequency distributions of hourly values of wind speed and direction at four places in Britain are shown in Figure 38. It can be seen that the wind regimes at these four major centres of population differ significantly from one another. Both Glasgow and London lie in valleys which run roughly west-east, and this is reflected in the frequency distribution of wind direction. There is a deficiency of winds from both north and south, especially at Glasgow. During about one-quarter of the hours the wind is very light at Glasgow, with speed below 1·5 m/s. In London, the proportion of light winds is also high, over 17 per cent.

At Birmingham, on its Midland plateau some 150 m above sea-level, the distribution of wind direction is that usually regarded as typical for Britain, with a peak between south and west. The proportion of light winds is only 10 per cent. At Manchester there is yet another distribution, with a peak at south. Presumably its position on the Cheshire plain, to west of the predominantly north to south alignment of the Pennine range, is responsible for this. The proportion of light winds is rather higher than at Birmingham.

It would seem desirable, therefore, to examine in more detail than is usual the wind regime at a site before deciding on the relative siting of buildings.

It should be borne in mind in this context that the 'prevailing' wind, whatever this may be, may not be the significant one for producing severe pollution. When an appreciable wind is blowing, the air is usually more or less turbulent. Thus pollution emitted by individual sources becomes diffused through a large volume of air, and no one place is exposed to a high concentration of the pollution. But when the wind is light, there is little turbulence to disperse a plume emitted by a chimney. Moreover, there is likely to be a 'temperature inversion' when the wind is light, so that the temperature of the air increases with height up to 100 or 200 m above ground, instead of decreasing as is usual. Under such conditions, any pollution emitted into this lower layer of air will be unable to escape to the upper air.

Thus, under an inversion it can be expected that levels of pollution will be high. At the same time, the direction of the air movement may not be that of the 'prevailing' wind. Indeed, Lawrence (1966b) has shown that at Crawley, Sussex, temperature inversions are most likely to be associated with light easterly winds. This result is probably broadly true for the whole of south-eastern England, but is unlikely to be so for the whole of Britain.

When there is a very light wind at night, and a temperature inversion has formed some 100 metres above the ground, smoke trapped beneath the inversion may travel a long way.

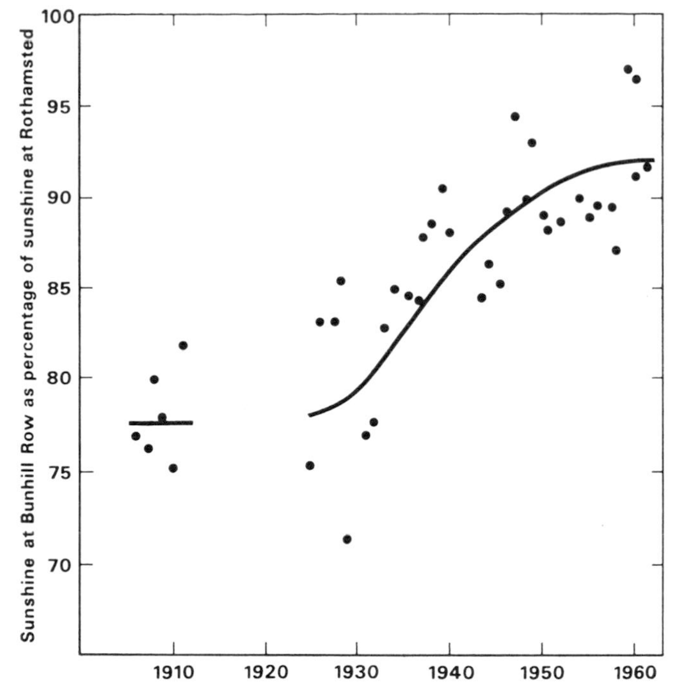

Figure 75 Hours of bright sunshine at Bunhill Row, in the City of London, as percentage of the sunshine at Rothamsted during 1906–63 (whole year). (From Craxford *et al* 1964. By permission of the Director of Warren Spring Laboratory)

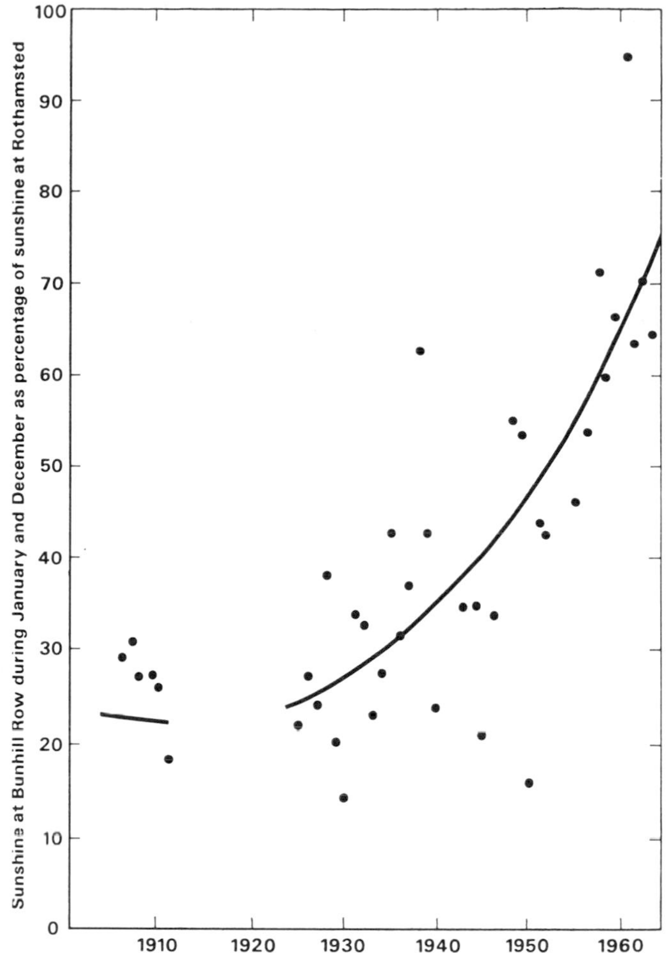

Figure 76 Hours of bright sunshine at Bunhill Row for January and December as a percentage of those at Rothamsted for the same months, 1906–62. (From Craxford *et al* 1964. By permission of the Director of Warren Spring Laboratory)

At Garston, Hertfordshire, for example, the distinctive sulphurous smell of the smoke from brickworks making Fletton bricks has been noted on more than one occasion, just after sunrise. On some of these mornings the trajectory of the air was such that the smoke must have come from brickworks south of Bedford, some 43 km to the north-north-west of Garston. On another, the source was estimated to have been the works at Peterborough, about 95 km north of Garston. Before sunrise, this smoke would have been concentrated in a layer just below the inversion, but after sunrise, as the ground becomes heated, convection currents form in the air below the inversion and the smoke layer becomes mixed into the air between the inversion and the ground. Thus the smoke reaches the ground for a short while, before the convection becomes vigorous enough to destroy the inversion, after which the concentration of the smoke decreases rapidly.

Scorer (1968) has emphasized the necessity of making a case study for each locality, pointing out the importance of local topography in determining airflow when there is an inversion. The difficulty of calculating the rise and dispersion of plumes was also stressed at a discussion reported by Scorer and others (1968). Local meteorological and topographical conditions were important and past experience should be used as a guide.

Not only does the pollution do damage to buildings and their contents, but it cuts down the amount of sunshine reaching the ground, and so causes a marked loss of amenity. In the past the greater cities lost much of the short winter sunshine because their air was so dirty, but things are now improving. There is evidence of a considerable increase in the amount of sunshine reaching the centre of London during the last few years, which may be attributed to a decrease in the amount of smoke in the air (Figures 75–77). In fact, at High Holborn in London the radiation reaching a well exposed roof-top is now at all times of the year about the same as that reaching the ground outside the town.

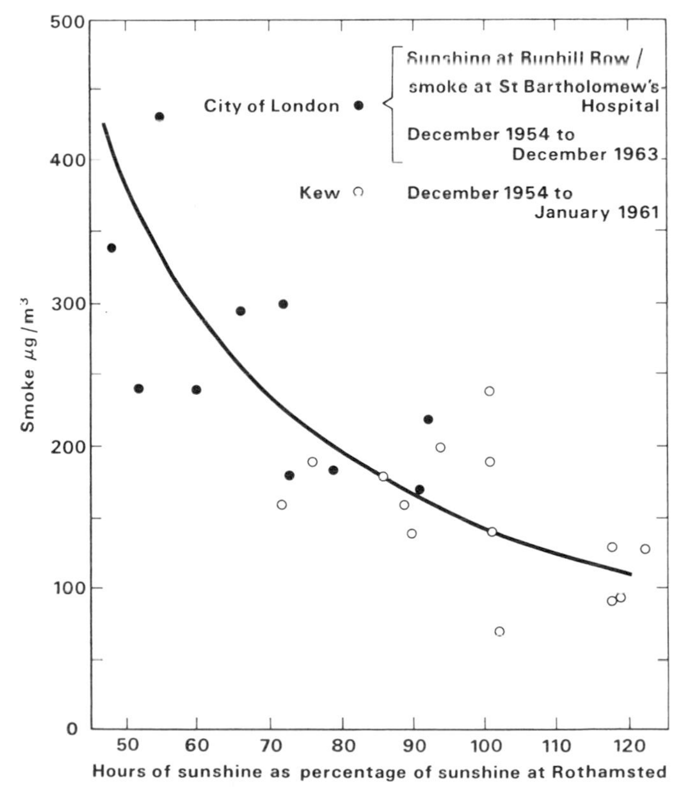

Figure 77 Relation between ground-level smoke concentration and duration of bright sunshine during January and December in the City of London, the sunshine being expressed as the percentage of that outside the city. (From Craxford *et al* 1964. By permission of the Director of Warren Spring Laboratory)

Atmospheric humidities in towns

Observations of the humidity of the air in towns are relatively uncommon, but it appears that in general the air in the town is somewhat drier than that outside in the open country. Kratzer, for example (1956), quoted the mean vapour pressure in six cities in Germany and Austria as being 0·3 to 0·7 mb below that in the surrounding country. He stated that the difference was almost zero in winter, and about double the annual average difference in summer. Relative humidities will differ correspondingly, but are also dependent on air temperature. Chandler, on the other hand (1965, page 202), concluded that there is only a 'weak overall tendency to drier air in the built-up area' of London, after a study of extensive measurements of humidity made in traverses of the town.

Less information is available about short-term and diurnal variations of humidity in towns, but it is to be expected that major changes in the general humidity (for example with changes in air-mass) will be the same in the town as in the country around. However, Chandler has shown that on clear, calm, summer nights the absolute humidity of the air in towns may be higher than that in the country around. He has reported measurements made in London (Chandler 1965) and in Leicester (Chandler 1967) which clearly demonstrate this. The maximum vapour-pressure differences amounted to around 3 mb. He suggested that pockets of warm, moist daytime air may remain enmeshed among the buildings during calm, clear nights; on these occasions the air in the country loses moisture because it cools down so much that part of its moisture is deposited as dew on vegetation and the ground. In the town the air does not cool so rapidly and therefore retains its moisture.

Now in fact it is possible that some of the extra moisture may arise from evaporation of water which has been absorbed by the structures in the town. It is frequently stated that the surface of a town is more or less completely impervious to water, so that rain drains away completely and rapidly. Following on this it is assumed that there is little or no water available for evaporation into the atmosphere. Both Kratzer and Chandler stress this point. In fact, measurements of run-off in towns have shown that this is not so. For example Watkins (1963) gives the results of measurements in a number of English towns. One of the test districts comprised 200 acres of Kensington, London, described as 'old dense development', of which 95 per cent was paved. The run-off from this paved area was about 50 per cent of the rainfall. Six other districts, ranging in area from 4·8 to 59·5 hectares, were described as 'housing' or 'old mixed development'. The proportion of paved ground varied from 20 to 42 per cent, averaging about 30 per cent. From this paved ground the run-off in storms was about 75 to 80 per cent of the rainfall, that is some 23 per cent of the rainfall on the whole of the district.

Most towns consist partly of dense and partly of mixed development, and in the following calculations a run-off of 40 per cent has been assumed. It is assumed that the remaining 60 per cent of the rainfall has been absorbed by unpaved ground and by porous building materials (bricks, tiles, concrete, etc) although a proportion may lie as puddles on tarmac and flat roofs, etc. Usually the materials are not saturated with moisture so that the water can only escape to the atmosphere by diffusion through the pores of the material (as is the case with soil). The rate of diffusion will be low compared with the rate of evaporation from a free water-surface (probably usually in the range 1×10^{-3} to 1×10^{-2} $g/m^2 s$,) and will be determined largely by the properties of the material and not by meteorological factors. The subsoil is taken to be impervious, so that there is no percolation, and all water absorbed by topsoil in the town is available for evaporation.

Consider a city the size of London, some 40 km across. The mean annual rainfall is about 625 mm, so that 375 mm is available for evaporation, ie $3{\cdot}75 \times 10^5$ g/m^2. Thus the annual mean rate of evaporation will be $1{\cdot}2 \times 10^{-2}$ g/m^2 s. If we suppose that water vapour escaping at this rate becomes uniformly mixed into a layer of air 100 m deep, travelling at 3 m/s from the edge of the city to the centre (a distance of 20 km), the amount of moisture taken up by a column of air of 1 m^2 cross-section is about 80 g. The increase in moisture-content is therefore about 0·8 g/m^3, which is some 10 per cent of the annual average moisture-content of the air over London.

There will be other sources of moisture, from the combustion of fuel, for example, but the quantities of fuel used in London (quoted by Craddock 1965) yield moisture equivalent to only 1 per cent of that from the evaporation of rain.

It is noteworthy, though presumably only a coincidence, that the amount of rainwater estimated to be absorbed in the town, about 375 mm, is nearly the same as the mean annual evaporation from the tank at Camden Square, 408 mm. This was a standard evaporation tank, about 2 m square, sunk in the ground in the back garden of a house in Camden Square, some 4 km from the middle of London. It was in use from 1885 to 1955. The situation was very sheltered, and probably typical of an average inner suburban exposure.

Chandler described (1967) measurements in and around Leicester on the nights of 19, 23 and 31 August 1966. On these nights there were strong inversions of temperature, and hence presumably of water-vapour pressure, in the air over open country. Thus the air could gain no water vapour from the ground, and indeed it may have lost some by deposition of dew. In the town, on the contrary, the higher thermal conductivity of the materials of streets and buildings probably permitted the release of heat at a rate sufficient to delay or even inhibit altogether the formation of an inversion of the temperature and vapour-pressure gradients. Thus in a large enough built-up area, evaporation from damp materials could continue, and a significant increase in the moisture-content of the air could occur. The weaker the wind, the greater the increase in vapour pressure near the ground, and the greater the chance of pockets of moister air occurring in exceptionally closely built-up areas as described by Chandler. On these occasions he recorded vapour pressures of 1 to 2 mb above those in the open country. The calculated increase of 0·8 g/m^3 is equivalent to about 1·2 mb. This calculation was for London. Leicester is a smaller town, only 7 km or so across, but if the wind speed was lower, the increase in moisture-content of the air might be little different from the value estimated for London.

This excellent agreement may be fortuitous, for the rate of loss of water by building materials does vary, by a factor of ten or more, depending on their moisture-content, while the depth of the 'mixing layer' in the air is uncertain and no doubt variable. The rate of loss in a fairly prolonged dry spell would be expected to be 1×10^{-3} g/m^2 s or less (Granum 1965), but August 1966, when Chandler made his measurements in Leicester, was exceptionally wet. Indeed, on the 29th about 20 mm of rain fell in the Leicester area, and an evaporation rate of about 1×10^{-2} g/m^2 s on the 31st is not unlikely, especially as there might have been many pools of water in streets and on flat roofs.

It appears, therefore, that the slow release of moisture, by diffusion through the pores of building materials, may be sufficient explanation of the increased humidity of the air in towns on certain calm nights, although clearly more information is needed.

It would be of interest to be able to set up a complete heat and moisture balance for a town, but it is clear from the figures quoted that generally speaking the difference in humidity between town and country air is not great enough to be significant when considering comfort and air-conditioning.

Trees in towns

It is often said that trees help to ameliorate the harsher aspects of town climate, in particular to cool streets during hot weather and to deaden the noise of traffic (see also page 57).

In fact, there is no evidence that trees in streets cause any measurable reduction in the air temperature. Even in a green area as large as Hyde Park, some 1·5 km across, Chandler (1965) found that air temperatures were never as much as 2 deg below those in the London streets around the park. In fact the differences were least in the hottest part of the day. However, although the trees may not lower the mid-day temperatures, they will make conditions more agreeable for the pedestrian, by shading him from the sun's rays, and from solar radiation reflected from buildings, as well as from long-wave radiation which is re-radiated from the hot surfaces of buildings. For this reason alone the use of trees is justified.

On the other hand, measurements have shown that the absorption of noise by the narrow belts of trees used in towns is quite insignificant. This can easily be demonstrated by walking into a wood alongside a busy road. It will be found that the noise level decreases at about the same rate as if the wood were not there. A shelter belt of trees needs to be about 50 m wide at least to give a useful reduction in noise level. Furthermore, the types of trees used would need to be selected with care to give maximum effectiveness at all times of the year. In any case it is clear that it might take many years to grow an effective shelter belt for this purpose.

The influence of weather on the propagation of sound

In perfectly still air, with no temperature gradients, sound travels in straight lines and the intensity of a sound at a given distance from a source will be the same in all directions. In practice there are usually gradients both of temperature and of wind speed, which bend the path of the sound waves.

Thus, since wind speed normally increases with height above ground, the path of the sound downwind from a source tends to be deflected downwards, and at a given distance the sound is louder than it would be in the absence of the wind. Conversely, upwind of a source the path of the sound is deflected upwards, so that the intensity heard at ground-level is reduced. The decrease in intensity upwind is proportionately greater than the increase in intensity downwind. (Turbulence in the wind scatters the sound in a random manner, but the effect is small compared with that caused by the wind-gradient.)

The effect of a lapse of temperature (ie the temperature decreasing with height above ground) is to bend the path of a sound upwards, because the speed of sound decreases as the air temperature is reduced. Thus near the ground there is a uniform reduction in noise in every direction, with a tendency to form a symmetrical 'shadow zone' all round the source. The effects due to wind and temperature are additive, but usually the wind effect is the most important, even at quite low speeds of 2 m/s or so (Scholes and Parkin 1967).

Temperature decreases with height in this manner on most days. At night, however, it is common for the wind to be light, and for the temperature to *increase* with height (the so-called temperature inversion) up to a height of perhaps 100 metres. The inversion is shallow when it first forms around sunset, the depth increasing through the night to a maximum around sunrise (although on a still, foggy day the temperature inversion may persist all through the daylight hours). While the inversion lasts, the sound rays from a source will be bent downwards, so increasing the noise perceived in all directions. Winds are always light when there is a temperature inversion, in fact there is usually no more than a gentle drift of air, less than say 0·5 m/s.

Absorption of sound by the ground has a significant effect on noise transmission, wet ground being less absorbent than dry ground (Scholes and Parkin 1967). For a more exhaustive discussion of meteorological effects on outdoor sound see Ingard (1953).

In a typical fairly open inland site in the UK the annual average wind speed, measured at a height of 10 m above ground, is about 4 m/s and for about three-quarters of the time the speed is greater than about 2 m/s. At this speed the effect of the temperature gradient on the path of the sound can be neglected, and we can say that the only important meteorological parameters are the speed and direction of the wind relative to the line between source and observer. It is of course tempting to suggest that sources of noise should be kept downwind of living and office areas, with respect to the 'prevailing wind'. In fact, as is clear from Table 17 and Figure 38, wind direction is much more variable than is generally believed. The wind actually blows from the so-called prevailing direction for only a few per cent of the time (eg at Manchester, it blows from south, 170° to 190°, for some 15 per cent of the time during an average year). Indeed, the frequency of light wind speeds, of 1·7 m/s or less, may be greater than the frequency of winds from the 'prevailing' direction – as can be seen for data from Renfrew in Figure 38.

According to some figures for Glasgow quoted by Bilham (1938, page 235), on about 10 per cent of nights the grass-minimum temperature fell below the screen minimum by at least 5·5 deg C (10 deg F). These may be assumed to be nights with a strong and deep inversion of temperature. They were commonest in the months September to November (about 18 per cent of nights) and least common in January–March (about 6 per cent of nights). Significant inversions probably occur on some 30 per cent to 40 per cent of all nights at inland places. On these occasions there will be some intensification of sound from noisy sources, in all directions from the source.

On the basis of these statistics it seems that although on the whole it may be desirable to have noise sources to eastward of offices and dwellings, the difference is perhaps not enough to override other planning considerations. This is especially so when conditions at night are important, for then winds are usually lighter and temperature inversions most common.

Chapter 3 The effects of climate on the design of buildings

In this chapter some of the ways in which the weather affects the design of buildings are discussed. In some cases it has been possible to include data which can be used directly in design. In others the way in which the weather influences the building is not known with sufficient precision and one can do little more than indicate the existence of a problem and suggest that it should be given some consideration.

Wind loading on buildings

Britain is one of the windiest countries of the world. Figure 31 shows that even in the southern parts of England the mean wind speed is some 4 to 5 m/s, while in exposed western and northern parts it is 7 m/s or more. Gales, during which the mean wind speed exceeds about 17 m/s continuously for a period of at least 10 minutes, occur on about two days a year on average in southern England away from the coasts and on 30 days or more each year on the most exposed coasts (Figure 40). Even in the least exposed parts of the country a gust with a speed of 40 m/s lasting for about three seconds may be expected on average once in 50 years (Figure 86). In the most exposed parts the gust speed may be over 55 m/s.

It has therefore been traditional to erect buildings stout enough to resist such conditions. However, current developments in building design and construction are making buildings more susceptible to the wind and are tending to reduce the safety margins that have enabled older buildings to survive. There has been a considerable advance in knowledge of wind loading in recent years, as a result of research and the study of buildings on which there have been failures from the action of the wind, but there is still need for a wider appreciation of the problem by designers.

This Section of Chapter 3 largely reproduces parts of BRE Digests. Experience has suggested some changes in the advice given in the earliest publications on this subject. The main changes are to the Table which is numbered 24 in the present work, which now allows for a greater reduction of wind speed at heights below the general level of obstructions. In this height zone there is, however, the probability of some local higher wind speeds due to eddies and the funnelling of wind between buildings. Moreover it should be borne in mind that local exposure can change as new buildings are put up or old ones demolished. The corresponding factors of Table 24 should therefore be used with caution; they are enclosed within a box as a reminder of this.

The wind-speed averaging times have been adjusted and the use of the 3-second gust is restricted to loads on cladding and glazing and their fixings. For the structural loading of all buildings having a maximum overall dimension of less than 50 m, the 5-second gust is used.

It had been intended to give guidance in this Section on the use of the wind data in design, quoting typical pressure and force coefficients. However, knowledge and experience in this field are increasing so rapidly, and the sheer bulk of material is so great, that it is felt to be preferable to refer readers to the relevant Code of Practice (British Standards Institution 1972) and to the *Wind loading handbook* (Newberry and Eaton 1974). Between them these give the full information needed by designers.

The generation of pressures and suctions

When the wind blows more or less square-on to a building, it is slowed down against the front face with a consequent build-up of pressure against that face; at the same time it is deflected and accelerated around the end walls and over the roof with a consequent reduction of pressure, that is suction, exerted on these areas. A large eddy is created behind the building which exerts a suction on the rear face. All these pressures and suctions act in a direction perpendicular to the surface adjacent to them. These effects are illustrated in Figure 78. The frictional drag on a surface over which the wind is blowing is, in most cases, insignificantly small.

The greater the speeding up of the wind, the greater will be the suction. Thus the channelling of wind between two buildings can produce severe suction loadings on the sides facing the gap between them. Access openings through and under large slab-like blocks are usually subjected to high wind speeds through them due to the pressure difference between the front and rear faces of the building; the facings of such openings are particularly prone to high suction loadings which may damage the glazing and cladding.

On the windward slope of a roof, the pressure is dependent on the pitch. It is at maximum negative value, that is maximum suction, at pitches below 15° but diminishes with increasing roof pitch until it is virtually zero at about 30°. Roofs steeper than about 35° generally present a sufficient obstruction to the wind for a positive pressure to be developed on their windward slopes, but even with such roofs there is a zone near the ridge where suction is developed and roof coverings may be dislodged, as shown in Figure 79, if they are not securely fastened.

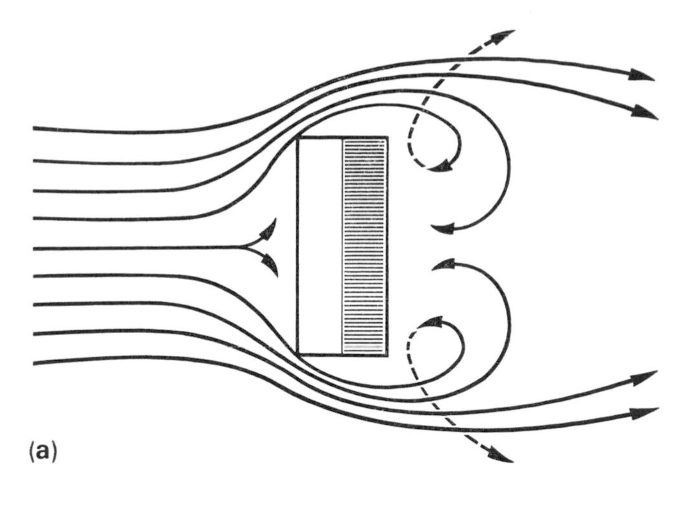

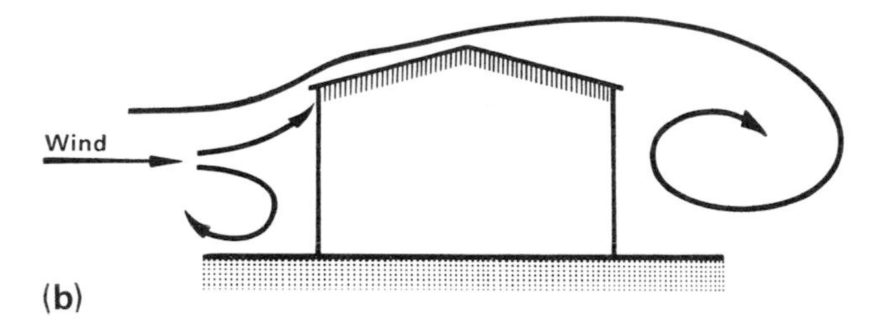

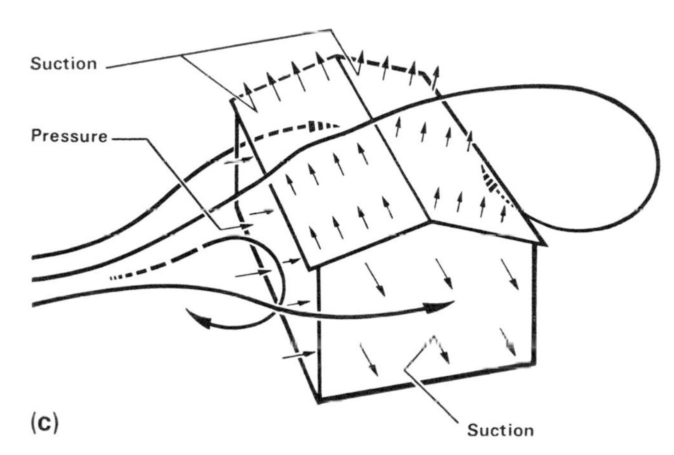

Figure 78 Airflows and resulting pressures and suctions on an isolated building

Figure 79 Tiling lifted off by suction forces. Wind direction from right of picture

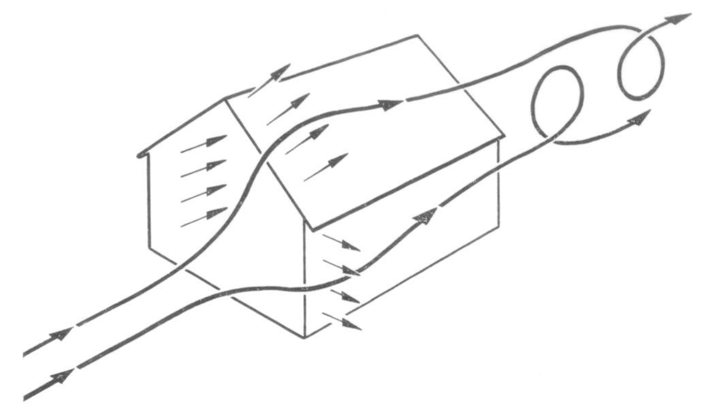

Figure 80 Wind flow and forces with wind blowing along direction of ridge

Lee slopes are always subject to suction. Roofs of all pitches are affected by suction along their windward edges when the wind blows along the direction of the ridge (see Figures 80–82).

The suction over a roof, particularly a low-pitched one, is often the most severe wind load experienced by any part of a building. Under strong winds the uplift on the roof may be far in excess of its dead weight, requiring firm positive anchorage to an adequate foundation to prevent the roof from being lifted and torn from the building.

Figure 81 Effect of wind suction on a steep roof. Wind direction from right of picture

Figure 82 Tiles ripped off at leading edge by suction forces

Figure 83 Suction over a low-pitched roof combined with pressure below it from the inset balcony gives an increased uplift force on the roof over the balcony

Variation of pressure over a surface

The distribution of pressure, or suction, over a wall or roof surface is generally far from uniform. Pressure tends to be greatest near the centre of a windward wall and falls off towards the edges. The most severe suction is generated at the corners and along the edges of walls and roofs; careful attention to the fixings at these locations is always worth while. Any projecting feature such as a chimney stack or a bell tower will generate eddies in the airflow which increase suction locally in their wake. The roof cladding around projections therefore needs special attention. Notice the effect on tiling in the lee of the bell tower in Figure 79.

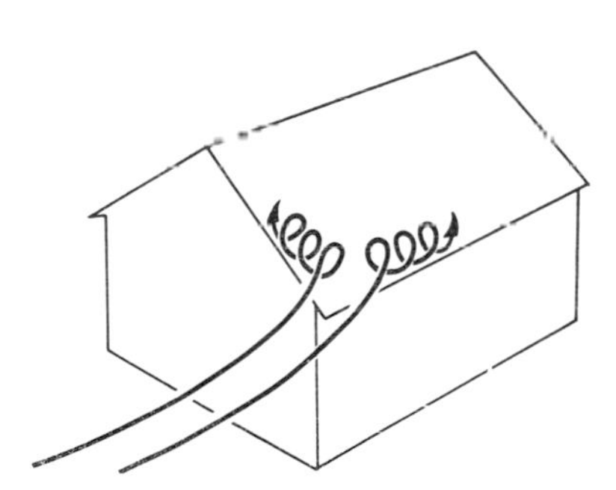

Figure 84 Vortices produced along edge of roof when wind blows on to a corner

In the assessment of uplift on roofs the effect of roof overhangs is sometimes overlooked. The flow diagrams in Figure 78 show how a windward wall deflects the airflow downwards and upwards. The upward element gives rise to a pressure on the underside of the roof overhang and any other section of the roof, such as the cover to a balcony, to which it has access (see Figure 83). This upward pressure on the underside reinforces any suction there may be over the roof and must be taken into account in assessing the roof uplift.

When the wind blows obliquely on to a building it is deflected round and over it. The pressures on the walls are generally less severe than in the previous cases, but strong vortices may be generated as the wind rolls up and over the edges of the roof as in Figure 84. These give rise to high suctions on the edges of the roof which must be resisted by especially firm fixing of the roof structure and covering in these areas. Since at most sites the wind can blow from any direction, all edges and corners need special attention (see Figure 85).

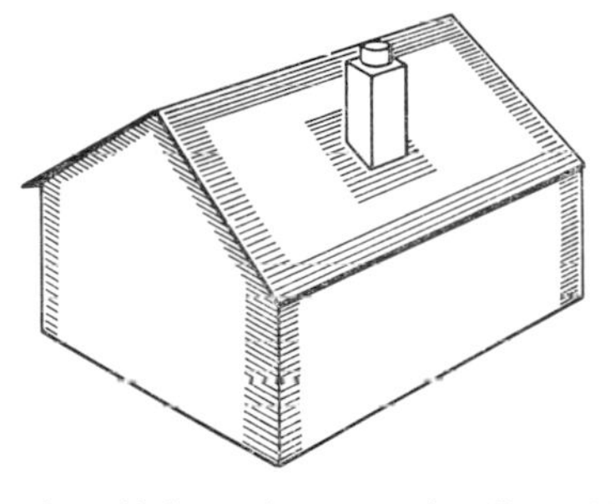

Figure 85 Areas where high suctions must be allowed for on the cladding

The response of buildings to the wind

Tall chimneys and towers and especially flexible structures such as suspension bridges often have natural periods of oscillation of several seconds. For example, the Post Office Tower in London has a natural period of about 6 seconds; but most ordinary buildings are relatively stiff with natural periods of less than 1 second. An ordinary building is thus responsive to short-duration loadings and so will respond to short gusts if these are of sufficient spatial dimension to encompass the building. It is therefore important to ensure that deflections under the loading due to a short gust, do not exceed the permissible limits. Trouble from excessive deflection has been rare in the past but the situation is being reached where blocks of flats rising to about 30 storeys have been subject to sway and wind vibration which, although not structurally dangerous, has been beyond the acceptable limits for residential occupancy.

Gust loadings are not normally sufficiently frequent or regular to excite a resonant oscillation but it is possible that eddies shed by the structure in a regular pattern may give rise to such a resonant oscillation.

Wind speeds in the United Kingdom

The gusty nature of the wind and the methods of measuring it were explained briefly in Chapter 1 (page 30). The Meteorological Office records the hourly mean wind speeds and the maximum gust speed each hour, the standard measuring height being 10 metres above ground, at open level sites. Gust speeds are used as the basis of wind load calculations. Maximum 3-second gust speeds, at a height of 10 m above ground, likely to be exceeded not more than once in 50 years in open level country have been calculated (see, for example, Shellard 1965) and are shown by isotachs (lines of equal wind speed) on the map in Figure 86. It should be assumed in calculations for structural loading that the maximum wind may come from any horizontal direction, though near the coast there is a tendency for the strongest winds to blow from the sea.

This map should not be considered as definitive in the sense that the values given on it will be valid for all time. It is based on records from only about 100 places (for comparison, rainfall maps are derived from records taken at about 6000 places in Britain), but the various records have been carefully considered, and the map is unlikely to be changed much in the near future by the acquisition of extra data. However, the map needs to be used with care, and the following paragraphs show how wind pressures at any place can be estimated by making corrections for local effects to the values read from the map. It would be quite impracticable to show local variations on the map itself, since the speed may vary by a large amount in a short distance, especially in areas with a rugged terrain. It is possible to make estimates of the likely extreme wind speed which may occur locally, based upon experience elsewhere, although clearly even the best advice is likely to be somewhat subjective.

Design wind speed V_s

The first step in the assessment of wind load is to determine the maximum wind speed appropriate to the structure. This is based on the maximum gust speed V for the locality, as given in the map, but to convert this to the design wind speed V_s three factors must be applied:

- S_1 for local topographic influences
- S_2 for surface roughness of the environment, gust duration appropriate to the size of the building, height of the structure
- S_3 for the design life of the building.

For a full explanation of the derivation of the values of the S factors, readers are referred to BSI Code of Practice CP3: Chapter 5 '*Loading*, Part 2 *Wind loads*' (1972) and to the *Wind loading handbook* (Newberry and Eaton 1974).

The basic wind speeds given on the map take no account of the local effect of hill and valley configuration. They must, therefore, be adjusted in accordance with the factors S_1 shown in Table 23 and described below.

Examination of the wind speed records has shown that the height of the site above sea-level does not by itself affect the basic speed, so unless there are special local effects, the value of S_1 should be taken as 1·0.

Exposed hills rising well above the general level of the surrounding terrain may give rise to accelerated winds; so may some valleys, particularly those shaped so that funnelling occurs when the wind blows up the valley. Sites so affected are often well known locally for their abnormal winds. For any of these situations, S_1 may be taken as 1·1.

On the other hand, there are some steep-sided, enclosed valleys where wind speeds will be less than normal. Caution is necessary in applying a reducing factor but for such sites a value of 0·9 may be used for S_1.

More extreme values may be necessary at some especially abnormal sites but until more data are available it is recommended that values outside the range 0·85 to 1·2 should not be used. Local knowledge may help the designer to select the S_1 value, but if he is in doubt the advice of the Meteorological Office should be sought. See Appendix 2 for a list of the main advisory offices.

In conditions of strong wind, the wind speed usually increases with height above ground. The rate of increase depends on ground roughness, and also on whether short gusts or mean wind speeds are being considered.

Ground roughness is divided into four categories as follows:

1. Long stretches of open, level or nearly level country with practically no shelter. Examples are flat, coastal fringes, fens, airfields and grassland, moorland or farmland without hedges or walls around the fields.
2. Flat or undulating country with obstructions such as hedges or walls around fields, scattered wind-breaks of trees and occasional buildings. Examples are most farmland and country estates with the exception of those parts that are well wooded.
3. Surfaces covered by numerous obstructions. Examples are well-wooded parkland and forest areas, towns and their suburbs, and the outskirts of large cities. The general level of roof-tops and obstructions is assumed to be about 10 m, but the category will include built-up areas generally apart from those that qualify for category 4.
4. Surfaces covered by large and numerous obstructions with a general roof height of about 25 m, or more. This category covers only the centres of large towns and cities where the buildings are not only high, but are also reasonably closely spaced.

A number of photographs, taken from the air, showing areas typical of these four gradings, are printed in the *Wind loading handbook* already mentioned.

In towns, of course, the local exposure to wind may be changed sharply by the demolition of buildings, perhaps over quite a large area. When this happens, buildings which were

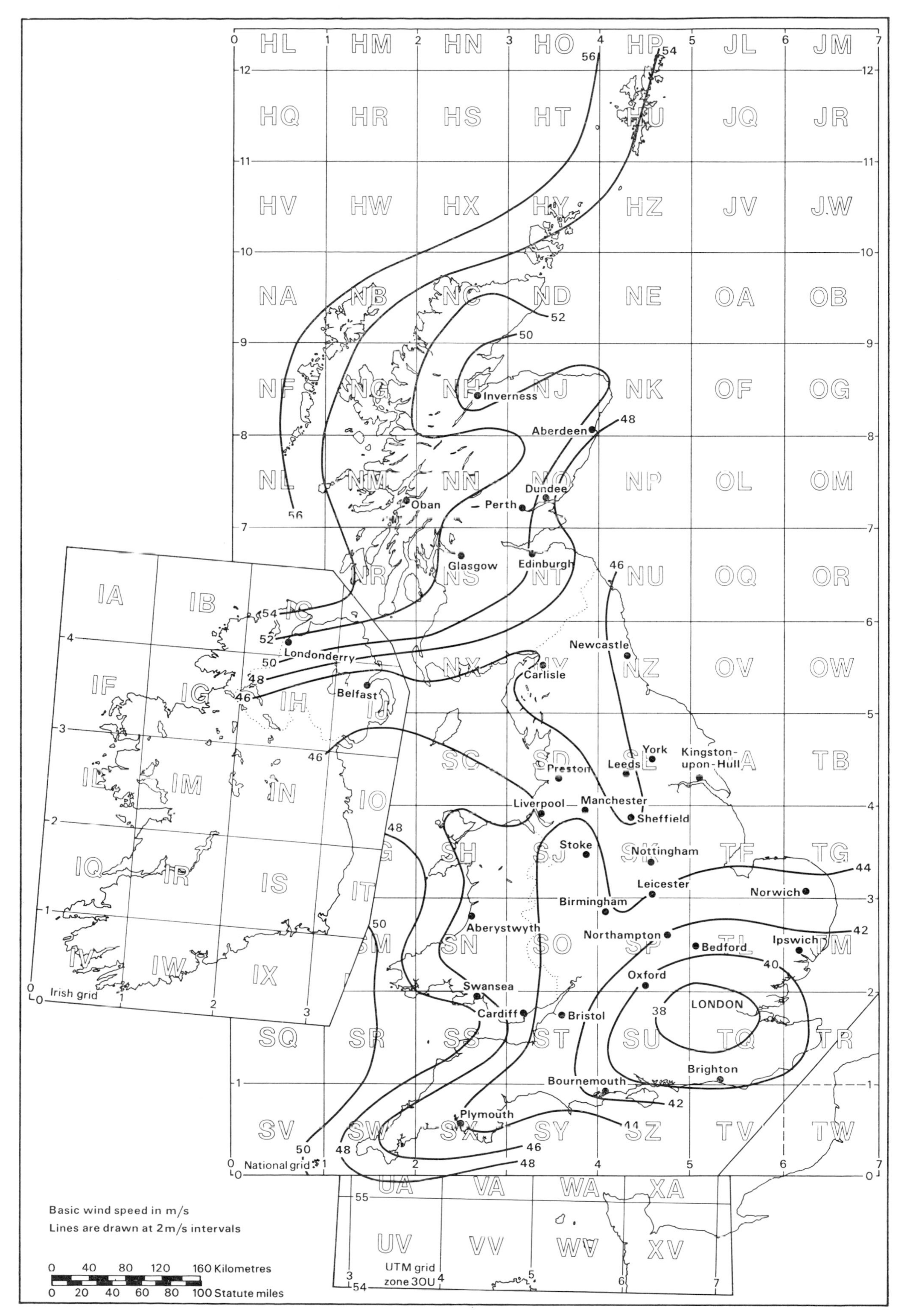

Figure 86 Map of United Kingdom showing basic wind speeds. (Adapted from BS CP3, Chap 5, Pt 2)

Table 23 Factor S_1 for topography

Topography	Value of S_1
All cases, except as below	1·0
Very exposed hill slopes and crests where acceleration of the wind is known to occur Valleys shaped to produce a funnelling of the wind Sites that are known to be abnormally windy due to some local influence	1·1
Steep-sided, enclosed valleys, sheltered from all winds	0·9

previously sheltered may subsequently have a much freer exposure to the wind. Whether there will be any consequent risk of failure will depend upon what factors of safety were incorporated in the design. As was remarked on page 82, the values of S_2 in Table 24 below for lower heights of building are enclosed in boxes as a reminder that the possibility of changes in exposure should be considered.

Although a classification of ground roughness has been made, it should be recognised that the change from one category to another is necessarily a gradual process. The wind must traverse a certain ground distance before equilibrium is established in a new velocity profile. The change starts first in the layers of wind nearest the ground, and the new profile extends to an increasingly deep layer as the distance increases.

It may be assumed that a distance of a kilometre or more is necessary to establish a different roughness category, but within the actual roughness layer, ie below the general level of buildings or obstructions in the windward direction, a lesser distance may apply as follows, depending on the density of buildings and other obstructions on the ground:

Ground coverage:	*Required distance*
not less than 10 per cent	500 m
not less than 15 per cent	250 m
not less than 30 per cent	100 m

For a site where the ground roughness varies in different directions, the most severe grading should be used, or, exceptionally, appropriate gradings may be used for different wind directions. For example, the sea-front of a coastal town would generally rank as ground roughness category 1.

A method of estimating the S_2 factor at various distances downwind of the boundary between two areas of different roughness is given in the *Wind loading handbook*, this being based on the Australian wind loading code.

The variation of wind speed with height is, however, also dependent on the size of gust that is being considered, ie on the wind-speed averaging time. Thus the table gives values of the factors for averaging times of 3, 5 and 15 seconds.

The recorded gust (a 3-second average) should be used in the design of all units of glazing, cladding and roofing, whatever the size or proportion of the whole building.

For structural design, a 5-second gust should be used for buildings whose largest horizontal or vertical dimension does not exceed 50 m; a 15-second gust should be used for buildings whose largest horizontal or vertical dimension is greater than 50 m.

The factor S_2 may be taken as appropriate to the total height of the structure above the level of the surrounding ground; alternatively, the height of the structure may be divided into convenient parts and the wind load calculated on each part, using a factor S_2 which corresponds to the height of the top of that part.

Table 24 Factor S_2 for roughness of environment, gust duration and height of structure. See text for note on values enclosed in boxes

Surface category	1. Open country with no shelter			2. Open country with scattered windbreaks			3. Country with many windbreaks; small towns, outskirts of large cities			4. Surface with large and numerous obstructions, eg city centres		
Height above ground-level m	3-s gust	5-s gust	15-s gust	3-s gust	5-s gust	15-s gust	3-s gust	5-s gust	15-s gust	3-s gust	5-s gust	15-s gust
3 or less	0·83	0·78	0·73	0·72	0·67	0·63	0·64	0·60	0·55	0·56	0·52	0·47
5	0·88	0·83	0·78	0·79	0·74	0·70	0·70	0·65	0·60	0·60	0·55	0·50
10	1·00	0·95	0·90	0·93	0·88	0·83	0·78	0·74	0·69	0·67	0·62	0·58
15	1·03	0·99	0·94	1·00	0·95	0·91	0·88	0·83	0·78	0·74	0·69	0·64
20	1·06	1·01	0·96	1·03	0·98	0·94	0·95	0·90	0·85	0·79	0·75	0·70
30	1·09	1·05	1·00	1·07	1·03	0·98	1·01	0·97	0·92	0·90	0·85	0·79
40	1·12	1·08	1·03	1·10	1·06	1·01	1·05	1·01	0·96	0·97	0·93	0·89
50	1·14	1·10	1·06	1·12	1·08	1·04	1·08	1·04	1·00	1·02	0·98	0·94
60	1·15	1·12	1·08	1·14	1·10	1·06	1·10	1·06	1·02	1·05	1·02	0·98
80	1·18	1·15	1·11	1·17	1·13	1·09	1·13	1·10	1·06	1·10	1·07	1·03
100	1·20	1·17	1·13	1·19	1·16	1·12	1·16	1·12	1·09	1·13	1·10	1·07
120	1·22	1·19	1·15	1·21	1·18	1·14	1·18	1·15	1·11	1·15	1·13	1·10
140	1·24	1·20	1·17	1·22	1·19	1·16	1·20	1·17	1·13	1·17	1·15	1·12
160	1·25	1·22	1·19	1·24	1·21	1·18	1·21	1·18	1·15	1·19	1·17	1·14
180	1·26	1·23	1·20	1·25	1·22	1·19	1·23	1·20	1·17	1·20	1·19	1·16
200	1·27	1·24	1·21	1·26	1·24	1·21	1·24	1·21	1·18	1·22	1·21	1·18

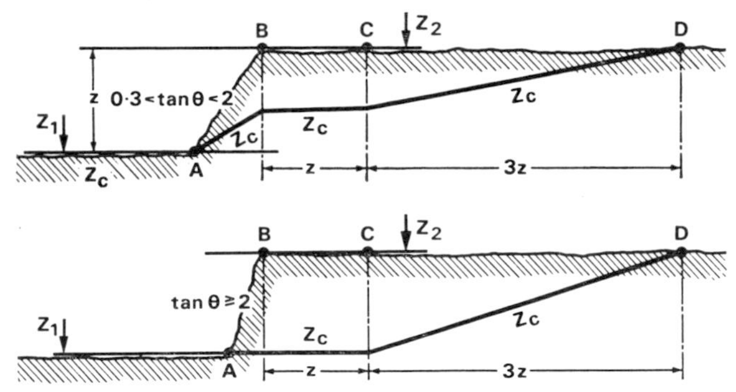

Figure 87 Effect of a hill or escarpment on the factor S_2, the inclination of the slope of the escarpment being θ:
(*a*) when tan θ exceeds 0·3 but is less than 2·0
(*b*) when tan θ is 2·0 or greater

If the structure is on or near to a cliff or escarpment, its effective height H for the purpose of determining the value of S_2 should be measured as follows:

First, find the angle θ, the inclination of the mean slope of the escarpment to the horizontal.

1 If tan θ does not exceed 0·3, the height of the structure should be measured from the natural ground-level immediately around the building.

2 If tan θ exceeds 0·3, the height of the structure should be measured from an artificial base Z_c (see Figure 87), which is set out by one of the following rules in which
  Z_1 is the general level of the ground at the foot of the escarpment
  Z_2 is the general level of the ground at the top of the escarpment
  z is the difference of level $Z_2 - Z_1$.

(a) If tan θ exceeds 0·3 but is less than 2·0:
  Set out the following points:
  A, at the intersection of the level Z_1 with the mean slope of the escarpment
  B, at the intersection of the level Z_2 with the mean slope of the escarpment
  C, such that BC = z
  D, such that CD = $3z$
  In front of A, $Z_c = Z_1$
  from B to C, $Z_c = Z_1 + \dfrac{2 - \tan\theta}{1{\cdot}7} z$
  behind D, $Z_c = Z_2$
  between A and B, and between C and D, Z_c is obtained by linear interpolation.

(b) If tan θ is 2·0 or greater:
  Set out the points B, C and D, as in (a)
  In front of C, $Z_c = Z_1$
  behind D, $Z_c = Z_2$
  between C and D, Z_c is obtained by linear interpolation.

This method of defining the effective height of a building is taken from the French code of practice *Règles NV 65*.
Note that at some distance back from the top of the cliff, the wind speed over the plateau will revert to the open country value.

Whatever the wind speed adopted for design purposes, there is always an element of risk that it may be exceeded in a storm of exceptional violence; the greater the intended life-span of the structure, the greater is this risk. It is in the nature of things that the recorded maxima of natural phenomena tend towards more extreme values with the passing of time, and statistical methods have to be used to estimate the trend of the extremes. For this reason it is not possible to state

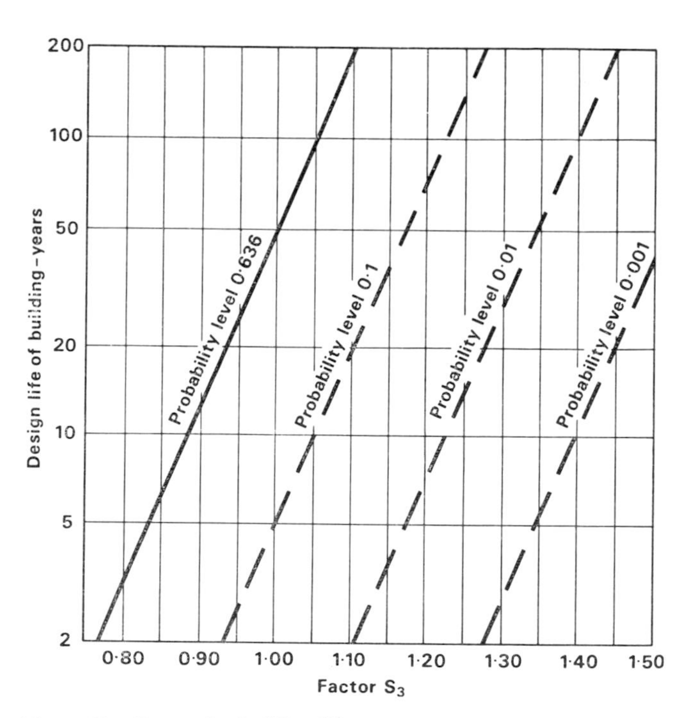

Figure 88 Factor for building life

categorically that a certain maximum value will never be exceeded: probability levels have to be used. The map wind speed is that which is likely to be exceeded on average only once in 50 years. This implies that in any one year there is a 1 in 50 (0·02) probability that the map speed will be exceeded. However, in any period longer than one year there is an increased probability of its being exceeded. It follows statistically that in any one period of 50 years there is a 0·63 probability that it will be exceeded at least once.

Figure 88 shows values of S_3, a factor which takes account of the intended life of the structure and the degree of security that is required. Normally, wind loads should be calculated using $S_3 = 1{\cdot}0$, but there are exceptions, as follows:
(a) temporary structures
(b) structures for which a life of less than 50 years is appropriate
(c) structures with an abnormally long intended life
(d) structures where additional safety is required.

For these special cases both the intended lifetime and the probability level may be varied according to circumstances.

The design wind speed V_s can now be calculated from:

$$V_s = V \times S_1 \times S_2 \times S_3 \quad \ldots\ldots\ldots\ldots (9)$$

The dynamic pressure of the wind

If the wind is brought to rest against the windward face of an obstacle, all its kinetic energy is transformed to dynamic pressure q, sometimes referred to as the 'stagnation pressure'. This can be calculated from the formula:

$$q = \frac{\rho V_s^2}{2} \quad \ldots\ldots\ldots\ldots (10)$$

where ρ = density of air
V_s = wind speed

In SI units, and at a temperature of 15°C and the standard atmospheric pressure of 1013·25 mb,

$$q = 0{\cdot}613\ V_s^2\ \text{N/m}^2 \quad \ldots\ldots\ldots\ldots (11)$$

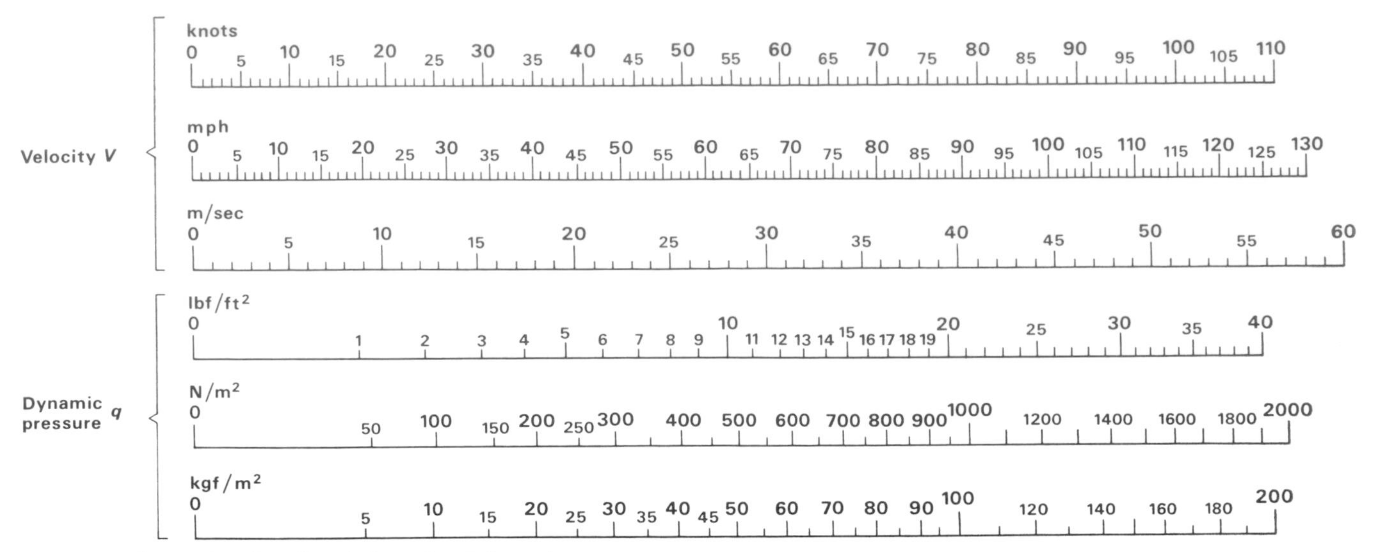

Figure 89 Conversion chart for wind speed and dynamic pressure

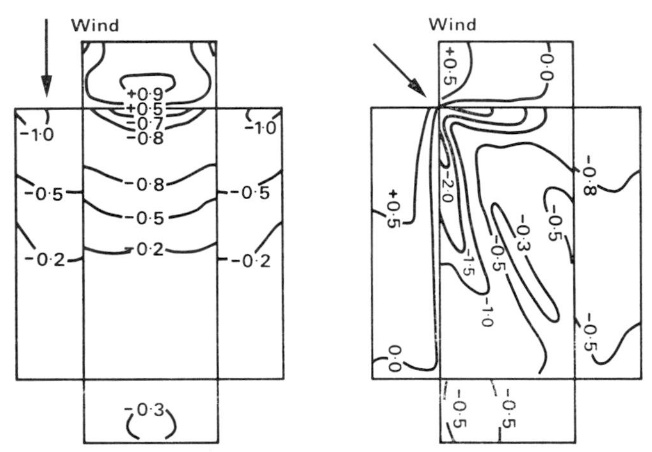

Figure 90 Distribution of pressures over a rectangular building for two directions of wind

Because information may be available in various units, a conversion chart covering V in knots, mph and m/s, and q in lbf/ft², N/m² and kgf/m² is given in Figure 89.

The pressure on any surface exposed to the wind varies from point to point over the surface, depending on the direction of the wind and the pattern of flow (Figure 90). The pressure p at any point can be expressed in terms of q by the use of a pressure coefficient C_p. Thus:

$$p = C_p . q \quad \text{.......................... (12)}$$

A negative sign prefixing the coefficient C_p indicates that p is negative, that is a suction rather than a positive pressure. The pressure or suction at every point acts in a direction normal to the surface.

In the calculation of wind load on any structure or element it is essential to take account of the pressure difference between opposite faces (Figure 91). For clad structures it is therefore necessary to know the internal pressure as well as the external, and it is convenient to use distinguishing pressure coefficients:

C_{pe} = external pressure coefficient

C_{pi} = internal pressure coefficient.

Values of C_p have been determined by experiment on models for a number of building shapes and wind directions.

The pressure coefficients which apply to a completed building will generally not be appropriate at all stages of the construction, and care is necessary to guard against adverse conditions. The most common variation is due to extremes of internal pressure when the building is partially clad. This risk should be borne in mind when the construction work is programmed, so that the most vulnerable conditions of partial cladding or dangerous structural shapes are avoided.

It is reasonable to assume that the maximum design wind speed V_s will not occur during a short construction period, and a reduced factor S_3 can be used to estimate the probable maximum wind. For normal construction, it is suggested that the S_3 factor for a 2-year building life should be used. The graphs of Figure 88 should not, in any circumstances, be extrapolated to a period of less than 2 years.

Detailed information on the coefficients to be used in a range of situations will be found in the *Code of Practice* and *Wind loading handbook* to which reference has already been made.

The pressure on a window from wind-driven raindrops
When the wind is sufficiently strong during rainfall, raindrops will impact on vertical surfaces. Their energy due to their motion perpendicular to a surface, will, on impact, be transformed to a pressure on the surface. The motion of both

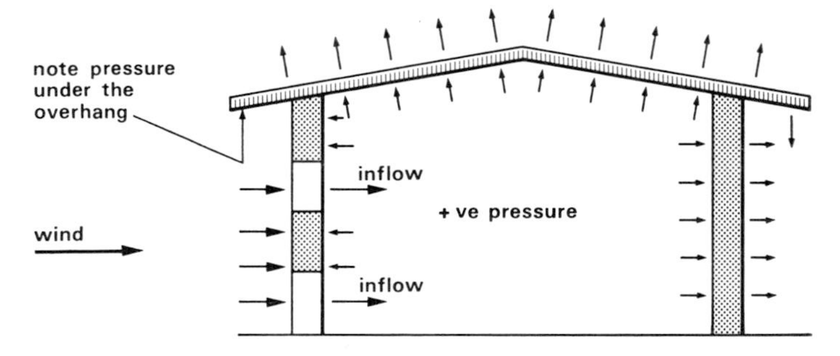

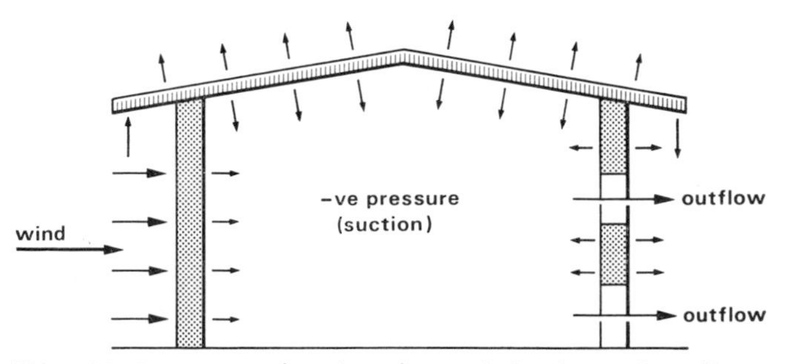

Figure 91 Pressures and suctions due to wind acting on the walls and roof of a building:
(*upper*) with openings on the windward wall
(*lower*) with openings on the leeward wall

Table 25 The pressure on a vertical surface caused by the impact of wind-driven raindrops.

Rate of rainfall r_h mm/h	Wind speed V m/s	Rate of impaction of wind-driven raindrops r_v kg/m^{2}h	Density of air and raindrops kg/m^3	Increase in pressure on surface by impact of rain Per cent
—	0	0	1·250	0
0·4	10	1	1·25003	0·002
2	20	8	1·25011	0·009
35	10	50	1·2514	0·11
75	10	100	1·2528	0·22
90	18	200	1·2531	0·25

Air density assumed is a typical one at a temperature of 10°C.
See page nos 97–98 for an explanation of the derivation of r_v from r_h and V.

the air and the raindrops becomes complex near a building, but the assumption that all raindrops approach a surface on straight paths perpendicular to the surface, and with the speed V of the free air, will ensure that the resulting pressure is not underestimated.

Let the rate of impaction of raindrops on a window or other vertical surface be r_v kg/m^{2}h, which is equivalent to $r_v/3600$ kg/m^{2}s. Then with a wind speed of V m/s normal to the surface, a volume of V m^3 (that is the column V m long with 1 m^2 cross-section) of air contains $r_v/3600$ kg of raindrops. Therefore the rain-content of one cubic metre of air is given by

$$c_r = r_v/3600V \text{ kg/m}^3 \quad \ldots\ldots\ldots\ldots\ldots(13)$$

We can then treat the mixture of air and raindrops as a fluid with a density which is the sum of the two individual densities. Taking the density of air as 1·250 kg/m^3, which is a typical value at a temperature of 10°C, the density of the mixture is

$$\rho_m = 1{\cdot}250 + r_v/3600V \text{ kg/m}^3 \quad \ldots\ldots(14)$$

Putting this value in the relation for the dynamic pressure due to the wind (eqn 10) leads to the increase in pressure due to the raindrops shown in the last column of Table 25 for a range of rainfall intensities and wind speeds. In strong winds with normal rainfall intensities, such as can be experienced in prolonged rain, the increase in pressure is no more than about 0·01 per cent. Even the heaviest thunderstorm rain with a strong wind gives an increase of pressure of only about one-quarter of one per cent. It is evident that this effect can be ignored as far as strength of structures is concerned.

Accumulation of ice on structures

Small, unfrozen water droplets can exist in the atmosphere at temperatures well below 0°C. When such supercooled droplets touch an object which is at a temperature below 0°, they freeze and adhere to the object. The ice-formation which results from the continued impact of droplets in this manner is called rime. Most of the ice forms on the windward side of an object, although some droplets may be carried round by small eddies in the current of air and freeze on the leeward side. The rate of accretion depends *inter alia* on the 'catching efficiency' of the object. In low-lying country a thick deposit of rime is relatively uncommon. One notable occasion in south-eastern England was during December 1962, when fog persisted for some days with an air temperature in the region of −2 to −6°, and with a very light wind. After about four days of continuous fog and frost, rime on trees had built up to a thickness of some 25 mm. It had even formed a deposit on parts of some buildings (Figure 92).

On high land, suitable conditions for the formation occur more commonly, and structures will often be in cloud when the temperature is below 0°. Because the amount of ice deposit will depend, in part, on the number of water droplets which strike an object, a larger thickness of ice can be formed on an object in a cloud blown along by the wind than in a fog which forms in calm weather (Figure 93). Banner-like projections of as long as 0·6 m have been reported.

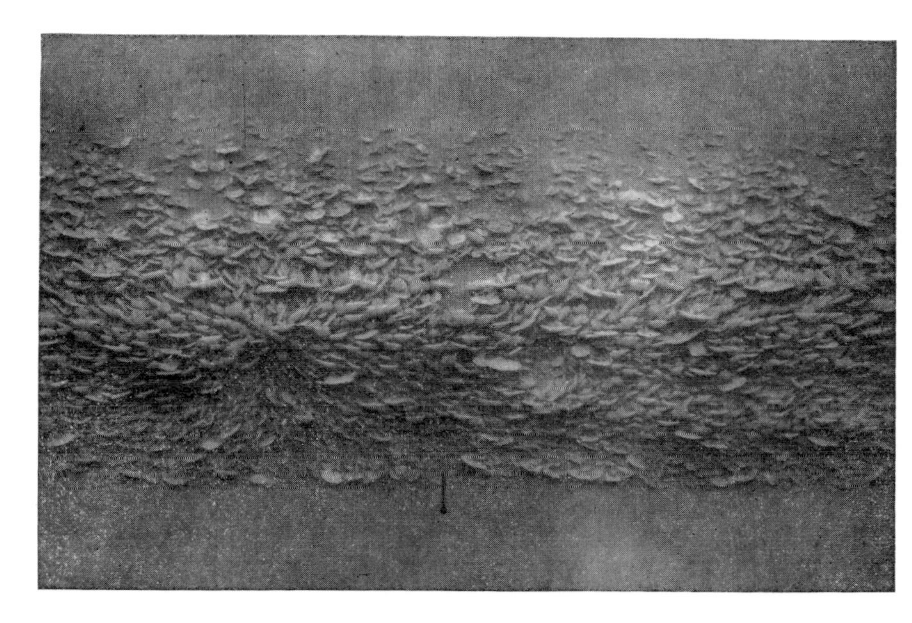

Figure 92 Rime deposited on a wall in a supercooled fog, 6 December 1962, at Garston. The ice has formed on the coldest part of the wall where the hardboard sheet is backed by a wooden frame

Figure 93 Rime on a post. Even small objects like twigs, grass and wire netting will acquire deposits of rime. This photograph shows a slender post at Great Dun Fell, 840 m above sea level, and illustrates the effect of wind speed. The deposit is least near the surface of the snow, where the speed of the wind is lowest, and increases with height, like the wind speed. (Photo by W Scriven)

Figure 94 shows a lattice radio-tower on Great Dun Fell, Cumbria, at a height of about 840 metres above sea-level, with a deposit of rime some 0·3 metre thick. It will be seen that as the ice builds up it changes the shape and increases the effective cross-section of the tower. This increases the collecting area for more rime, while snow can also collect, still further increasing the area. It is clear that the increase in area will also increase the wind loading on the tower, at a time when the vertical loading caused by the weight of ice has itself increased.

Figure 94 Rime on a lattice tower, accumulated to a thickness of 0·3 m or more. Great Dun Fell, altitude 840 m. (Photo by W Scriven)

Evidently the tower shown in Figure 94 was strong enough to withstand these severe conditions. However, during prolonged and severe icing conditions during March 1969, a guyed tubular mast, 380 m high, on the 260 m high Emley Moor, collapsed, perhaps through failure of a guy (Page 1969). The guys became coated with ice of density approaching 600 kg/m^3, to a thickness of 75 or even 100 mm in places. It was estimated that the ice on the cables weighed altogether some 37×10^3 kg.

At the site the temperature was continuously below freezing-point for some days, with the mast continually in cloud. The wind direction was approximately east-north-east. It is thought that in this case some at least of the icing was due to the accretion of freezing rain (see below).

Although such severe conditions are believed to be rare in Britain, except perhaps on the Scottish mountains (see below), it is clear that icing should be considered in the design of tall towers, and especially those mounted at high altitude. See Caspar and Sandreckzki (1964) for European experience in this respect.

During March 1969 there was also severe icing of trees, and there were even deposits to a thickness of 25 mm or more on heated, thick-walled buildings (Figure 95) at high altitude. The formation of ice on such buildings requires not only a prolonged spell of supercooled fog with sufficient wind to impact the water droplets or freezing rain (see below) on the surfaces, but a preceding spell of weather cold enough to chill the outer layers of the walls to well below 0°C.

A second form of icing is the so-called 'glazed frost'. This consists of a deposit of clear ice, formed when rain falls onto chilled objects – ground, vegetation and building. It requires a rather unusual combination of circumstances – a relatively shallow layer of cold air which has persisted for some days (so that the ground and other objects are cooled to below 0°C), overlain by milder air from which rain or drizzle is falling. The raindrops become supercooled when they pass through the lowest, cold layer of air (so becoming 'freezing rain') and freeze after impact on the cold surfaces.

The designer needs to know what extra loads are put on to his structure when it is iced. These will be vertical loads caused by the weight of ice, and horizontal ones caused by wind forces on the increased surface area. Because the deposit of ice will normally be non-uniform (as the ice forms mainly on the windward side), the extra weight-load will be asymmetric, so that it will be necessary to know the shape and density of the various parts of the deposit. Because the wind speed may be varying, it is desirable to know what wind speeds will be experienced whenever there is ice on the structure. The precise form of the deposit will presumably determine the aerodynamic drag and thus the loading for a given speed of the wind.

Reports of freezing precipitation are often localised and occur once every few years in some parts of the country, being mainly confined to England and Wales. The most widespread glazed frost of recent years, perhaps of this century, occurred in January 1940 and was described by Brooks and Douglas (1956). They reported deposits of 1488 g of ice on a spray of beech twigs weighing 100 g and deposits of 50 mm diameter on telegraph wires. Another widespread glazed frost occurred 11–15 March 1947, but on this occasion the thickness of the deposit was generally less well reported.

Vertical surfaces exposed to freezing precipitation are generally coated with ice on the side facing the wind. If the surface is flat and broad (eg a house side) the deposit has the form of a sheet of ice of more or less uniform thickness, rarely 40 to 50 mm thick. If the surface is curved and relatively narrow laterally (eg trees, telegraph poles) the deposit tends to build smoothly but with a thicker deposit directly into wind and thinner deposits at the sides of the object, producing a change of curvature. Thicknesses of up to 50 mm have been observed. If the surface is markedly curved and very narrow (eg cables, wire mesh fences) then the deposit, before freezing, may run to encase the exposed surface completely, often embedding the support in the centre of the ice section, but just as frequently building an oval section ice deposit with the wire roughly at the focus on the major axis furthest from the oncoming wind. If a cable or wire is at an angle to the vertical or if it is horizontal, the asymmetry of the load may induce a twisting moment. The coating of ice may then exhibit spiral effects with a very uneven surface, but the absence of spirals cannot be taken to imply that twisting has not taken place. Glazed frost adheres strongly to most metallic, mineral or organic surfaces and is relatively dense; in the absence of measurements it must be assumed that the density is 0·92 g/cm^3, the density of pure water ice at 0°C, because little or no air is trapped in the deposit (when air is trapped the deposit is white).

Figure 95 Ice on wall of a house. A layer of ice about 25 mm thick on the wall of the Fox House Inn, Derbyshire, 340 m above sea level, in March 1969. The ice was apparently deposited in part from a super-cooled cloud, in part from rain falling through cold air and impacting on the chilled surface. (Photo by J K Page)

There are two types of meteorological situation in which the conditions for glazed frost may be met. One is a steady situation in east or south-east winds with a narrow band of warm air overlying a very cold surface layer of air and with very cold air above so that snow falling from above is melted in the warm layer and then the drops or droplets are super-cooled as they pass through the cooler underlying layer. This situation may persist for days and is mainly reported in England and Wales. The hourly mean wind speeds during formation are usually in the range 6 to 10 m/s and exceptionally speeds of 15 m/s may be experienced. The glazed frosts of January 1940, March 1947 and March 1969 are typical examples in which deposits of 50 mm or so were recorded on trees, cables and house sides. More frequently, the meteorological conditions are met for only a short time after a cold spell. On these occasions a period of two or three hours' freezing rain may be followed by very strong warm winds with mean hourly speeds of 20 to 25 m/s, but melting of the ice takes place rapidly. Such an occasion was reported on 4 March 1970. Because the deposit is formed in a short period of rain, thicknesses of 25 mm or so are unlikely to be exceeded except on rare occasions. This type of glazed frost but with variations in deposit thickness is reported in some part of the United Kingdom every year, with wind speeds up to about 15 m/s or so, mainly in Scotland but particularly on hills. While no detailed observations are available it is reasonable to assume that temperature and precipitation conditions at heights up to 200 m or so above ground will not vary greatly from conditions at the surface. However, it may also be assumed that wind speeds will increase with height according to a power law with exponent 0·16 for hourly mean speeds. Thus a design criterion of 15 m/s for wind speed at the surface becomes 23 m/s or so at 200 m.

It is suggested that for glazed frosts in England and Wales design criteria might be as shown in Table 26. These two

Table 26 Design criteria for glazed frost conditions

Criterion	Values in England and Wales (lowland)	Values in Scotland and on hill tops in England and Wales
Maximum ice thickness	25 mm to windward	15 mm to windward
Ice density	0·92 g/cm^3 (920 kg/m^3)	0·92 g/cm^3 (920 kg/m^3)
Hourly mean wind speed during formation	10 m/s at the surface and increasing according to the power law to about 17 m/s at 200 m	15 m/s at the surface increasing according to the power law with exponent 0·16
Hourly mean wind speeds after formation	15 m/s at the surface and increasing with height to 25 m/s or so at 200 m	20 m/s at the surface increasing according to the power law with exponent 0·16

Note: The two sets of criteria are equally likely.

sets of criteria spring from different meteorological situations and the probability of their occurrence cannot be assessed at the present time.

The least dense ice deposits are formed when fog or cloud is blown, usually at steady hourly mean speeds of 3 to 8 m/s on to cold surfaces, freezing on impact to form a loose aggregate feather ice or rime ice which builds into wind in the shape of pointed icicles. Air is trapped between the frozen droplets which are white in appearance. The density of rime ice has been measured in controlled conditions by Macklin (1962) who studied the influence of wind speed, ambient air temperature, droplet diameter and liquid water concentration, but no densities have been determined so far as is known from actual atmospheric samples. Macklin found densities of from 0·1 g/cm³ to 0·9 g/cm³ being dependent directly on droplet diameter and wind speed, but inversely proportional to temperature of the accreting surface. For practical purposes a value of 0·5 g/cm³ may be adopted because the very dense cases are associated with large droplet diameters. The ice is usually fragile, disturbances in the airflow and movements of the depositing surface being of sufficient force to result in lumps of ice breaking off. In steady wind conditions, on fairly solid objects like masts, rime ice will form a banner *into wind.* If the object is thin enough, say wires, there will be only one banner but flat narrow objects may acquire two banners, one at each edge facing the wind. The December 1969 issue of *Weather* shows some photographs of rime ice on page 497 and also in this issue there is an article by Page (1969). These suggest that banners may grow to lengths of 600 mm. Occasionally the banners may fill the spaces between adjacent upright members of a structure, the strength of the ice deposit being increased by the additional support and the tendency to fracture being much reduced.

During the period November to March in the British Isles, rime ice may be experienced on structures rising from ground at 200 m or so above sea-level particularly on the windward slopes of hills. Usually banners of less than 200 mm length are recorded at lower levels, but banners of 600 mm or more may occur at 1000 m or so above sea-level every year and once every few years at lower altitudes. However, the length of the banner will depend on the size of the depositing surface, its shape, the steadiness, particularly in direction, of the wind and the duration of the conditions of icing. Table 27 presents suggestions for suitable design criteria for this type of deposit. These criteria should be applied for all structures on hills at heights greater than 200 m above sea-level.

Ice may also be deposited as melting snow on cold surfaces. The resulting deposit is whitish in appearance and is midway between glazed frost and rime ice both in appearance and density, this latter being about 600 kg/m³ if the one measurement made on ice from the cables of the television mast at Emley Moor, Yorkshire, is accepted (Page 1969). Because it is a midway case, no design criteria are offered, but hourly mean wind speeds of 15 m/s are fairly common particularly in Scotland.

Combinations of the three methods of deposition in any order are possible but perhaps the sequence at Emley Moor during March 1969 will serve to illustrate the complexities. During the period 12–19 March 1969 hourly mean speeds varied from 3 to 9 m/s. Throughout the period fog was reported at temperatures near 0°C to about −2°C. Superimposed on this more or less continuous accretion of rime ice were periods of melting snow and freezing rain. On the 19th, ice up to 75 mm thick was measured on cables of about 30 mm diameter. On this occasion, however, hourly mean wind speeds did not exceed 10 m/s even on the 19th when rapid thawing took place with air temperatures between 1 and 3°C, the higher temperatures being at 200 to 370 m above the ground.

It is essential to assume that the interstices between members of lattice masts will be filled with a wall of ice. Because the ice will have been deposited by at least two processes, the density will be about 600 kg/m³ in hourly mean wind speeds of 15 m/s. The thickness of the wall of ice will depend largely on the breadth of the surface facing into wind but values of 75 mm are not uncommon. Such icing will usually only be experienced at ground level which is 200 m or more above sea-level. Hourly mean wind speeds in the free air above the ground may be assumed to increase according to the power law with exponent 0·16. This increase should be limited to heights up to 400 m above the ground because the gradient wind level should be assumed to be 400 m and above this level there should be no increase of hourly mean wind speed. In very hilly terrain or on exposed hill tops rising 600 m or more above sea-level, the hourly mean wind speed should be assumed constant at 25 m/s for reasons just given. Table 28 presents suggested design criteria for these conditions on lattice masts.

At the present time knowledge of the combined frequency of wind speed, ice deposit thickness and ice density is fragmentary. Much work is required by collecting observations of ice deposits on structures, by relating these to the meteorological conditions and by designers who must decide what particular combinations of wind and ice loads are appropriate for code purposes.

Table 27 Suggested design criteria for banner ice

Maximum length into wind	600 mm
Density of ice	0·5 g/cm³ (500 kg/m³)
Hourly mean wind speeds during formation and after formation	10 m/s at 200 m rising to 30 m/s at 1000 m

Table 28 Suggested design criteria for icing on lattice masts

Location of ice	Wall of ice filling the spaces between upright members
Density of ice	0·6 g/cm³ (600 kg/m³)
Hourly mean wind speeds during and after formation at heights of 200 m to 600 m above sea-level	15 m/s rising according to power law to about 25 m/s and constant at all heights above 600 m

Icing of gutters

It is worth mentioning as an addendum to the preceding notes on the icing of structures, that overhanging eaves gutters may be damaged by the weight of snow or ice. This is not a riming effect, but is caused by excessive quantities of snow. At one time it was common to see a snow guard at the foot of the slates or tiles, which prevented accumulated snow from sliding off the roof. Although perhaps the intention was mainly to stop the snow falling on to people below, it also helped to reduce the risk of damage to the gutter.

However damage is perhaps more likely to be caused in freezing weather by slow melting of snow on a roof, mainly by heat from within the building, so that the resultant water freezes in the gutter. When this is full, the overflowing water freezes into icicles so that the total load may be much greater than that caused by a gutter full of water or snow. Figures 96 and 97 show the enormous icicles which may form given the right conditions.

Figure 96

Figure 97

Figures 96-97 Icicles formed by the melting of snow from a roof during freezing weather. Great Dun Fell. (Photos by W Scriven)

Rain penetration

Perhaps one of the commonest failures in building practice, and one which causes many complaints, is the penetration of wind-driven rain into houses and other buildings. The trouble arises in two main ways. Firstly, porous materials such as bricks and lightweight concretes may become wetted, even saturated, with rainwater. The result may be dampness on the inner surfaces as a direct result of the water in the materials. Perhaps more important, the damp material has a higher thermal conductivity than a dry one, so that heat losses are increased, while more heat has to be used to evaporate the moisture. Moreover, because of this the inner face of a wall will be cooler when it is wet, thus increasing the risk of condensation of atmospheric moisture on and in it (this condensation will in its turn tend to make the material even wetter – or will appear as liquid water if the material is already saturated). Thus a house so affected will tend to be colder and damper than a dry one, with increased risk of mould growth on structure and furnishings.

Secondly, rainwater may be driven through cracks in the structure, gaps around windows, and so on. Generally gaps down to 0·1 mm wide will permit water to be driven through by air pressure. With smaller gaps, capillary forces are the most important, and wind pressure is relatively ineffective. Thus the strength of the wind has no effect on the rate of absorption of water by masonry structures. Rainwater coming through openings is more obvious than the gradual accumulation in masonry, so that complaints of rain leakage usually refer to penetration through cracks.

Driving rain is defined as rain which is carried along by the wind so that it impinges on to vertical surfaces. As will be shown later, the intensity of driving rain depends both on the intensity of the rainfall and on the speed of the wind. Thus it may be expected that driving rain will be most intense and most frequent in the wettest and windiest parts of the country. In the first Section of this Chapter it was stressed that there could be considerable local variations in the speed of the wind – it is to be expected that there will be equally dramatic variations in the intensity and amount of driving rain, caused by local variations in degree of shelter. These may be caused by local topography, by shelter belts of trees, or by other buildings.

Reliable statistics of rain penetration are hard to come by, but a survey in 1969 by the Building Research Station for the Scottish Working Party on Component Performance has produced some interesting results (Figure 98). The survey of over 1200 tenants in local-authority housing in Scotland showed that 33 per cent of those living in areas with severe exposure complained of leakage of rain through windows, while in areas of moderate exposure the proportion of complaints was only 17 per cent. Severe and moderate exposure were defined by the annual mean driving-rain index, as described on page 106. There was also a high correlation between number of complaints and height above ground, as measured by the storey number. In Figure 98 the results of the enquiry are shown by circles for tenants living in the zone of moderate exposure and by crosses for those in the severely exposed zone. Where tenants lived in two-storey houses or maisonettes the results are plotted at the mean height of the unit; for instance a maisonette on fourth and fifth storeys is counted as one at four and a half storeys' height. The increase in frequency of rain penetration with height above ground is most marked.

It appears that the proportion of tenants complaining of rain penetration through windows can be related to height

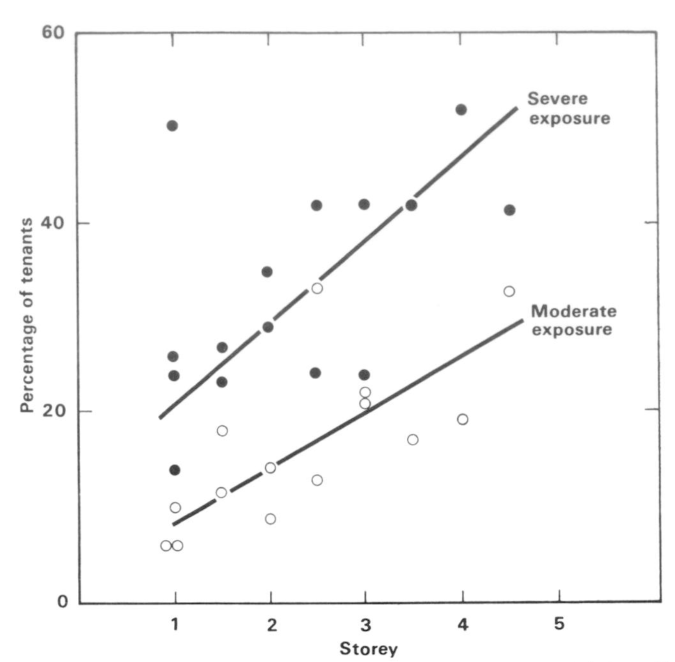

Figure 98 Relation between proportion of tenants complaining of rain penetration through windows, the degree of exposure of the district, and the height of the dwelling above ground

above ground (S_h measured in storeys) by a simple equation such as

$$P_c = A + BS_h \text{.....................(15)}$$

where A and B are constants. In this particular investigation, in the area of moderate exposure to driving rain, A and B had values respectively of about 2 and 6, so that the proportion of tenants complaining was given by

$$P_c = 2 + 6S_h \text{ per cent.............(16)}$$

In the area of severe exposure the values of A and B were respectively about 12 and 9. The values of the constants would be expected to vary to some extent with local exposure, so that these values would not necessarily apply to other sites (nor to other types of construction).

Driving rain

In perfectly calm weather, raindrops fall vertically with a velocity that depends on their size. By the time they are nearing the ground, all drops will be falling at their terminal velocity – Figure 99 shows the relationship between raindrop diameter and velocity that was determined by Best (1950). It will be noted that the curve flattens out towards the top. This is because the very biggest drops deform and their aerodynamic drag increases, so that they fall at speeds little different from those of smaller drops. Bigger drops still deform even more and break up into smaller ones.

When a wind is blowing (assuming for the moment that it is a steady, horizontal wind), all the raindrops are carried along horizontally at approximately the speed of the wind. Since they are also falling vertically, relative to the air, at the terminal velocity corresponding to their diameter, it follows that they are moving at an angle with the vertical relative to the ground (Figure 100). Hence, in principle it is possible to calculate the incident angle of the raindrops to the vertical, if the size of the raindrops and the wind speed are known.

In practice the problem is not so simple, for rain normally contains drops in a wide range of sizes, while the wind may be deflected by buildings. Although there may be a great variation from one rainstorm to another, on average the distribution of drop sizes takes the form shown in Figure 101 (from Laws and Parsons 1943). Three distribution curves are shown, one

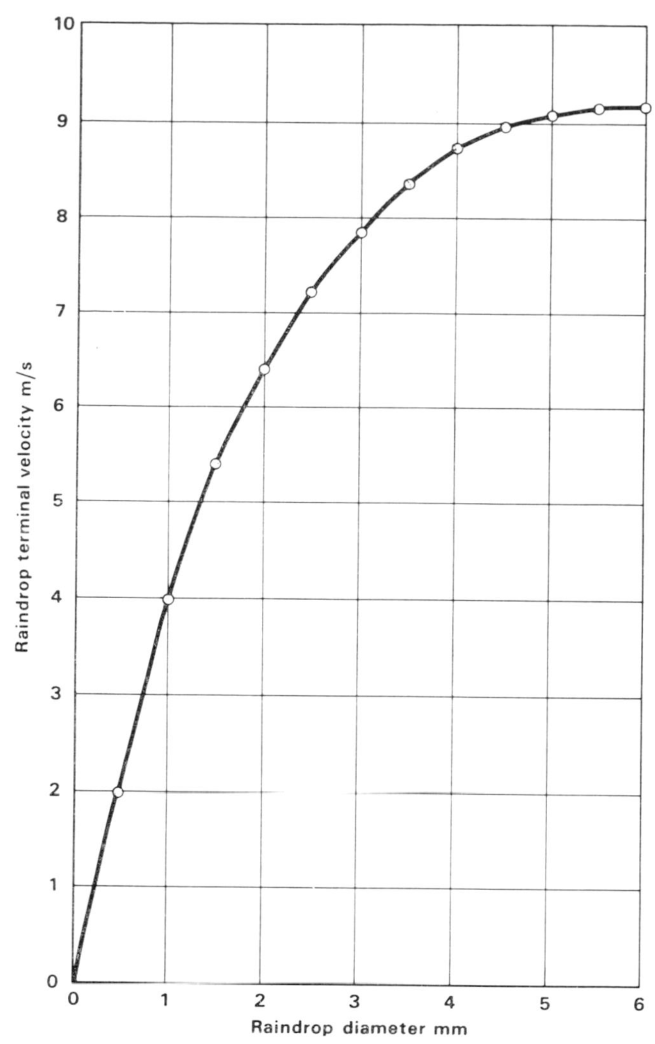

Figure 99 Relation between the diameter of a freely-falling raindrop and its terminal velocity. (After Best 1950)

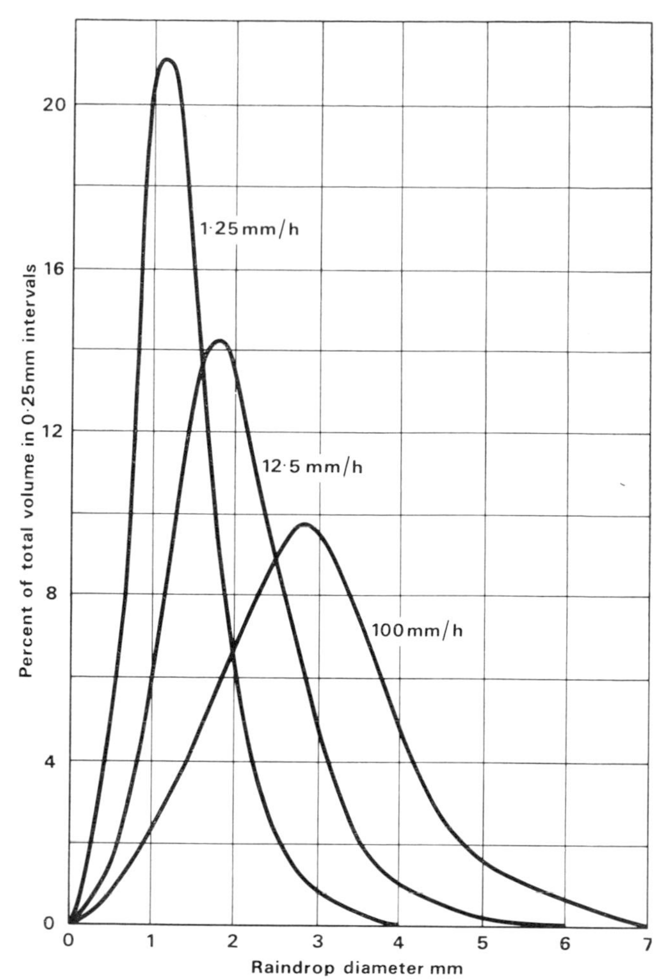

Figure 101 Distribution curves showing typical proportions of raindrops of different sizes in rainfall of three rates. (After Laws and Parsons 1943)

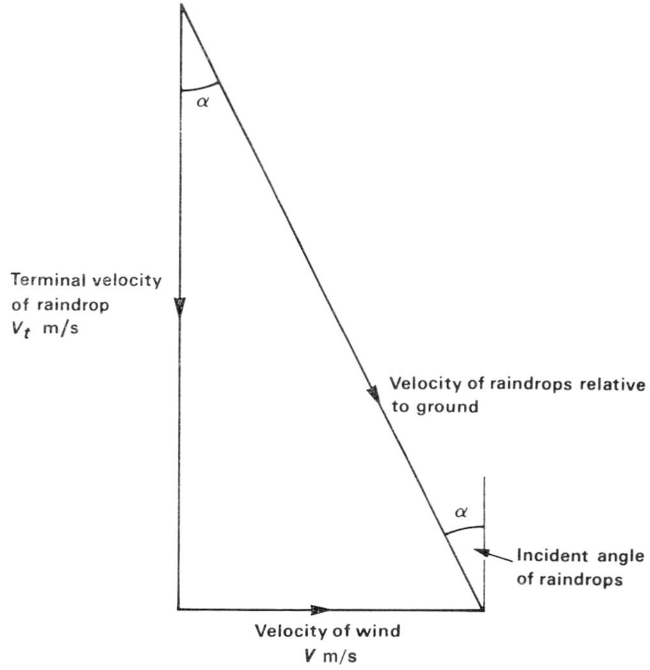

Figure 100 Diagram showing path of raindrops in relation to raindrop terminal-velocity and wind velocity

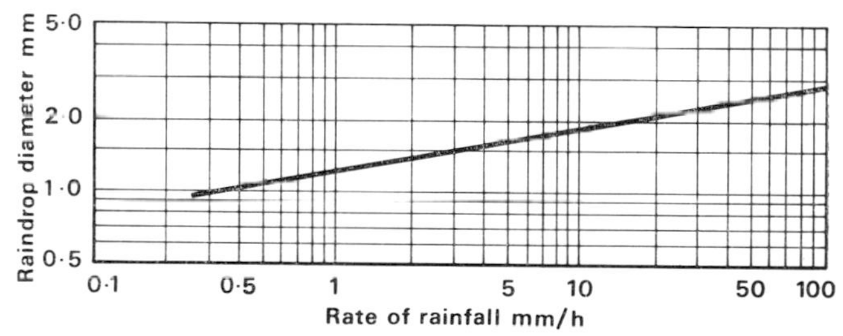

Figure 102 Relation between the rate of rainfall and the median raindrop diameter D_{50}. (After Laws and Parsons 1943)

for a rainfall with a rate of 1·25 mm/h (corresponding to ordinary warm-frontal rainfall), and two for shower rainfall, 12·5 mm/h (moderate) and 100 mm/h (heavy) respectively. There is in each case a marked peak in the curve corresponding to a dominant size of raindrops. The more intense the rainfall the bigger the dominant raindrops.

Laws and Parsons (1943) developed an empirical relationship between the rate of rainfall on the ground (r_h) and the median drop size (D_{50}),

$$D_{50} = 1{\cdot}238\, r_h^{\,0{\cdot}182} \qquad (17)$$

where drop diameter is in millimetres and the rate of rainfall is in mm/h (Figure 102).

(Note: Half the volume of the rain is made up of drops whose diameter exceeds the median value D_{50}.)

Now clearly the ratio of the amount of rain driven onto a vertical surface (perpendicular to the wind) to that falling on the ground in the same time is the ratio of the velocity of the wind V to the raindrop terminal velocity V_t; that is to say,

$$r_v/r_h = V/V_t = \tan i \qquad (18)$$

which is the tangent of the incident angle i (see Figure 100). This assumes no deflection of the wind or of the raindrops by the vertical surface, and that the rain behaves as if all the drops were of the median size D_{50} appropriate to the intensity. Combining the relationship between median drop-size D_{50} and rainfall intensity r_h found by Laws and Parsons, and the relationship between raindrop diameter D and terminal velocity V_t found by Best (Figure 99), we obtain the equation

$$V_t = 4{\cdot}505\, r_h^{\,0{\cdot}123} \quad \ldots\ldots\ldots\ldots\ldots\ldots (19)$$

Combining this with the relation $r_v/r_h = V/V_t$ we obtain

$$r_v = 0{\cdot}222\, V r_h^{\,0{\cdot}88} \quad \ldots\ldots\ldots\ldots\ldots\ldots (20)$$

which relates the rate of driving rain r_v to the rate of rainfall r_h and wind speed V. This relationship is conveniently displayed on a family of curves in Figure 103, from which the required rate of driving rain may easily be interpolated.

It must be stressed that this is a theoretical rate, assuming no deflection of the wind or of the raindrops. Although the flow of particles near objects has been studied in wind tunnels, most of the published results refer to circular obstacles, and little or no research has been done on objects shaped like buildings. Thus we do not know in detail how raindrops behave in the vicinity of buildings. Observation shows that edges of buildings, where the wind sweeps around corners, or over cornices, collect more rain than the rest of the face of the wall. But every detail of the building affects the local air-flow and thus the way raindrops move. This too can be observed in the wetness patterns on buildings (Figure 104). (See also pages 111–112).

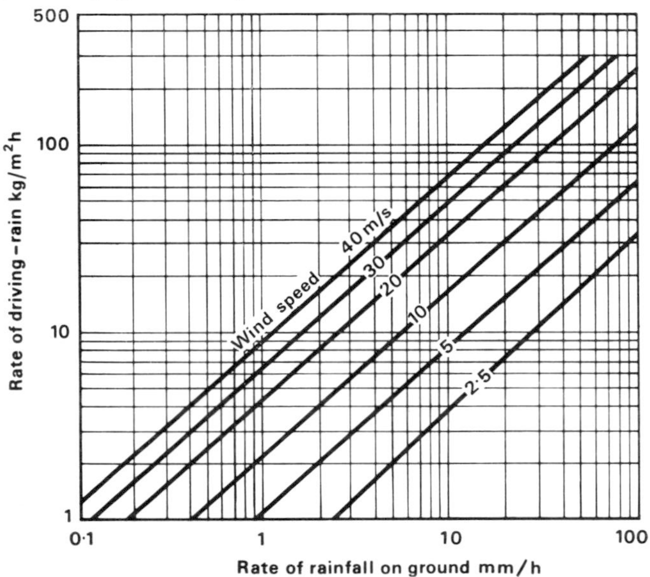

Figure 103 Relation between rate of driving rain on a vertical surface, rate of rainfall and wind speed

Figure 104 Wetness patterns on a building, showing how variable these can be from one part to another

Examples of driving-rain measurement

From considerations of rainfall alone, it would be expected that buildings in the west and north of Britain would experience most trouble from rain penetration. These wettest areas are also those which experience the strongest winds, so that there is more driving rain in northern and western Britain than in the drier south and east. Inland low-lying regions especially have relatively little driving rain, but with local differences caused by topography.

The differences between the amounts of driving rain experienced at a relatively exposed place in South Wales and at a sheltered place in the south-east of England (Garston, north-west of London) are demonstrated by the measurements reported in the following paragraphs.

Tredegar is an old coal-mining and iron-working town, the oldest part of which lies at an altitude of 290 to 310 m above sea-level, near the head of the Sirhowy valley. Like most of the other valleys of South Wales, this one runs almost due south, to the sea some 30 km distant.

The district is notably wetter than south-east England, Tredegar having about 1600 mm of precipitation annually as against about 600 mm in the London region. It is also much more subject to strong winds than the London region. However the old town centre of Tredegar, being in the bottom of the valley, is fairly sheltered from westerly winds. The houses at Cefn Golau were built in about 1960 on the western slopes of the valley, the highest houses lying at an altitude of 400 m, some 90 m above the town centre and 700 m away.

These new houses (mostly of brick construction) have suffered severely from condensation, from rain penetration, and even from frost damage to the outer leaves of the walls. It was decided to install driving-rain gauges – a special form of raingauge with the aperture in a vertical plane, so that it collects only rain falling at an angle to the vertical. It can be either fixed to the wall of a building or free-standing. Three driving-rain gauges were placed on a pair of the new houses (Nos 93 and 95 on the plan, Figure 105) which back on to open moorland. Gauges No 2 and 3 were on the west-facing wall, about 1 m above the level of the back garden, and No 4 on the south wall about 2 m above ground. Gauge No 5 also faced west, but was on house 78 on the other side of the road, and about 3 m lower down the hillside. A further driving-rain gauge (No 1) was placed, facing west, on a post in the back garden of house 95, 1·4 m above the ground. A cup-contact anemometer was mounted 2·8 m above ground near this gauge, and gave the daily mean wind speed.

Another driving-rain gauge was mounted 2 m above ground, facing west, on a post at the town's climatological station in Bedwellty Park. This is at an altitude of 315 m above sea-level and 100 m from the town centre.

Table 29 summarises the results for the period 24 November 1967 to 30 April 1968, omitting a few days with snowfall. The column headed A contains data for all the 78 days on which rainfall amounted to 0·2 mm or more. Column B contains data for those 12 days which had the greatest amount of driving rain – on which 10 mm or more was collected in the free-standing, west-facing gauge No 1. It is remarkable that most of the driving rain occurred on these few days. Indeed, the south-facing wall-gauge No 4 collected 131·7 mm (61 per cent of the winter total) in only 3 days. The figures in brackets are daily mean values.

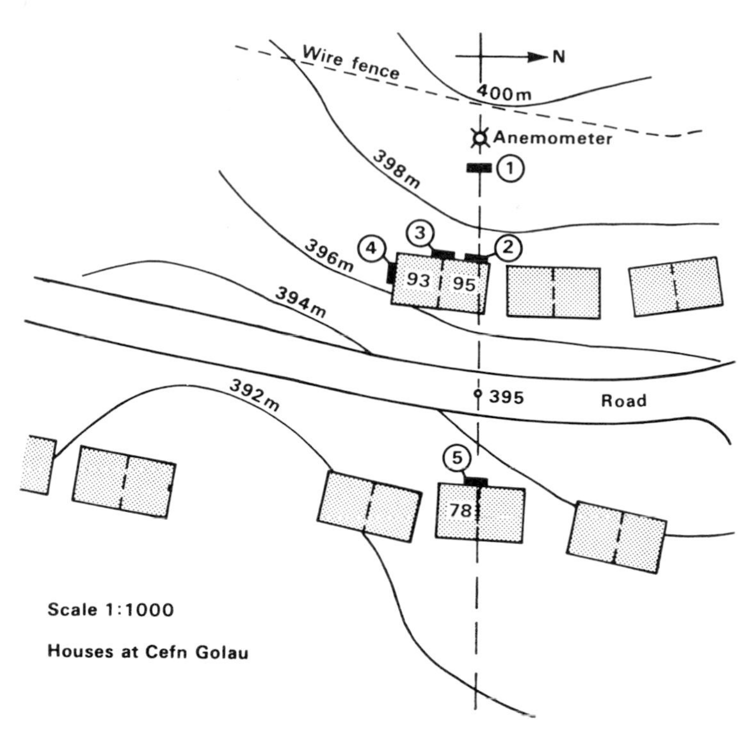

Figure 105 Plan of houses at Cefn Golau, Tredegar, showing positions of five driving-rain gauges and anemometer

Table 29 Measurements of rainfall and driving rain at Tredegar and Garston, winter 1967–68

		A All days with rain	B 12 days with most driving rain	B/A Per cent
Tredegar				
No of days with rain (0·2 mm or more)		78	12	15
Rainfall	mm	575·4 (7·4)	254·0 (21·2)	44·1
Duration of rainfall	h	—	123·4 (10·3)	—
Mean rate of rainfall	mm/h	—	2·06	—
Mean wind speed (at 2·8 m)	m/s	4·3	7·4	172
Estimated speed at 15 m	m/s	5·7	9·9	
Driving rain: Bedwellty Park	mm	76·8	32·3	42·1
Cefn Golau 1	mm	509·4	334·7	65·7
Cefn Golau 3	mm	238·4	174·7	73·3
Cefn Golau 4	mm	232·7	193·9	83·3
Cefn Golau 5	mm	120·0	81·8	68·1
Computed driving rain, west	mm	—	298	—
Computed driving rain, south	mm	—	298	—
Garston				
No of days with rain		76	12	16
Rainfall	mm	224·9 (3·0)	81·0 (6·7)	36·0
Duration of rainfall	h	186·5 (2·5)	46·7 (3·9)	25·0
Mean rate of rainfall	mm/h	1·21	1·73	143
Mean wind speed (at 15 m)	m/s	3·9	4·3	110
Driving rain (free-standing octagonal guage) west	mm	47·8	20·6	43·1
south	mm	55·0	29·1	52·9

Values in brackets are daily means.

The quantities of driving rain intercepted by the walls on several of these days were considerably more than that required to saturate the walls (Table 30).

The lowest parts of Tables 29 and 30 contain some corresponding data for the same period of time at Garston.

Comparing the readings at Cefn Golau and Garston, the main points are:
Rainfall at Cefn Golau was about 2·5 times that at Garston, and wind speed on rain-days about 50 per cent higher, but total driving rain was nearly 11 times as much. On the 12 days with most driving rain, Cefn Golau received about 15 times as much as on the 12 worst days at Garston. Rate of rainfall at Tredegar was not much higher than at Garston, but the duration was greater.

There was a very marked difference between the exposure to driving rain at Cefn Golau and at Bedwellty Park near the town centre, with on average a ratio of about 6·5 between the catches in the free-standing gauges. On some days this ratio reached 15. Presumably the degree of exposure of buildings varied in the same ratio.

Calculations of the 'theoretical' driving rain have been made for the 12 worst days listed in Table 30, using Tredegar rainfall and the wind speed and direction during the rain. The method used was that described on page 98. As only 24-hour mean winds were available at Tredegar, the speeds used in the calculation were the hourly values recorded at Rhoose Airport, about 40 km SSW, near the coast. As can be seen from Figure 106 there is quite a good correlation between calculated and measured driving-rain values on these 12 days.

Table 30 Rainfall and driving rain during most severe storms, Tredegar and Garston, 1967–68

Tredegar			Driving rain mm					
				Cefn Golau				
Date	Rainfall mm	Rain duration h	Bedwellty (west free-standing)	1 (west free-standing)	3 (west wall)	4 (south wall)	5 (west wall)	Mean wind speed m/s
23. 3.68	73·7	23·3	7·0	76·0	45·0	78·0	16·0	10·3
22.12.67	33·0	14·0	3·0	44·7	28·1	11·5	17·1	6·6
13. 1.68	15·2	18·7	3·5	35·0	9·0	5·5	6·0	8·0
14. 1.68	7·6	11·8	2·2	33·0	11·4	3·0	7·5	9·7
16. 1.68	20·8	7·3	2·0	25·0	14·8	6·4	6·0	8·3
19. 3.68	17·3	6·2	2·6	24·5	15·1	25·5	5·8	9·8
16. 3.68	9·9	6·9	4·2	22·5	13·0	5·5	7·5	7·4
23.12.67	19·8	8·9	2·5	22·0	15·8	4·1	7·0	5·8
22. 3.68	14·0	8·3	1·0	21·0	14·0	8·0	5·0	7·1
30. 1.68	2·5	2·8	1·0	11·0	2·3	1·2	1·3	6·3
4. 2.68	26·7	6·0	1·3	10·0	2·2	38·2	0·6	4·5
15. 1.68	13·5	9·2	1·0	10·0	4·0	7·0	2·0	5·2

Garston			Driving rain mm		
Date	Rainfall mm	Rain duration h	west free-standing	south free-standing	Mean wind speed m/s
22.12.67	5·0	4·1	3·2	2·5	5·5
20. 3.68	4·1	2·2	3·0	1·1	4·9
16. 3.68	1·6	0·9	2·5	0·1	7·5
18. 4.68	11·7	3·5	2·3	5·1	3·3
15. 3.68	3·6	1·0	2·1	0	4·5
24.12.67	5·2	4·3	1·9	0·7	2·9
13. 1.68	7·7	4·3	1·7	5·0	4·6
18.12.67	16·5	12·1	1·7	1·8	1·8
4. 2.68	6·0	4·3	1·5	5·3	4·3
22. 3.68	4·0	2·4	0·5	2·3	5·4
28. 4.68	8·2	4·3	0·2	1·9	2·9
5. 2.68	7·4	3·3	Trace	3·3	3·9

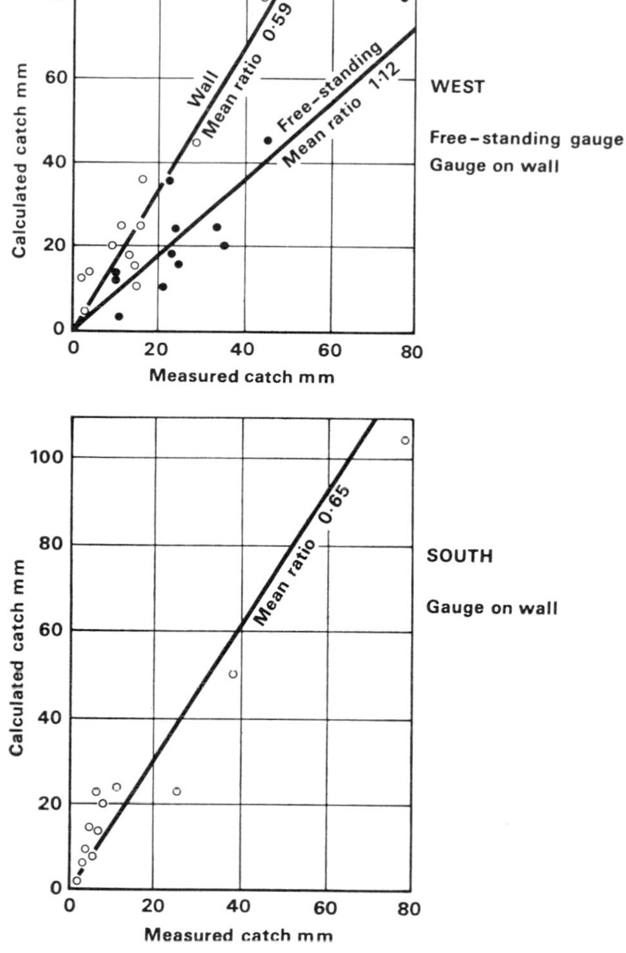

Figure 106 Comparison between calculated and measured driving-rain at Cefn Golau, Tredegar

Prolonged spells of driving rain

It appears that most driving rain in Britain is produced in more or less prolonged frontal-rain, although there may be high intensities of driving rain in showers lasting for only a few minutes (see page 103). There is at present little information on the duration of spells of driving rain, except for data from Garston and Tredegar. Table 30 gave the durations of rainfall during the 12 worst storms of driving rain in the winter 1967–68 at these two places. The total duration in these 12 storms at Garston was 46·7 h, which was 25 per cent of the total duration of rain in that winter. The durations of rainfall in the individual 12 storms ranged from 0·9 to 12·1 h, an average of 3·9 h. Only one exceeded 4·3 h.

At Tredegar the 12 worst storms of driving rain had rain lasting in all for 123·4 h, ranging from 2·8 to 23·3 h with an average duration of 10·3 h. Only one lasted less than 7·6 h. Thus the duration of the worst storms at Tredegar averaged about 2·6 times that of the worst storms at Garston, although the mean rate of rainfall was only 1·2 times as high. Thus the rainfall in the worst storms at Tredegar was about 3·1 times as high as at Garston, compared with an annual mean ratio of about 2·5 for all rainfall.

On the other hand, as described on pages 98–100, driving rain at Cefn Golau, Tredegar, was some 15 times as high as at Garston averaged over the 12 worst storms. These data are based on observations at two places during only one season, but they are probably typical of the differences between exposed and sheltered parts of the country.

Rainfall data from Garston alone have also been analysed to determine the proportion of driving rain which was attributable to longer storms. For this purpose, records of continuous spells of rain have been examined. A continuous spell has been defined thus:

(a) The long spell is deemed to be continuous if the aggregate of any breaks is not more than 10 per cent of the whole.
(b) No single break is to be more than 5 per cent of the time.
(c) The total duration is defined as the time from beginning to end of the spell, *including* the dry breaks.

The bulk of the records used were of spells of rain lasting for 10 hours or more – all available data for the 22 years 1948–69 being used. In addition, records of all storms giving more than about 5 mm of rain in the single year 1963 were used to give information on driving rain from shorter storms. Because there were so many of them it was sufficient to use data for one year only. For each spell the driving-rain index was calculated, using the mean wind speed during the spell. (The driving-rain index is defined on page 106.) Finally each individual storm index was expressed as the percentage of the annual mean driving-rain index at Garston. The results are displayed in Figure 107, with the individual values plotted against the corresponding durations of the rain-spells. It appears that the greatest amount of driving rain in individual continuous spells is produced by storms lasting about 15 h, with both shorter and longer ones producing less. Continuous spells lasting more than 24 h are scarce at Garston. In the 22 years there were only two, one lasting 26·5 h (giving very little driving rain) and one lasting 46·0 h, the driving-rain index for which was 7·5 per cent of the annual average. Such a prolonged spell of rain with strong wind is exceedingly rare in south-east England, and no return period can be estimated at present.

Indeed, it is believed that such prolonged rain-spells are distinctly uncommon anywhere in Britain. For example, at Eskdalemuir during the 21 years 1922–42 the longest truly continuous spell of rain lasted 33 hours, although there was one of 66 h including gaps amounting to 5 h. In this same period there were 85 spells during which it rained continuously for 12 hours or more. At Kew, on the other hand, there were only ten such spells in these years, although the longest spell lasted 36 h, 3 h more than the longest at Eskdalemuir (information from the *Observatories' Year Books* of the Meteorological Office).

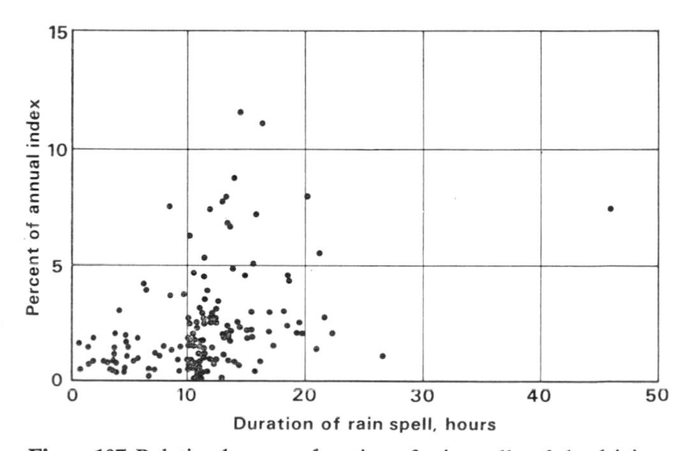

Figure 107 Relation between duration of rain-spell and the driving-rain index attributable to the spell. Index expressed as percentage of annual mean index. Data from Garston: spells with duration of 10 h or more for 22 years 1948–69; shorter spells for 1963 only

Table 31 Relation between duration of rain-spells and rate of driving rain in the spells at Cefn Golau, Tredegar. Data from December 1966 to October 1967, and 12 worst spells of 1967–68 winter

	Driving rain intensity mm/h										Total frequency
Duration hours	0·1–0·9	1·0–1·9	2·0–2·9	3·0–3·9	4·0–4·9	5·0–5·9	6·0–6·9	7·0–7·9	8·0–8·9	13·0–13·9	
1·0–1·9		3	2	2	1	1	1			1	11
2·0–2·9	2	4	3	1	1				1		12
3·0–3·9	1	4		1			2				8
4·0–4·9	1	2	3	2	2			1			11
5·0–5·9		4	1	3	1						9
6·0–6·9			2	2	1	2	1	1	1		10
7·0–7·9	1	2	2	1	2						8
8·0–8·9	1	3	1		2						7
9·0–9·9		1	1	1	1	1					5
10·0–10·9		3									3
11·0–11·9	1	1	1								3
12·0–12·9			1		2						3
13·0–13·9			1								1
14·0–14·9				2							2
15·0–15·9			2								2
16·0–16·9				1							1
17·0–17·9							1				1
18·0–18·9		1									1
19·0–19·9											
20·0–20·9						1					1
21·0–21·9											
22·0–22·9											
23·0–23·9		1				1					2
Total freq	7	29	20	16	13	6	4	3	2	1	101

Rates of driving rain calculated using wind speeds from Cardiff, Rhoose.

It appears from the measurements at Tredegar that there, unlike at Garston, the longer the rain-spell, the more driving rain it produces. It is true that an occasional spell lasting 15 h or more produces relatively little driving rain, but on the whole it seems that long spells give a lot of driving rain (Figure 108).

The limited evidence suggests that long continuous spells are much more frequent in wetter areas – the number with a duration exceeding 10 h increasing perhaps three times as fast as the rainfall. The rate of rainfall in the longer spells may also be higher in the wetter areas, while as wind speeds are also generally higher in these places, the amount of driving rain in individual spells is greatest in the wetter parts of the country. However, although it seems that everywhere in the country the largest part of the driving rain may be provided by a few severe prolonged storms, an individual severe storm may produce a higher proportion of the annual driving rain in a dry area than in a rainy one. Figure 107 shows some storms giving as much as 12 per cent of the mean annual value at Garston. The highest value noted at Tredegar was about 8 per cent, though admittedly in less than two years of observation.

The rates of driving rain in the prolonged spells at a high-altitude place like Tredegar may be quite high. Table 31 shows the relation between duration of spell and the calculated rate of driving rain (using winds from Cardiff, Rhoose) for all significant spells in the period December 1966 to May 1967, as well as for the worst storms in the winter 1967–68. Although the very highest rates were generally confined to short spells (of showers), the rates in spells lasting six hours or more ranged up to 8 mm/h. The two longest spells, both lasting 23 hours, had mean rates of driving rain of 1·1 and 5·5 mm/h respectively. The latter, especially, gave a lot of driving rain, and was rivalled by two other spells, one of 20 h at 5·4 mm/h and the other lasting 17 h at 7·1 mm/h.

No corresponding data are available for lowland sites, but it is likely that the mean wind speed during prolonged spells of rain does not exceed about 5 m/s, so that the rate of driving rain will be about the same as the rate of rainfall, commonly about 1 to 2 mm/h. The greater rates of driving rain at the rainier places arise mainly from the higher wind speeds, and the higher total amount of driving rain from a combination of this factor and the longer duration of rainfall. Figure 109 shows the relation between annual mean driving-rain index

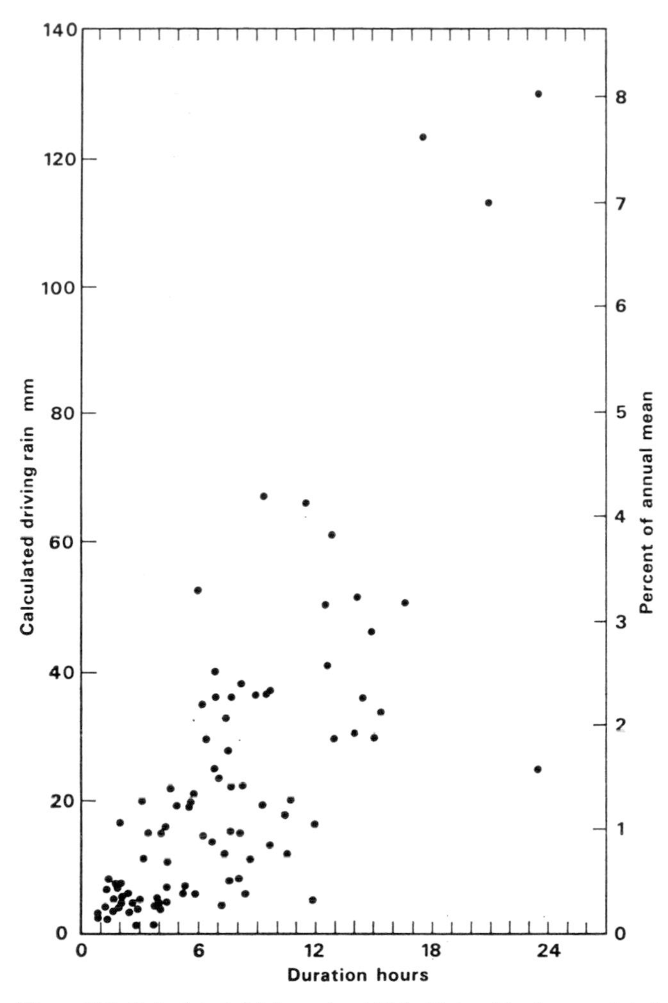

Figure 108 Calculated driving rain at Cefn Golau, Tredegar, related to duration of storm. 1966–67 data, with 12 worst storms of 1967–68, and others to 20 October 1967

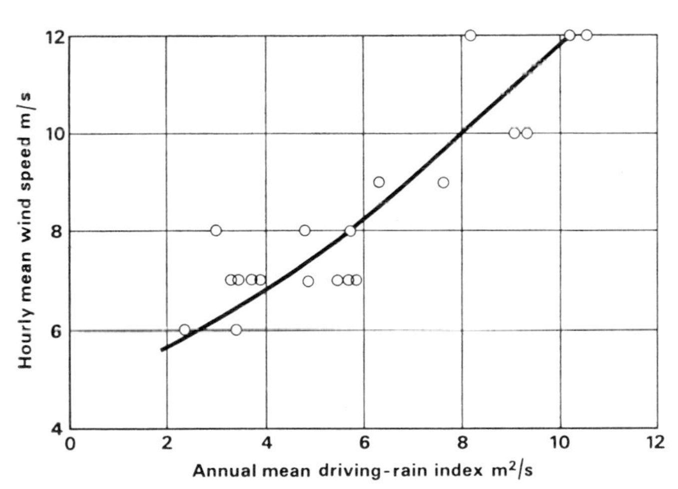

Figure 109 Relation between annual mean driving-rain index and hourly mean wind speed (during rain) above which half the driving rain occurs (1957–66)

(see page 106) and the hourly mean wind speed during rain, above which half the driving rain occurs. The wind speed where the annual mean driving-rain index is 9 m^2/s (about that at Tredegar) is about twice that for an index of 2 m^2/s (the value for Garston). On some occasions, however, there may be in upland country prolonged spells of rain of an intensity of 5 mm/h or more with strong winds. The increased intensity is caused by orographic uplift of the moist air, as explained by Pedgley (1967).

High-intensity driving rain

Driving rain may also occur in showers lasting only a few minutes. Whereas the depressions (lows or cyclones) which give the prolonged rains described in the previous Section may be upwards of 1000 km across, shower clouds are usually no more than a few kilometres in diameter. Because the showers move with the wind, often at speeds of 10 m/s or more, the duration of the rain at any place is usually only a matter of a few minutes. But because the intensity of the rainfall may be so great in showers, up to 100 mm/h or even more, the intensity of the driving rain may be high if the wind is strong enough. Table 53 (page 141) shows the frequency of occurrence of rainfall of high intensity.

Generally the intense rainfalls are produced by convectional storms of the types described by Byers and Braham (1949) and by Pedgley (1962). Briefly, when the storm is young, in the first few minutes of the life of a new cell, the intensity of the rainfall increases rapidly. As the rain falls, it cools the column of air through which it passes, mainly by evaporation from the raindrops, so increasing its density. In addition, the frictional drag of the raindrops imparts an acceleration to the air.

The two effects produce a down-draught, which spreads out on approaching the ground to develop the characteristic ring of squally winds round the base of the storm. According to Byers and Braham the divergence of this outflow is linearly proportional to the rate of rainfall, so that presumably the wind speed should be similarly proportional to the rainfall rate. However, this simple picture is complicated by at least two factors, firstly by the velocity of the storm relative to the ground, so that the winds ahead of the storm are stronger than those to its rear, and secondly by the structure of the air near the ground, which may modify the pattern (Pedgley observed that a temperature-inversion prevented the out-draughts from reaching the ground in some of the storms that he studied).

It was also noted by Byers and Braham that the out-draught spreads more quickly than the area of rain, so that in a more mature storm the most intense squally winds arrive at a given place before the rain. It follows that only a relatively young shower, in which the rain has been falling for no more than a few minutes, is likely to produce a high rate of driving rain. On the other hand, if the general wind speed of the air is high, as may well happen in showery airstreams in this country, especially in the west and north, quite high rates of driving rain may occur throughout the life of a shower.

Individual shower clouds of this type die out after about half an hour, although in a showery type of weather new ones are continually forming. In certain weather situations, fortunately rare in this country, exceptionally severe, self-perpetuating storms may form and last for some hours before decaying. These severe local storms, as they are called, move quite rapidly and produce not only intense rain but large hailstones (see page 29). Other intense storms are formed on occasion during the summer and may persist for quite prolonged periods giving 25 or 50 mm of rain, but there are no records of the winds experienced in such storms.

Because of the short duration of most of the storms it is difficult to get reliable statistical information on the driving rain from them. Even those meteorological stations which determine hourly values of the elements have at best hourly means of the wind speeds on record, and the speed of the wind during the shower may be appreciably different (usually higher) from the hourly mean. Furthermore, the intensity of rainfall averaged over the hour, or over that part of the hour

Table 32 Annual mean driving-rain index (DRI) at 23 places, with maximum hourly driving-rain indexes in 10 years, 1957–66* and corresponding wind direction

				Maximum hourly driving-rain index			
Place	Altitude m	Map reference	Annual mean DRI m²/s	Amount m²/s	As per cent of annual mean DRI %	Direction of wind degrees	Month of occurrence
Lerwick	82	HU 453397	10·60	0·117	1·1	180	December
Wick	36	ND 364522	6·37	0·169	2·6	030	August
Stornoway	3	NB 459332	9·34	0·104	1·1	240	August
Kinloss	5	NJ 069625	2·99	0·099	3·3	030	June
Edinburgh, Turnhouse	35	NT 159739	3·68	0·096	2·6	060	September
Tiree	9	NL 999446	10·25	0·181	1·8	150	September
Glasgow, Renfrew	5	NS 480667	5·51	0·205	3·7	060	June
Prestwick	16	NS 369261	5·69	0·125	2·2	030	August
Dishforth-Leeming	32	SE 305890	3·39	0·211	6·2	240	June
Waddington	68	SK 988653	3·46	0·107	3·1	120	August
Mildenhall	5	TL 683779	2·33	0·104	4·5	270	August
Birmingham, Elmdon	97	SP 171837	3·88	0·094	2·4	360	November
London, Heathrow	25	TQ 077769	3·37	0·140	4·1	090	July
Kew	5	TQ 171757	2·75	0·078	2·8	030	July
Manston	44	TR 335666	4·45	0·086	1·9	240	January
Thorney Island	3	SU 758026	4·42	0·110	2·5	210	September
Boscombe Down	126	SU 172403	4·82	0·147	3·1	300	August
Manchester, Ringway	76	SJ 818850	4·87	0·167	3·4	090	September
Holyhead Valley	10	SH 310758	8·18	0·178	2·2	210	September
Aberporth	133	SN 242521	7·67	0·146	1·9	240	December
Cardiff, Rhoose	62	ST 060670	5·82	0·199	3·4	360	August
Plymouth, Mt Batten	27	SX 492529	9·08	0·404	4·5	090	July
Aldergrove	69	IJ 147798	5·74	0·112	2·0	090	December

*Records from Manston are for about 2 years only (November 1964 to December 1966).

for which it rained, may grossly underestimate the rate which is most significant; often the most intense rain falls for only a few minutes, and is followed for the remainder of the period by more gentle rain. Since the size of the raindrops is related to the rate of rainfall, such errors in estimating the significant rate of rainfall may also cause errors in estimating the rate of driving rain. Obviously it would be possible to extract manually the required data from autographic charts of rainfall, wind speed and wind direction, but this would be a formidable task and has not been attempted.

An analysis of hourly driving-rain indexes

The first approach to the problem of estimating frequencies of driving rain from showers has therefore been to use hourly data, although undoubtedly it must underestimate the quantities of driving rain. The following notes are based on as yet unpublished Meteorological Office analyses of hourly observations of rainfall, rate of rainfall, wind speed and wind direction from about 20 places in Britain. The data for periods of up to 10 years were available on magnetic tape so that analyses could be made by high-speed computer. For each hour, the driving-rain index was computed (that is the mean wind speed multiplied by the rainfall in the hour), and frequency tables prepared with the results divided into groups according to wind speed, wind direction and rate of rainfall at the time.

The analysis shows that in the west and north of Scotland some 95 per cent of the annual driving-rain index is attributable to rain with an intensity of 5 mm/h or less, while in south-east England the proportion falls to about 80 per cent. The remainder occurs in showers with a rainfall intensity of over 5 mm/h, the proportion at any station correlating rather well with the frequency of thunder (Figure 110).

Although the total amounts of driving rain from showers are relatively small, the intensities in individual hours can be quite large. The highest driving-rain index in any one hour at each station in the 10 years was generally in the range 0·1 to 0·2 m²/s. At relatively sheltered places this represents about 5 per cent of the mean annual index, and at the more exposed places about 1 per cent.

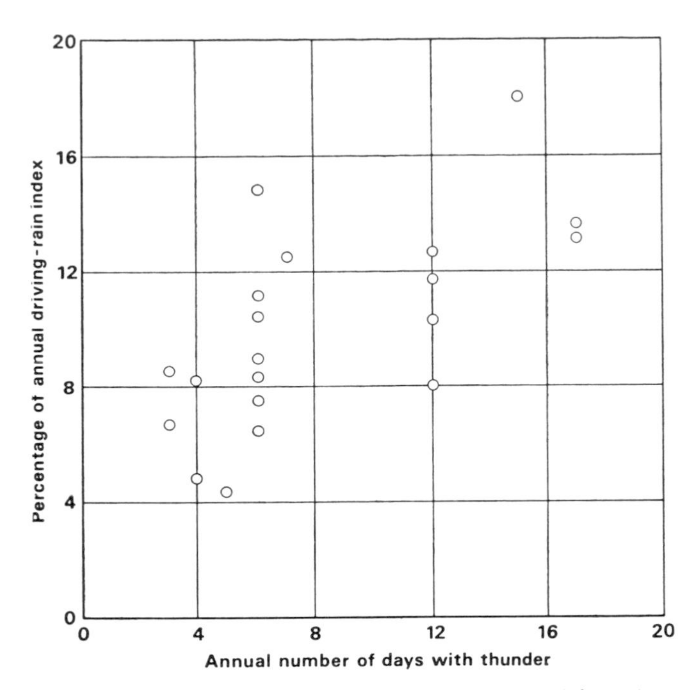

Figure 110 Relation between average annual number of days with thunder (1901–30) and percentage of annual mean driving-rain index (1957–66) occurring with rainfall rates more than 5 mm/h

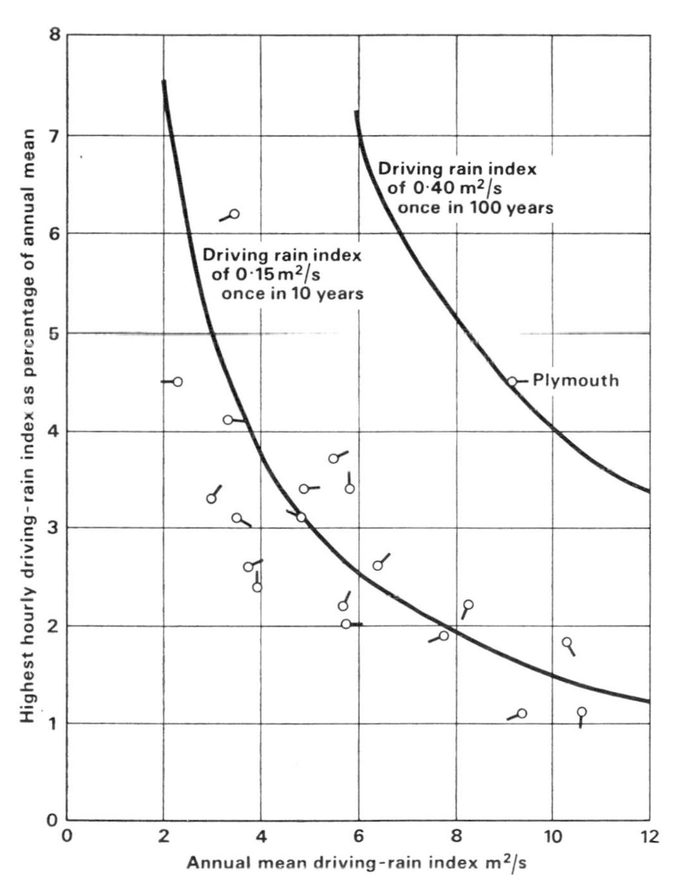

Figure 111 Relation between annual mean driving-rain index and highest hourly index (expressed as percentage of annual mean) in 10 years (1957–66)

Some of the results are summarized in Table 32, and in Figure 111. It seems that it may be reasonable to adopt a value of 0·15 m^2/s as the hourly driving-rain index which may be expected anywhere about once in 10 years, and one of the curves on Figure 111 shows this value. The maximum value at Plymouth was found to be 0·40 m^2/s, much higher than at any other place. It occurred with a wind of about 9 m/s and a rainfall of 46 mm in 0·9 h. Such a fall in this period of time is expected to recur not more than once in about 100 years. It is likely that the frequency of occurrence with a wind speed as high as 9 m/s is lower still, but assuming that once in 100 years is in fact the correct frequency, the upper curve has been drawn in Figure 111. Obviously both these curves must be regarded as extremely tentative.

The direction of the wind when the highest hourly index occurs is shown in Figure 111 by the 'tails' on the points. As will be expected with shower rain, the wind can blow from any point of the compass, because the wind blows outward from the centre of the storm. Thus the direction of the wind depends on the position of the storm relative to the measuring point.

It is not certain what intensities of driving rain are represented by these indexes, but 0·1 m^2/s probably represents about 20 litres/m^2h (5·6 $\times$ 10^{-3} kg/m^2s).

Other analyses have been made using the original data – rainfall, wind speed, etc, rather than the driving-rain index. A more elaborate analysis, calculating the actual amount of driving rain in each hour, rather than the index, could have been made, but it was thought that the extra expense of doing this was not justified.

Calculation of driving rain from records of high-intensity rainfall

An alternative method of using existing records is to consider data on exceptionally heavy showers, using, for example, an analysis such as that provided by Holland (1964). He has produced tables showing the frequency of occurrence of high rates of rainfall during short periods, these being based on observations at stations in England and Wales. The data are reproduced here in Tables 52 and 53 (page 141).

The statistics of rainfall rate produced by Holland do not contain any reference to the associated wind speeds. It might be possible to obtain wind data for a few of the showers by reference to original records, but it is believed that for most of them no wind data exist.

It is, however, reasonable as a design value to assume for each shower a definite wind speed. In the absence of any exact data on the variation of out-draught velocity with rainfall intensity, a constant value of 18 m/s has been adopted for shorter showers, lasting for no more than 15 minutes. This is the value which has been adopted for driving-rain test apparatus, after consultation with the Climatological Services Branch of the Meteorological Office. For showers lasting for 30 to 60 min a wind speed of 10 m/s has been adopted. Using the rainfall intensities shown by Holland in his rainfall-intensity-frequency graphs and tables, intensities of driving rain have been computed by the method described above (page 98), and these are plotted in Figure 112 against return periods in years.

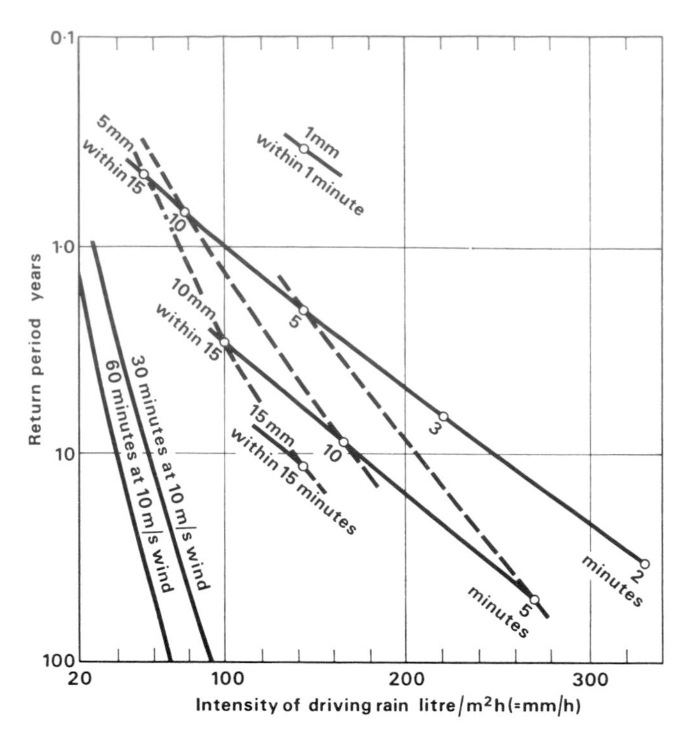

Figure 112 Estimated frequencies of driving-rain intensities from intense showers, assuming wind speed of 18 m/s from Holland's (1964) rain intensity frequency graphs for Britain, with additional data for storms lasting 30 min and 60 min with wind speed of 10 m/s

These curves can only be regarded as approximations, and certainly indicate frequencies higher than those actually experienced. This is because the computation assumes that the high wind speed (18 or 10 m/s respectively) always occurs at the time of extreme rain intensity. Thus the once in 10 years maximum hourly driving-rain index corresponding to the 60-minute curve in Figure 112 is 0·25 m^2/s, with a rainfall rate of 25 mm/h. In fact at the 23 stations whose data were analysed only one had an hourly value exceeding 0·21 m^2/s (this being the once in 100 years value at Plymouth) and the mean of the other extreme maxima was about 0·15 m^2/s.

It seems, therefore, that the driving-rain intensities shown by the 60-minute curve in Figure 112 may be about twice those found in practice on average. There are no data available to check the other curves for the short, very intense spells. However, it may be reasonable to take the diagram as a guide to the frequency of driving rain at the most exposed parts of higher buildings, in the height range of about 30 to 100 m.

An index of exposure to driving rain

Although the amount of driving rain in the free air can be estimated fairly accurately by the method mentioned above, the method requires a knowledge of the rate of rainfall and of the wind speed during the rain. These quantities are usually not known, but other measurements of driving rain on the faces of buildings in Scotland suggested that a simple index of exposure to driving rain would be obtained by using the product of rainfall and wind speed. A good correlation was obtained between the amount of driving rain caught by a gauge on the wall of a building and the product of rainfall and of wind speed (resolved in a direction perpendicular to the wall) during the rain. It seemed that it would be reasonable to take, for the annual mean index, the product of annual total rainfall and annual mean wind speed, as statistics of these parameters are available for many places in the country. Lacy and Shellard (1962) have shown that in Britain there is an almost constant ratio between the annual mean wind speed and the speed during all rainfall – so that this expedient does not distort the overall picture.

Figure 113 shows how an index calculated in this way varies over the country. This map differs in detail from that published in BRS Digest 127 (1971) which has been in use up to now. Because frequent attempts are made by users to determine the local degree of exposure for particular places, it was felt worth while to refine the details. The main problem is that the details of local variation of wind speed are known only imperfectly compared with the good knowledge of rainfall distribution. For the new map, the map of mean wind speed shown in Figure 31 has been used, in conjunction with the rainfall map for 1916–50, to derive the contours of mean annual driving-rain index over England, Wales and Scotland shown in Figure 113. The corresponding contours for Ireland have been prepared by Murphy (1973). The wind data apply only to localities which are not over 70 m above sea-level, and are representative of open country at a height of 10 m above ground. This is higher than normal houses, but the relative values of the index are not affected by this.

The map indicates the amount of rain which would be driven onto a vertical surface in an average year. The numbered contour lines in Figure 113 represent the product of annual mean rainfall in metres and the annual mean wind speed in metres per second. Thus the annual mean driving-rain index is in units m^2/s y. It has been shown that a driving-rain index of 1 m^2/s y is equivalent to about 200 kg/m^2 of driving rain a year (Lacy 1965). However, for most purposes the units used are not important, it is the relative values that matter when comparing the exposure of one place with that of another.

The worst rain penetration of absorbent structures such as brickwork probably occurs in a relatively few severe storms, in each of which the product of rainfall and wind speed might be 10 per cent or more of the mean annual driving-rain index. So much is indicated by the measurements at Tredegar already described. However, the relative severity of worst conditions in different parts of the country seems to be much the same as the relative severity of average conditions, and the same map is applicable to both. An analysis of hourly values of the driving-rain index at a number of stations shows that the highest hourly index in 10 years varies from about 5 per cent to about 1·5 per cent of the mean annual index, as the mean annual index increases from 2 m^2/s to 10 m^2/s. No similar analysis is available for complete storms, but this would probably show a more nearly constant ratio between index for worst complete storms and the mean annual index.

In Figure 113 the indexes are proportional to the total amount of rain that would be driven in one year on to a vertical surface *always facing the wind.* In many parts of the country, most rain falls with winds blowing from directions between south and west or north-west but this is far from true in places near the east coasts of Britain. The 'driving-rain roses' in Figure 114 illustrate the amounts of driving-rain index attributable to winds from different directions. It is clear that in western parts of the country most driving rain reaches walls facing south, south-west or west, and little falls on north facing walls, so that it is desirable to pay special attention to making the former weather-tight. But in some eastern districts, north or north-east walls may be the ones most severely exposed to driving rain.

However, though most driving rain conforms to these patterns, short periods of intense driving rain may be experienced from any direction, as is shown in Table 32 (see page 104).

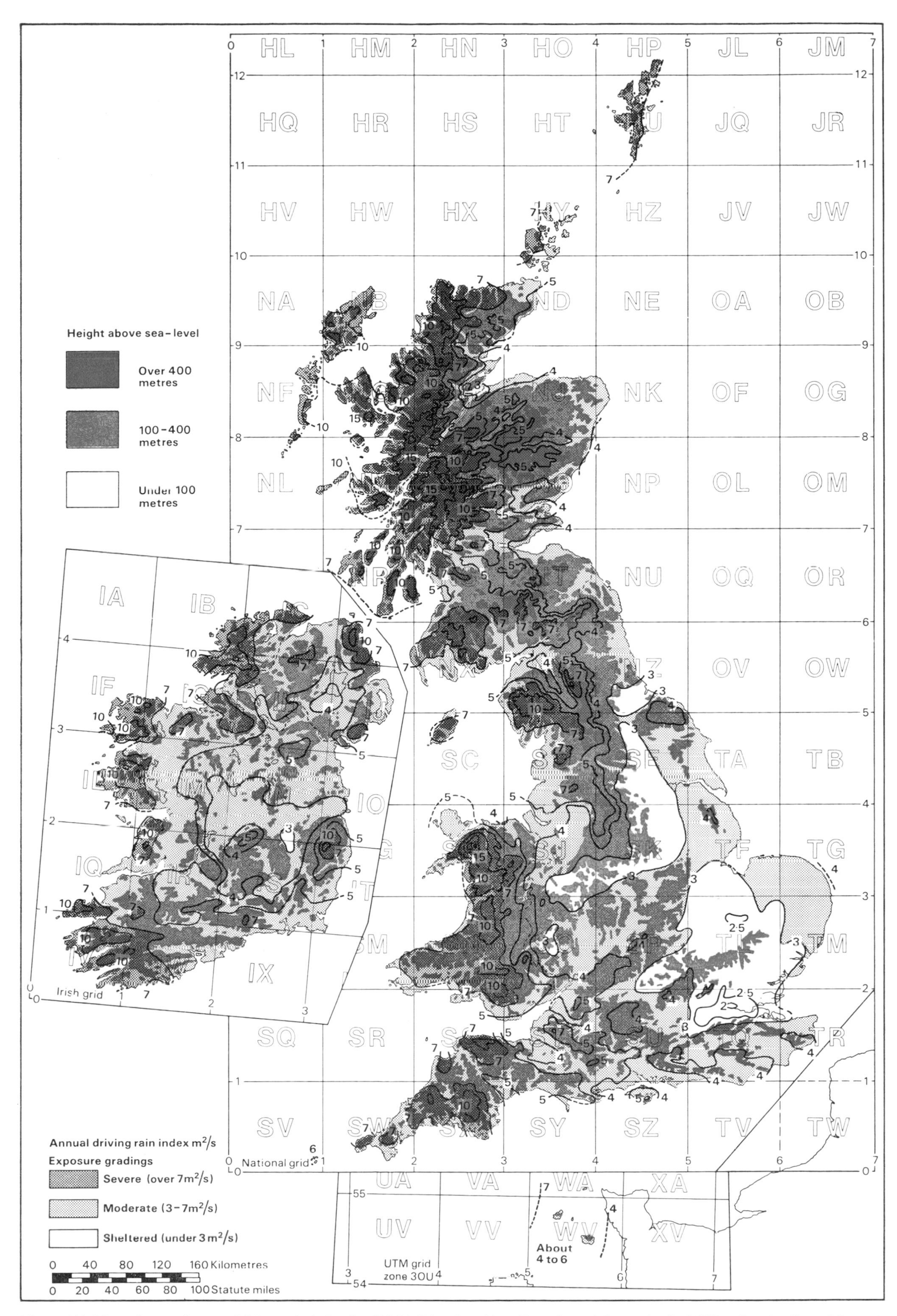

Figure 113 Map of annual mean driving-rain index for British Isles, in m²/sy. (Data from Meteorological Office, Bracknell, and from Department of Transport and Power, Dublin)

Exposure grading

The contour lines in Figure 113 join places having the same mean annual driving-rain index, based on wind speeds at a height of 10 m above open, level country. Shading distinguishes those areas where the map value of the index lies between 3 and 7 m^2/s, with darker shading for those with an index of 7 m^2/s or more. Until recently, the exposure grading of a site has been determined by the value read from the map, with a simple system of rules for allowing for local difference of exposure, as given in BRS Digests 23 and 127, namely:

a. *Sheltered* conditions obtain in districts where the driving-rain index is 3 m^2/s or less, excluding areas that lie within 8 km of the sea or large estuaries, where the exposure should be regarded as moderate (see also *d* below).

b. *Moderate* conditions obtain in districts where the driving rain index is between 3 and 7 m^2/s, except in areas which have an index of 5 or more and which are within 8 km of the sea or large estuaries, in which exposure should be regarded as *severe* (see also *d* below).

c. *Severe* conditions obtain in areas with a driving rain index of 7 m^2/s or more.

d. In areas of sheltered or moderate exposure, high buildings which stand above their surroundings or buildings of any height on hill slopes or hill tops, should be regarded as having an exposure one grade more severe than that indicated by the map.

As the present book went to press, new and more precise rules were being developed, which make more exact allowance for different heights of building and for local shelter. These new rules should be of value for all driving-rain problems, but are designed particularly for use in deciding whether it will be safe to improve the thermal insulation of cavity walls by injecting foam cavity-fill. The values of the local index derived by the application of the rules to the values read from the map are designed to correspond with the behaviour of filled cavity walls in practice. One consequence of the use of these new rules is that that the division of the map into 'exposure zones' does not necessarily correspond to the local exposure gradings found by application of the rules to the conditions at actual sites. The new rules recently became available in a BRE Report by Lacy (1976). The report includes a new version of the map to a scale of 1: 625 000, which allows the map value value of the index at a particular site to be determined more exactly.

The annual mean driving-rain index gives, it is believed, a reasonably precise method of comparing different sites with respect to total amounts of driving rain on walls. It enables a designer to compare the exposure of a place with that at another with which he is already familiar. It is doubtful if it will ever be possible to find an exact correlation between the weather conditions and every type of construction. Not only is there an infinite gradation in severity of exposure, but there can be a great variation in the performance of a given type of structure, because of slight variations in materials, in detailing, in sizes of components and in workmanship.

However, the exposure gradings defined by the driving-rain index do bear a relationship to the performance of buildings. It was noted above that complaints of rain leakage through windows are related to degree of exposure as defined by the index. The driving-rain index is used as a classification of exposure in the Building Standards (Scotland) Regulations 1963, and in the BS Code of Practice CP 121: *Walling*, Part 1: 1973: Brick and block masonry.

It should be recognised that there may be appreciable variations in the severity of exposure on different parts of the same building, as well as the variations between sites. Part of this variation may result from shelter by other buildings, or by projections from the wall, but even on a freely exposed face there may be significant differences. As the wind blows past the building the air is deflected from its normal course and is speeded up, especially around corners and over cornices, where its speed may be twice that of the undisturbed wind. The rain intensity is increased correspondingly, so that in such places the driving-rain index is abnormally high. While this need not affect the general assessment of the exposure of the building, the possibility that certain parts may be abnormally exposed in this way should be considered in the design. This variation can sometimes be observed in the pattern of wetting of a building. Some recent Swedish measurements with rain-gauges on a large building suggest that the range of variation in the amount of driving rain reaching various parts of a wall may be as high as a hundred to one. Measurements in Britain have never recorded a range of more than two or three to one, but the detailed design of the building may well play a part.

The wetness of a wall at any time will reflect the balance between the rate of gain of moisture from driving rain, and the rate of loss by evaporation from the surfaces. The rate of evaporation depends on a number of factors (see page 114), but broadly speaking the rate is rather lower in the wetter parts of Britain than in the drier parts. The net effect is to increase the risk of rain penetration in regions with a high driving-rain index, for here the walls have less chance of drying out between spells of rain. However, the range of variation of driving-rain index is so much greater than that of evaporation rate that the latter can be ignored when making assessments of the degree of exposure.

Estimating driving-rain catches on the various walls of a building

The statistics used in preparing the driving-rain roses in Figure 114 can be used to estimate the relative amounts of driving rain that will be expected to reach the different walls of a building. Such an estimate was made recently when investigating the causes of failures on a particular building, it being thought that wind-driven rain might have contributed to the trouble. The driving-rain data are available in the form of calculated values of the driving-rain index for each of 12 directions, respectively 0°, 30°, 60°, etc, from true north. The angles between these directions and the normals to the faces of the building are first calculated; then the respective values of the driving-rain index are multiplied by the cosine of these angles with normals and summed for each face.

For example, in this case one wall faced 048° true. Driving rain from 090° (east) has an azimuth (90 − 48)°, that is 42° from the normal to the wall, and the cosine of 42° is 0·74. The driving-rain index for winds from 090° at Elmdon (the nearest available site to the building) was 4·4 per cent of the total annual driving-rain index. Multiplying this by the cosine gives 3·3 per cent. Proceeding in the same way for all wind directions which would drive rain on to this face (0, 30, 60, 90, 120 and 330 degrees) and summing the results gives 20·0 per cent. Similarly for the other three walls, which faced 138, 228 and 318 degrees, gives relative driving-rain indexes of 40·8, 47·4 and 19·9 per cent. The local total annual driving-rain index is estimated to be about 3·0 m^2/s, so that the

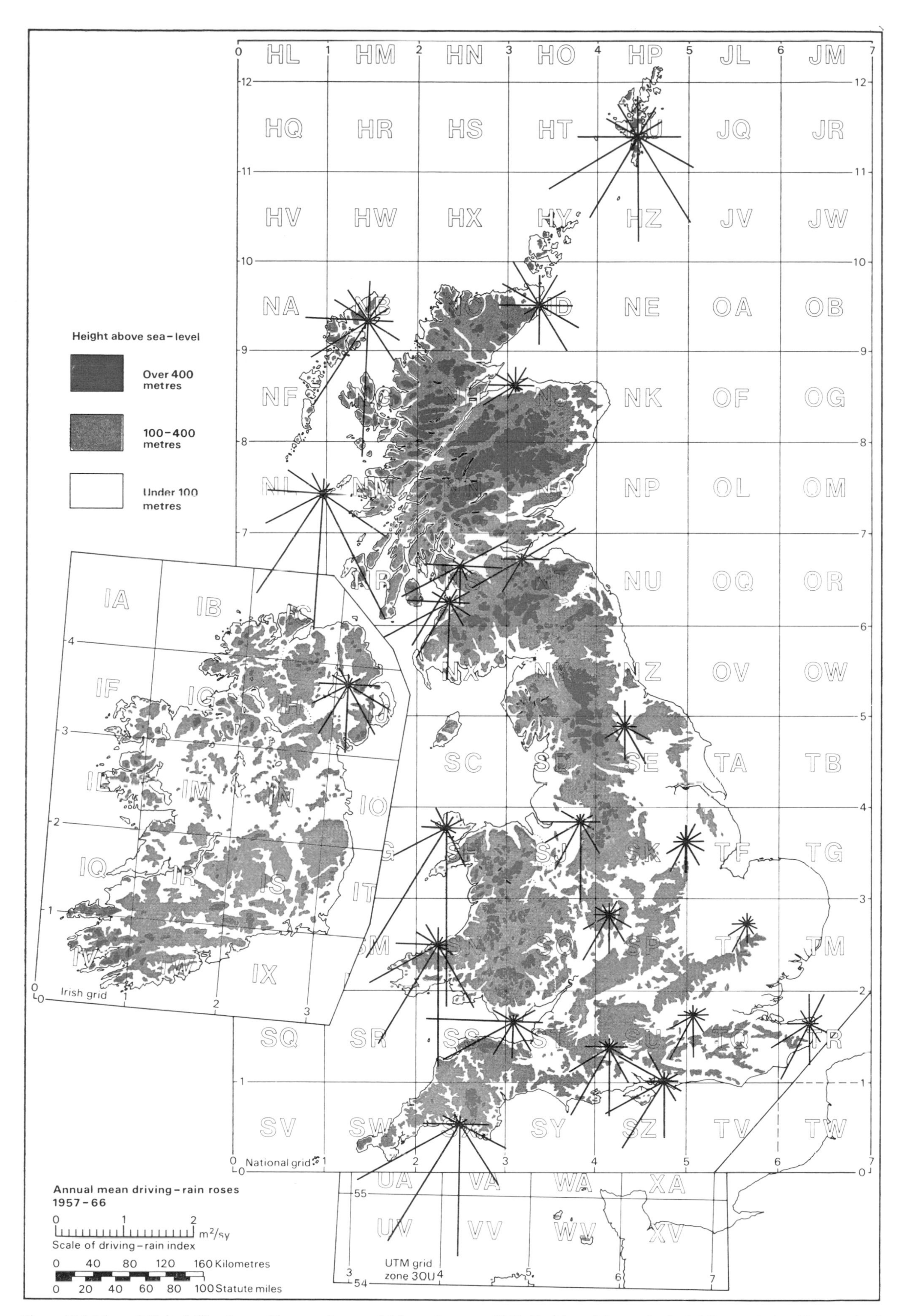

Figure 114 Map of United Kingdom with annual mean driving-rain roses, 1957–66. (From Meteorological Office analysis of hourly data)

indexes for the four walls were respectively about 0·60, 1·22, 1·42 and 0·60 m^2/s. As a driving-rain index of 1 m^2/s probably represents a driving rain amount of about 200 litre/m^2 on a vertical surface, the mean annual catch on these four walls is estimated to be respectively 120, 244, 284 and 120 litres/m^2.

These values do not allow for deflection of the wind by the building, nor for local shelter by any obstructions in the near vicinity, but give a rough indication of the amounts to be expected.

Run-off from walls

When a masonry wall becomes saturated with water, or if the rate of driving rain should be greater than the rate at which the wall can absorb water, water will run off the wall. (See page 113 for a note on the rate of absorption of rainwater by materials.) By definition, there will always be run-off from non-absorptive materials such as glass and metals.

The rate of driving rain on a vertical surface can be calculated from the rate of rainfall (on the horizontal) and the wind speed, using certain characteristics of rain as described on page 96. The calculation assumes that there is no deflection of the wind, or of the raindrops, by the vertical surface. Since such deflection must always occur, the calculation over-estimates the catch of driving rain over most of the surface. However, there is a pronounced tendency for a concentration of driving rain near the edges of a building, and the calculated rate is probably not very different from the actual rate at such exposed points. The rate over the greater part of a wall is half or less of this calculated rate.

The highest rate of driving rain that has ever been measured was somewhat greater than 100 litre/m^2h on a penthouse of some flats in Glasgow, about 24 m above ground. This high rate lasted for only about five minutes.

Observation of buildings during rain suggests that in fact there is little risk of large rate of run-off from buildings, except perhaps where the building is completely clad with ceramic tiles or other similarly smooth material. Most other buildings, although nominally smooth, have the walls broken by inset joints of one kind or another, or by metal sections of some kind. Small though these may be in comparison with the bulk of the building, they are sufficient to break the flow of water. Even a small horizontal member, such as a metal window-frame or sill, collects the water, and it drips off in relatively large drops, some of which may strike the building again lower down, but many of which fall to the ground. It is suggested therefore that it is not necessary to provide gutters on the face of a building, but that some form of canopy should be fitted, perhaps at first-floor level, to protect pedestrians from the 'curtain' of large raindrops.

This curtain of drips may be augmented by water splashing off the wall as raindrops break up on striking the surface, and perhaps also by raindrops being concentrated and carried down by the vigorous downcurrent of air near the foot of the building.

Although run-off has been discussed, it must be remembered that above the 'neutral point' on a building, the air will be flowing upwards (see Figure 60, page 63). If the wind speed is high enough, these upward air currents will carry raindrops with them, tending to drive them under flashings and over-hangs. This is especially likely near the cornice of a building, where the air currents will be especially strong, see for example Khlusov (1954). At such points, the film of water on a wall may even be driven upwards.

There is very little information on the frequency of high rates of catch on walls. The catch of over 100 litre/m^2h mentioned above was by far the highest value noted in a six-month period of measurement. It occurred during a spell of heavy showers, in a strong wind. Prolonged rainfall from depressions will often occur in combination with strong winds, but because the rate of rainfall is much smaller than in showers, the rate of driving rain is not so great. Typically, the rate of rainfall over low ground in a depression will be about 1 mm/h, and from data on extreme winds it is estimated that at a height of 10 m above ground the highest likely rates of driving rain will be as shown in Table 33. The table shows the estimated probable return periods in certain parts of the country, and also corresponding figures for a height of 100 m above ground. These are the rates of driving rain that may be expected at corners and edges of buildings. The rates over the faces may be half as great or less.

It would be expected that much higher rates of driving rain would be experienced from showers, as explained on page 103, so giving a greater likelihood of run-off from walls. However, observations of such run-off are entirely lacking, except for one case of a building at Bergen in Norway, on the very exposed western coast. Here it has been noted that the worst conditions for rain penetration arise if the design of the building is such that driving rain can be concentrated, for example in an angle between the face of the wall and a projecting mullion or column, so that a stream of water can run down the wall, much increasing the risk of penetration of water through the joints which usually exist in such angles.

Table 33 Estimated frequencies of rates of driving rain in strong winds during frontal rain, lasting for 30 seconds

Rate	Frequency
At a height of 10 metres above ground	
10 litre/m^2 h	not more than once in 20 to 50 years at exposed places in west and north-west Scotland.
5 litre/m^2 h	not more than once in 2 years in towns in Scotland, northern England and Wales.
	not more than once in 20 years in towns in midland and south-east England.
At a height of 100 metres above ground	
10 litre/m^2 h	not more than once in 3 years in exposed places in west and north-west Scotland.
	not more than once in 10 years on exposed western English and Welsh coasts.
	not more than once in 60 to 100 years in northern towns.
5 litre/m^2 h	not more than once each year in towns in northern England and Wales .
	not more than once in 3 years in midland and southern towns.

The above remarks have been concerned with the nuisance caused by the water itself. A further nuisance, which usually starts to be obvious on any building more than a few days old, is that of the transport of materials by rainwater, and its deposition in places where the water has evaporated. The material may be derived by the leaching out of salts from the structure, or may come from pollution in the atmosphere. A simple case of such staining is shown in Figure 115. The problem is widespread and takes many forms (see for example the book by Addleson 1972). Since rain and dirt will always be with us, it seems that it will be necessary to design buildings so that as far as possible water is not allowed to run down facades which are liable to stain. Arrangements must be made to throw it clear, as far as can be done, by string courses, lintels and the like.

The shelter afforded by a canopy or overhang

A question often asked is – how much shelter is given by a canopy or overhang? There is no simple answer, because the angle at which the raindrops fall depends on both the size of the raindrops and the speed and direction of the local wind. Although there is usually a range of raindrop sizes in any given fall of rain, on average, the median raindrop size is proportional to the rate of rainfall; the heavier the rain, the larger the raindrops (see Figures 99–101).

The larger the raindrop, the greater its terminal fall-speed, that is the vertical component of its velocity. From the relationship [equation (18)] given on page 97, the angle of fall of a raindrop can be calculated in any given wind speed. For example, if the drop diameter is 1 mm, the terminal velocity is 4 m/s. Thus in a wind of 6 m/s, the angle with the vertical of the path of the drop will be such that

$$\tan i = 6/4 = 1{\cdot}5.$$

Hence i, the angle with the vertical, will be about 56°. But even when the rate of rainfall is constant, there are many raindrops much bigger and much smaller than the median-sized ones (see Figure 101). Thus an appreciable proportion of the drops will fall at larger angles to the vertical than the value calculated for the median drop-size diameter. For example, at a rate of 1·27 mm/h, when the median diameter is somewhat over 1 mm (Figure 101), about five per cent by volume of the rain is composed of drops about 0·5 mm in diameter. Their terminal velocity is about 2 m/s and the angle of their path with the vertical is $\tan^{-1} 6/2 = 72°$.

These calculations apply to an open situation, or to open-sided flat-roofed shelters which do not deflect the wind. If the shelter is in the neighbourhood of buildings, local winds may be quite different from those in the open. In particular, in the case of an open-sided corridor connecting two buildings, there could be much stronger winds blowing between the buildings than in the open, so that raindrops would be blown further under the canopy.

Observations made with freely exposed driving-rain gauges can be used to estimate mean incident angles of rainfall (Lacy 1951). On a rather sheltered site at Garston it was found that daily mean values were less than 10° on about 35 per cent of rain-days, and less than 30° on 90 per cent of days. On the other hand, on a fairly well exposed site at Thorntonhall, south-east of Glasgow, the mean incident angle was less than 10° on only 9 per cent of days, and was over 30° on 50 per cent of days. These figures refer to mean incident angles of fall; it must be emphasized that half the volume of rain falls at greater incident angles than the mean.

In the case of canopies or overhangs attached to a building, it is not possible to make any estimate of the degree of shelter afforded without a detailed knowledge of the airflows close to the building. A useful approach is to observe the wetness patterns on existing buildings, see, for example, Figures 115 and 116. It should be remembered that snowflakes are carried more readily by the wind than raindrops, and may be driven up under overhangs which provide shelter from driving rain (Figure 117).

Some guidance for design can be obtained by calculating the likely frequency with which driving rain will strike the facade of a building such as a grandstand, so that an estimate can be made as to whether it is worth while going to much expense to extend the canopy beyond the seating. For example, consider a grandstand which is to be put up in a fairly exposed situation in the midlands of England, with the spectators facing approximately 130° (about south-east). First estimate the total duration of rainfall at site; from the map in Figure 146 it appears that it will be raining for about 6 per cent of the time, averaged over the whole year. Only rain which falls with winds blowing from directions between 040° and 220° will reach the front of the stand. Using driving-rain data from Elmdon Airport, near Birmingham (which was used to prepare the driving-rain rose in Figure 114) and assuming that the duration of rain from each direction is proportional to the amount, it is estimated that driving rain will reach the facade for about 4 per cent of the time. Probably about half this driving rain will fall with an incident angle to the vertical lying between 10° and 30°. Thus if the stand is in use for 4 hours on each of 20 days evenly spread through the year, there is a risk of driving rain reaching the spectators in the front seats on about 3 hours in each year. The risk will be less in summer and greater in winter. It is for the owner of the building to decide whether such a risk is tolerable, or whether he should go to the considerable expense of a wind-tunnel study.

Rain patterns on walls

When the wind blows it carries raindrops with it, so that they fall at an angle to the vertical. Thus it will be expected that any vertical surface will intercept these raindrops. However, the wind is deflected by any object in its path, and the resulting pattern of airflow may be quite complex, depending on the shape of the building, and its position relative to other objects in the vicinity. In consequence, the pattern of deposition of the raindrops is also complex. Indeed, the complexity is increased because raindrops exist in a range of sizes, each with its own rate of falling, and hence with its own angle of fall in a given wind speed.

Figure 115 Rain patterns on a wall caused by raindrops blown on to the surface and by run-off of water

Figure 115 shows a north-facing wall, shortly after the end of a spell of rain which fell with wind blowing from south-east to south. A variety of wetness patterns can be seen. On the extreme left-hand corner, and along the top edge, the wall has been wetted by raindrops blown over the edges and carried to the lee side of the building by wind-eddies. Much of the rest of the face has become wetted to a lesser extent by rain carried over the top of the building, but with parts on the right-hand (western) side of the projecting columns sheltered. This sheltering effect seems to disappear at the right-hand end, presumably because of a modification to the local airflow by the larger building on the right. There is also considerable evidence of water running down the face. Finally in several places there is a wet strip of wall below the pronounced horizontal joints between the concrete panels. This suggests that small local eddies, strong enough to cause deposition of raindrops, have been set up in these zones, by the joints.

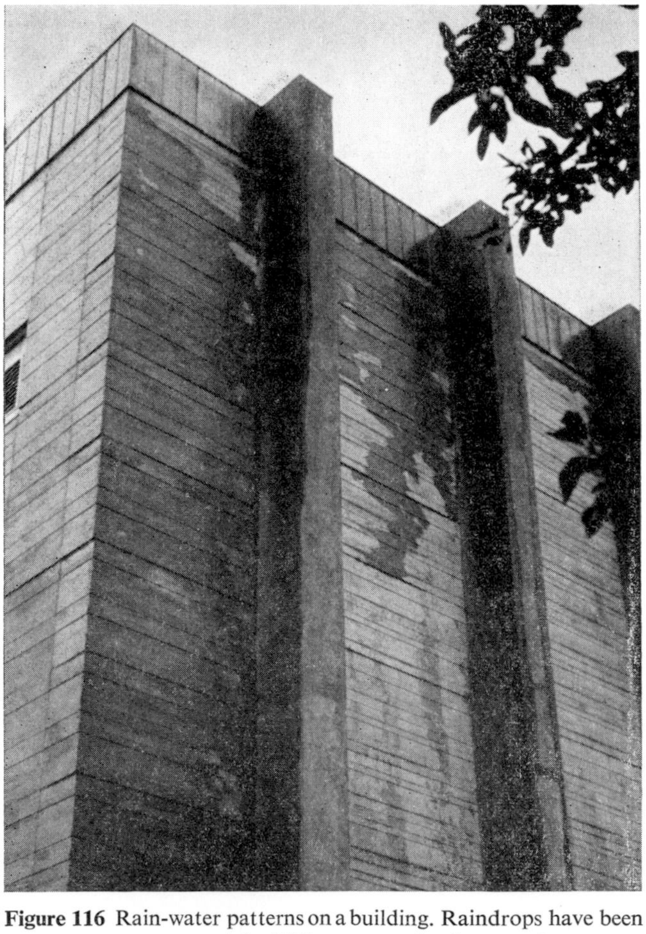

Figure 116 Rain-water patterns on a building. Raindrops have been swept round the NW corner, to impinge on the west face, and over the top of the parapet. The wetness between the first and second columns is in part caused by moist air emerging from the air-conditioning plant on the roof and being carried down by an eddy sweeping over the parapet

There is a considerable tendency for these patterns to be repeated in successive rain-spells, so that quite soon corresponding patterns of dirt are produced. These are conspicuous on this picture, which was taken when the building was less than two years old.

Figure 116 shows other examples of staining caused by water running down a wall, demonstrating the tendency for concentration at both external and internal angles. At one place the wet patch is partly caused by moist air emerging from the air-conditioning plant on the roof. This is gradually producing a permanent dark stain on the concrete.

Driven snow on walls

Snowflakes have a much lower density than raindrops, only one-tenth that of water, or even less. In consequence they are more responsive to the smaller fluctuations of air currents than raindrops. An overhanging roof effectively protects a portion of the wall below it from wind-driven raindrops, the effectiveness depending on the amount of the overhang and the speed of the wind. However snowflakes may be driven under the overhang and, if the temperature is a little above 0°, may stick to the wall. Figure 117 shows how snow has been plastered on the wall of a house, even to the top of the wall under the overhanging gutter. Thus when the snow melts, parts of the wall which are quite sheltered from rain may become wetted, although normally they are dry. It is clear that the pattern of snow-deposition on a wall may not be a guide to what happens to wind-driven rain.

Snowflakes have also been observed to stick on windows and soon to melt. The water ran down the glass to the metal framing, where it was blown through cracks at the corner of the window-frame.

Figure 117 Wind-driven snow adhering to house and garden walls at Tredegar on 2 December 1966. Note how snow is driven up under roof overhang

Moisture in walls

Walls of porous materials such as brick, stone or concrete, absorb rainwater which is driven on to them, and lose this water by evaporation to the atmosphere in the dry spells. Approximate estimates of the amount of driving rain reaching a wall can be made from meteorological data, as described on pages 108–10, but the amount of rainwater absorbed may depend also on the properties of the wall, as well as on the recent history (for example, if the wall has been saturated by previous rain, no more can be absorbed). Common bricks can, when relatively dry, absorb water at a high rate, in the region of $2{\cdot}5 \times 10^{-2}$ kg/m²s, which is much higher than any likely rate of driving rain. The rate of absorption falls off with the square-root of the time after the initial rain, but until the brick is nearly saturated it can absorb almost all the driving rain that can reach it. On the other hand, dense bricks, concretes and cement renderings can absorb only at a much lower rate, at perhaps one-hundredth of that for common bricks.

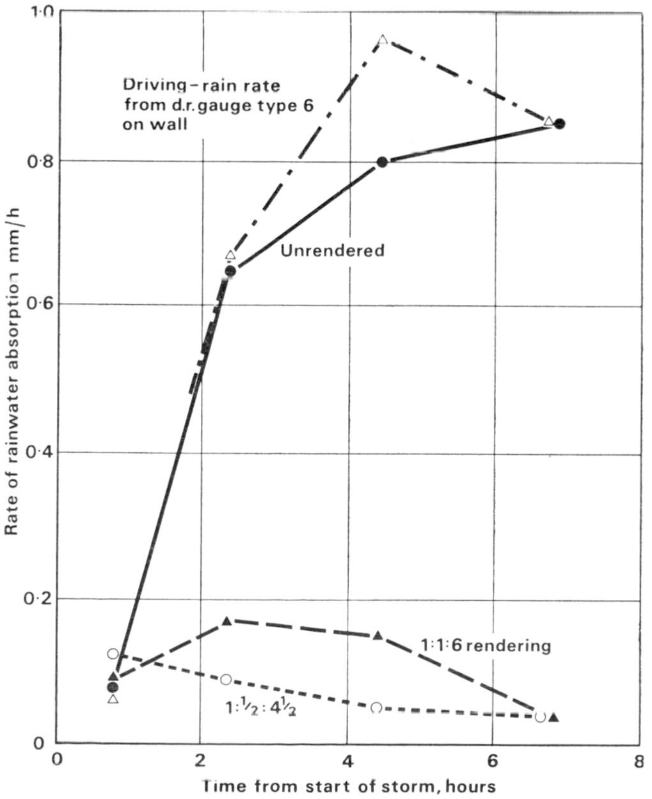

Driving-rain rate
from d.r. gauge type 6
on wall
Unrendered
1:1:6 rendering
1:½:4½
Rate of rainwater absorption mm/h
1·0
0·8
0·6
0·4
0·2
0
0 2 4 6 8
Time from start of storm, hours

Figure 118 Rates of absorption of driving rain by three walls, and rates of driving rain. Averaged during four consecutive spells on 9 December 1965

It follows that although a rendered wall may initially absorb all the driving rain reaching it, in an hour or two its rate of absorption may fall off to such an extent that excess rain runs down the face. Figure 118 shows the results of some measurements on wall panels which were mounted on weighing machines, so that the change in weight as water was absorbed could be followed. The weights of the panels, each about 0·9 m square, were taken 5 times on a day when there was continuous driving rain reaching them. The average rate of change of weight in each interval is plotted at the mid-point of the interval. The unrendered common Fletton brick panel continued to absorb water at approximately the rate of the driving rain as measured by the raingauge mounted on the face of the guard-ring round the panels. The two rendered panels, one with a 1 : 1 : 6 mix (proportions by volume of cement : lime : sand), the other with a $1 : \frac{1}{2} : 4\frac{1}{2}$ mix with pebble-dash finish, absorbed water at the same rate as the unrendered panel during the first interval of some $1\frac{1}{2}$ hours, but in the second interval, during which the rate of driving rain was about 8 times as high, their rate of absorption was little different from that in the first interval. The sensitivity of the weighing machines was not great enough to follow the changes in weight as closely as could be wished, but it seems clear that the rate of absorption of these two rendered panels gradually fell. As the rate of driving rain was much greater than the rate of absorption, water must have been running down these panels. There was therefore always an adequate supply of water, and the fall in the rate of absorption must have resulted from the particular pore structure of the rendering.

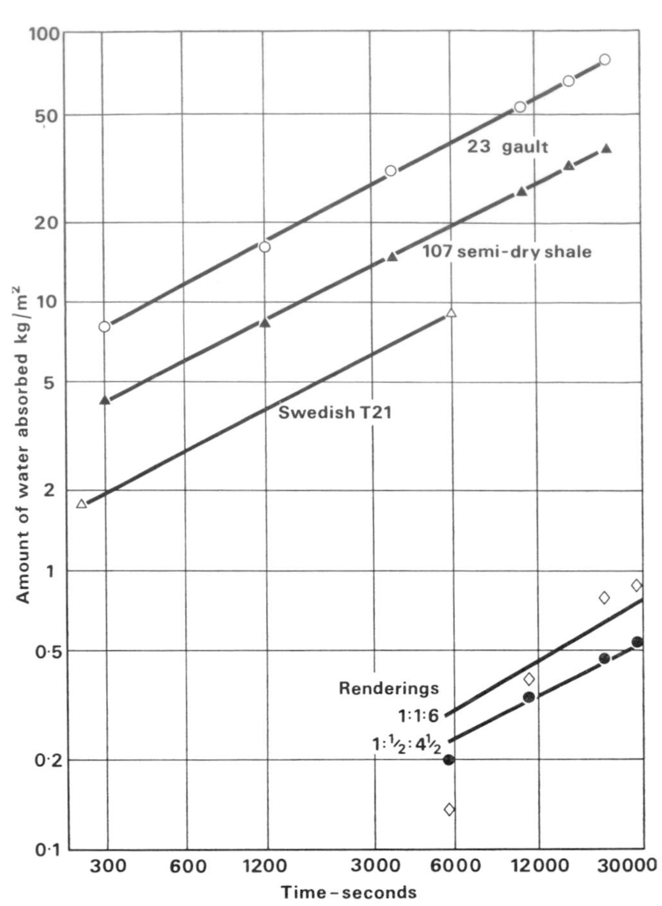

Figure 119 The absorption of water by bricks and renderings. Data for bricks from Butterworth (1947) and Jansson (1965), and for renderings from weighed walls at the Building Research Station, Garston, on 9 December 1965

Jansson (1965) has quoted the results of experiments on the absorption of water by different materials, measured in the laboratory, and the rate of absorption fell off with the square root of elapsed time, exactly as predicted by theory, when there was an unlimited supply of water. Figure 119 shows data from laboratory tests on three kinds of brick, one being taken from Jansson's paper, the other two being from measurements by Butterworth (1947).

The amount of water A_w absorbed up to time t is given by the relations:

Brick 23 – Gault $A_w = 4{\cdot}16 \times 10^{-1} t^{0.52}$ kg/m^2......(21)

Brick 107 – shale $A_w = 2{\cdot}62 \times 10^{-1} t^{0.49}$ kg/m^2......(22)

Brick T.21 – Swedish $A_w = 0{\cdot}01 + 1{\cdot}08 \times 10^{-1} t^{0.51}$kg/m^2(23)

Thus the rates of absorption of these three different types of bricks differ in a range of about 4 to 1.

The data at the foot of Figure 119 are taken from the weighed rendered wall panels described above. The lines through them are drawn assuming that they pass through zero at the beginning of the more severe driving rain on the occasion of the measurements. Although there is a good deal of scatter in the results, they lie approximately on lines representing the square root of time relation:

1: 1: 6 rendering $A_w = 1{\cdot}8 \times 10^{-3} t^{0.59}$ kg/m^2......(24)

1: ½: 4½ rendering $A_w = 3{\cdot}1 \times 10^{-3} t^{0.50}$ kg/m^2......(25)

Thus the renderings appear to absorb driving rain at a rate about one hundred times slower than ordinary bricks. In the first hour after the driving rain started, the 1: ½: 4½ rendering absorbed about 0·18 kg/m^2 of rain, about what would be produced by a light rainfall of 0·1 mm/h in a wind of 5 m/s.

Water loss from walls

The water contained in the pores of a wall, whether derived from driving rain or from water vapour in the internal air of the building, is gradually lost by evaporation from the surfaces, the rate of evaporation depending in part upon the humidity and speed of movement of the air in contact with the surfaces, and in part upon the nature of the materials of the walls and upon their moisture-content. If the surface material is completely saturated with water, the rate of evaporation is the same as that of a free surface of liquid water. It is then controlled by the wind speed and humidity of the air, if there is an adequate supply of sensible heat to make good the latent heat required. The relation between the rate of evaporation φ, the wind speed V, the water-vapour pressure e in the air, and the saturation water-vapour pressure e_w at the temperature of the surface, is of the form

$$\varphi = \mathrm{a}.V^{\mathrm{b}}(e_w - e) \qquad (26)$$

where a and b are constants.

In fact, building materials are usually only saturated for short periods, except perhaps in the most exposed areas of the country. Most of the time, the evaporation of water takes place at a rate which is determined by the resistance to flow of water, or of water vapour, through the pores of the material, rather than by the conditions in the air flowing over the surface. This is demonstrated by the results of measurements of the rates of loss of water from a number of wall panels, both naturally exposed and in laboratory tests.

The rates of loss of water from north-facing walls in the Wall Laboratory at Garston are shown in Table 34. They were rather sheltered by trees and by other buildings. The rooms behind the walls were heated in winter to 18·5°. The evaporation rates (deduced from measurements of the moisture-content of the walls at different times) were only about one-tenth of the computed wet-surface rates.

Losses of water from a weighed, south-facing, unrendered brickwork panel at Garston as high as $5{\cdot}8 \times 10^{-6}$ kg/m^{2}s have been measured, averaged over the first 4 days after driving rain has wetted the surface, and still higher rates (about $1{\cdot}9 \times 10^{-5}$ kg/m^{2}s) over periods of a few hours. These may be compared with calculated wet-surface rates of about $5{\cdot}5 \times 10^{-5}$ to 1×10^{-4} kg/m^{2}s. Much lower rates of evaporation were measured during long dry spells (Table 35). The rate of evaporation from the two rendered panels during these dry spells was appreciably less than that from the brickwork panel.

The rates of evaporation from similar panels exposed at Thorntonhall, near Glasgow, are of about the same magnitude as those at Garston, although Thorntonhall is appreciably more exposed to wind. For example, during an almost dry spell from 29 January to 4 March 1963, the mean evaporation rate from an unrendered panel of Fletton bricks at Thorntonhall was $1{\cdot}8 \times 10^{-6}$ kg/m^{2}s, and that from the panel with 1: 1: 6 rendering was 9×10^{-7} kg/m^{2}s. The mean moisture-contents were 20 and 8 per cent by volume respectively.

Granum (1965) has measured the evaporation from brickwork under controlled conditions and found that 10 to 20 days after the start of a test the rate averaged about 1×10^{-6} kg/m^{2}s, when the conditions were such that the surface layer became dry in that time.

It seems likely, therefore, that the drying of walls is influenced mainly by the frequency and duration of the dry spells between periods of driving rain, rather than by the actual weather during the dry spells. This is demonstrated by

Table 34 Water losses from north-facing walls in the Garston Wall Laboratory kg/m² s

Wall no	Description	Dates of test	Mean moisture-content, % vol	Net water loss	Estimated driving rain on wall	Overall water loss
35	20 cm 1:2½:7½ foamed-slag concrete	7/51 to 5/53	11 (in 2nd winter)	$3{\cdot}0 \times 10^{-7}$	$7{\cdot}9 \times 10^{-7}$	$1{\cdot}1 \times 10^{-6}$
36	20 cm 1:3:6 foamed-slag concrete	10/50 to 5/51	19	$4{\cdot}5 \times 10^{-7}$	$11{\cdot}1 \times 10^{-7}$	$1{\cdot}6 \times 10^{-6}$
37	20 cm 1:3:6 foamed-slag concrete	7/52 to 5/53	16	$6{\cdot}2 \times 10^{-7}$	$12{\cdot}1 \times 10^{-7}$	$1{\cdot}8 \times 10^{-6}$
38	20 cm 1:2:8 foamed-slag concrete	9/52 to 5/53	17	$5{\cdot}5 \times 10^{-7}$	$11{\cdot}8 \times 10^{-7}$	$1{\cdot}7 \times 10^{-6}$
41	23 cm perforated foamed-slag concrete blocks	10/52 to 3/53	7	$8{\cdot}3 \times 10^{-7}$	$9{\cdot}7 \times 10^{-7}$	$1{\cdot}8 \times 10^{-6}$

Note: All the walls were of foamed-slag concrete, rendered outside, plastered inside.

Table 35 Water losses from south-facing wall panels, at Garston and Thorntonhall, kg/m² s

Panel no	Description	Date	Mean moisture-content, % vol	Mean rate of loss of water
1	Fletton bricks, unrendered	17/1/65 to 14/3/65 (56 days)	19·0	$1{\cdot}4 \times 10^{-6}$
		15/5/65 to 15/6/65 (31 days)	13·5	$3{\cdot}9 \times 10^{-7}$
2	The same with 1:1:6 rendering	17/1/65 to 14/3/65	22·5	$6{\cdot}9 \times 10^{-7}$
		15/5/65 to 15/6/65	19	$2{\cdot}6 \times 10^{-7}$
3	The same with 1:2:9 rendering and pebble-dash	17/1/65 to 14/3/65	21·5	$5{\cdot}6 \times 10^{-7}$
		15/5/65 to 15/6/65	19	$2{\cdot}4 \times 10^{-7}$
4	Fletton bricks, unrendered	29/1/63 to 4/3/63 (34 days)	20	$1{\cdot}8 \times 10^{-6}$
5	The same with 1:1:6 wet-dash rendering	29/1/63 to 4/3/63	8	$8{\cdot}6 \times 10^{-7}$

All panels were of brickwork about 11·5 cm thick, the rendering being about 2 cm thick.
Panels nos 1, 2 and 3 were exposed at Garston, panels nos 4 and 5 at Thorntonhall.
All these results refer to periods during which no measurable rain reached the wall panels.

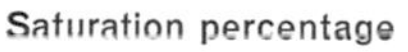

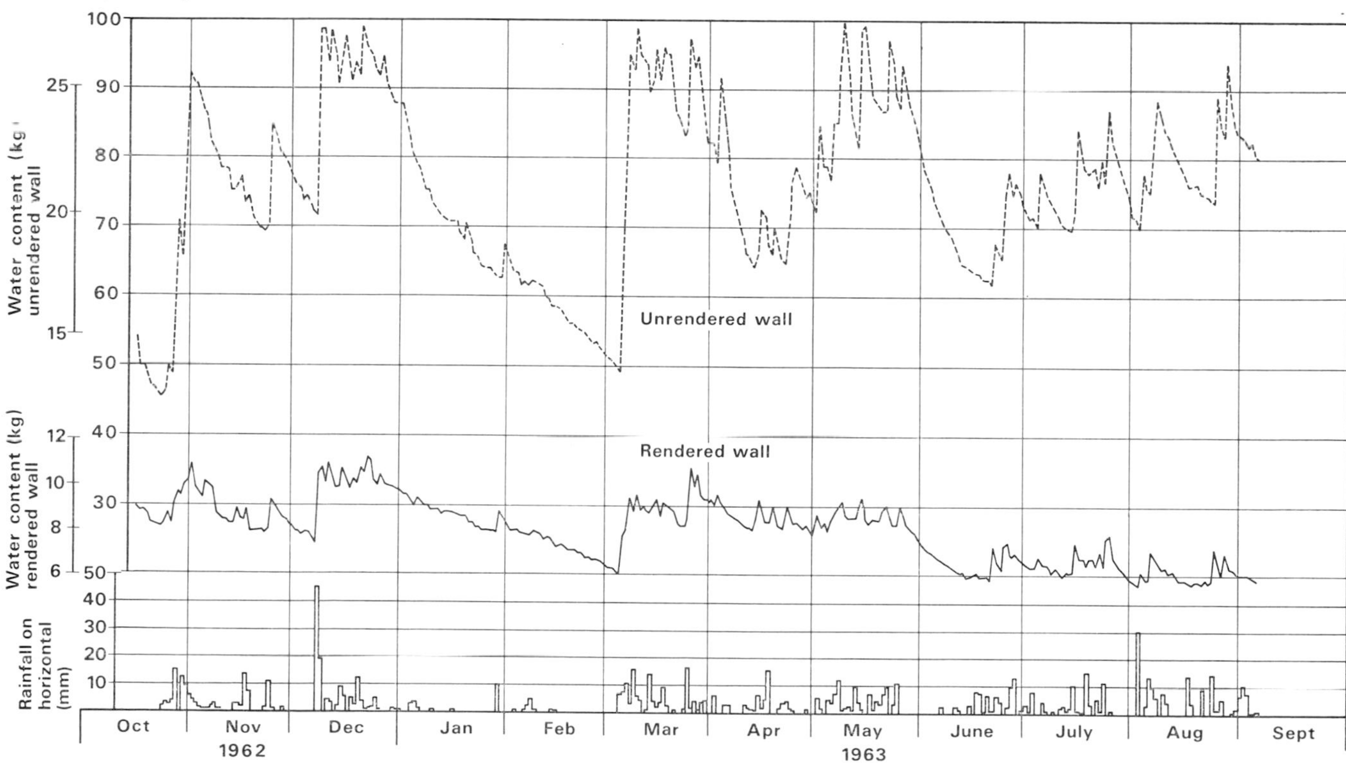

Figure 120 Daily variation in the water content of two walls exposed to the weather at Thorntonhall, facing south-west.
Above: Unrendered Fletton brick wall panel *Below:* 1: 1: 6 cement: lime: sand rendered panel

Figure 120, which is a plot of the daily weighings of two wall panels at Thorntonhall in the period October 1962 to September 1963. During dry spells in winter, the rate of drying was little different from that in spells during summer. In the wet spells, during December 1962, and in March and May 1963, the unrendered brick wall rapidly became saturated, and although after each individual fall of rain there was a rapid loss of water, the intervals were too short for much water to be lost, and each fall of rain brought the wall back to saturation.

In fact, since at even the wettest places the duration of driving rain does not exceed about 15 per cent of the total time, and on any given wall is unlikely to be much more than about half of this, the time during which a wall is not actually being wetted does not vary greatly from one place to another (from perhaps 90 per cent of the time in the wettest places to about 98 per cent of the time in the driest places in Britain).

Evidently the most important factor by far in assessing the likely wetness of walls is the probable amount of driving rain, in conjunction with the construction of the walls concerned. Where it is not possible to provide topographical shelter for walls, that is by building in places sheltered from strong winds and driving rain, it is necessary to provide local shelter. That is to say, the main masonry structure must be sheltered, by a suitable rendering or by a ventilated cladding. In this connection, it must be realized that certain combinations of masonry and rendering may be unsuitable, in that the relative pore structures are such that the passage of moisture from the masonry to the rendering is impeded, and the drying of the wall hindered, but this is not the place to discuss this problem.

Rain penetration through joints

In the very small pores which exist in porous materials (with diameters measured in micrometres) water is absorbed and moved around by capillary forces which are much greater than the pressures exerted by the wind. Thus the pressure of the wind does not influence the rate of flow of moisture through most brick walls. The capillary forces decrease as the diameter of a pore or crack increases, and when a hole is about 0·1 mm or more across, the wind pressure may be more than the capillary forces.

Joints are a feature of all types of building, whether they be nominally sealed ones between fixed members such as panels and frames, or partly sealed ones between movable members such as doors and windows. Water may travel through joints in a number of ways. If the raindrop is small enough relative to the size of the joint it may travel through it without touching the sides. Or rainwater may run along the faces of a joint, under the influence of gravity or of air blowing through the joint. Or water may bridge the gap between the faces of the joint, so that an air-pressure difference builds up across the water and forces it through the gap.

Direct penetration of raindrops through a joint is usually the least important, for only a few will have a trajectory which will avoid impaction on the sides of the joint (small, dry snowflakes on the other hand can easily penetrate open joints, for they will not stick readily to surfaces, and in addition will follow air currents through joints).

The penetration of joints by water running along surfaces, or being blown through joints, has been studied empirically, but is still not understood sufficiently. In particular, the effects of what may be called the micro-climate of the wind close to the building are quite unknown. All that can be said in this field is that observations of buildings during rain (Figures 115 and 116 for example) show great variations in the amount of rain striking a wall from one part to another, and that corners and edges receive more than the middle part of the faces. On an even smaller scale is the effect of small upstands or changes of level. For example, when the wind blows up the slope of a low-pitched roof, it will be expected that on average the air pressure acting on this section of roof will be less than atmospheric, so that there will be a pressure-drop across the cladding from inside to outside. In spite of this, water does find its way through joints between overlapping sheets or tiles, showing that locally there must be a fall of pressure along the joint from outside to inside, for part of the time at least.

On the contrary, a window recessed into an opening in a wall may be much more sheltered from wind than one placed flush with the surface. Experience shows that such windows are appreciably less liable to suffer leakage of driving rain. These two examples show that quite small changes in surface profile can have important effects on rain penetration.

Quite a large volume of water may need to be absorbed by a porous material before significant effects are noticed at the inner face. Joints between non-porous components however may show leakage with only a small amount of water present on the outer face. It was noticed on one occasion, for example, that in light snow showers the small amount of snow caught by the aluminium frame of a curtain wall was melted by the heat of the building, and the melt-water was forced through cracks by wind pressure. Even these small amounts may be significant if they drip on to documents or other valuable contents.

It appears likely therefore that for rain penetration of joints, wind pressure may be the most significant climatic parameter; if the pressure is high enough, serious penetration may occur even with small amounts of rain (or melted snow).

An as yet unpublished Meteorological Office analysis of hourly values of wind speed, rainfall amount and rain duration for a period of 10 years at each of about 20 stations in Britain has made it possible to estimate the proportion of the time during which rain fell at different wind speeds. The rainfall data were grouped according to wind speed and show clearly that as the speed of the wind increases, so does the chance that it will be raining. The rate of increase in the proportion of hours during which rain falls is greater at places with higher rainfall, and in Table 36 the results are grouped accordingly. One column includes values from eastern, drier places, the other those from western, rainier places. Because there are relatively few hours with high wind speed, there is less certainty about the exact proportions of rainy hours with the highest winds. This is part of the reason for the large spread of values in the lowest lines in Table 36. However, there is no doubt about the upward trend, although the amounts of rain in the windier hours tend to be less.

Table 36 Proportion of hours during which rain falls, in different wind-speed classes. British stations, means for period 1957–66

Wind-speed class (hourly averages) m/s	Eastern stations %	Western stations %
0·5–2·1	5–6	7–10
2·6–4·1	7–9	9–13
4·6–6·2	10	13–15
6·7–8·2	12–13	15–21
8·8–10·3	15–16	17–28
10·8–12·4	17–18	20–40
12·9–14·4	20–21	25–44
14·9–16·5		27–56
17·0–18·5		36–62
19·1–20·6		37–75

It is therefore thought to be reasonable to grade areas of the country by wind speed alone when considering the degree of exposure to which jointed structures will be subjected. However, there is still some doubt as to what averaging time is significant for the wind speed. If water is to be forced through a short passage, through material only a millimetre or two thick, it may be that gusts with a duration of only 3 or 5 seconds will last long enough to move an important amount of water. Longer and more intricate passages may require the wind pressure to be exerted for a greater period of time, perhaps 10 to 30 seconds.

In fact, although the longer the averaging time the less the wind pressure is likely to be in a given set of circumstances, one may assume a standard decrement in the pressure (see, for example, Table 24). Thus if a zoning scheme is drawn up, which divides the country into areas where given ranges of wind pressure will be experienced, the zone boundaries will be independent of the wind-averaging time that is chosen (see Figures 121 and 124).

Other methods of getting statistics have already been described on pages 103–6. Both lead to the tentative conclusion that the intensity and frequency of the most intense driving rain is the same all over Britain. This of course implies that a similar design of window or curtain wall could be used anywhere. However, this apparent discrepancy is perhaps resolved if, as suggested above, quite small quantities of water are all that is necessary to produce appreciable rain penetration in a strong wind. The high-intensity storms described on pages 103–6 in fact produce far more driving rain than is needed, while the air pressures associated with them are not as great as during the prolonged warm-frontal storms with strong winds, to which the statistics of Table 36 apply.

It must be emphasized that the design and test pressures quoted on pages 118–25 are based on meteorological data, using pressure coefficients which at the time of writing seem reasonable. Further experience may indicate that the pressure coefficients can be reduced, or that the factors which allow for local effects may be altered, thus altering the pressures quoted in the Tables. It may also be found advisable to include a reduction factor taking into account the way a window is set into the building – the factor perhaps having a value of unity when the window is flush with the surface of the wall, and a smaller value when the window is recessed into a window-opening. Such alterations will not affect the contour lines in the Figures, merely the values allotted to them.

It is also possible that developments in meteorological analysis, or the acquisition of further wind data, may alter the wind-speed contours (isotachs), but it is likely that such changes will be very slight.

Climatic data for tests on windows

The range of wind speed and driving rain over the country having been examined, it is now appropriate to look at the values which should be specified for making realistic tests on windows. Tests on windows which involve wind and rain are three in number:

The first is designed to test the mechanical strength of the window, and for this the test pressure corresponds to that of a design gust speed. A gust in this context is a brief increase in wind speed, lasting about 3 seconds, in which the window or panel will respond fully to the increase in wind pressure.

The second test is designed to discover the amount of water which leaks through the window when an air pressure is applied in the presence of water running down the outer surface, or sprayed on to the window.

The third test is designed to ascertain the rate of air-leakage through the window when an air pressure is applied.

In the strength test it is clear, as has been noted, that the 3-second gust that can be withstood by the window is the appropriate test criterion. It is not so obvious what should be the duration of the wind pressure which is applied in the other two tests. It is generally assumed by testing agencies that a steady pressure applied for about 5 minutes is appropriate, although there seems to be no scientific basis for this. As was suggested in the previous Section, rain might be forced through some kinds of gap in a few seconds, while in others a longer application of the pressure will be necessary. However, in the discussion which follows, 5-minute mean wind speeds will be adopted for the rain and air leakage test data, since the tests currently used by Agrément and other agencies specify mean pressures applied over this period.

Because there is such a wide range of wind speeds over Britain, it would be quite unreasonable to use the same design speed for all places. If windows had to be made sufficiently strong to withstand gusts of about 55 m/s in the Western Isles, they would be far stronger than is necessary to withstand the gusts of 38 m/s or so which might be experienced in the London area. Clearly it would be quite uneconomic to make them suitable for all parts of the country. It is desirable therefore to introduce a system of zoning, dividing the country into areas which experience a limited range of extreme wind speeds, so that a given type of window can be used anywhere in the zone, subject to consideration being given to any local effects caused by topographic shelter or extra exposure. Because of the way in which the statistics of gusts and mean speeds have been derived, it is possible to use a scheme of zoning common to the tests for all three requirements, strength, water-tightness and air-leakage.

The gust data used for the strength tests are, of course, the same as those used for building design (Figure 86), but a different coefficient is used in deriving the forces on a window, to take account of the difference in pressure that may develop across a wall because of wind effects on the pressure within the building. It may be pointed out here that the maps of wind speed used here are identical with that published in the current edition of British Standard Code of Practice CP3: 1972, Chapter 5, *Loading*, Part 2, *Wind loads*.

Design wind speeds for window strength testing

The strength test for windows uses a wind pressure corresponding to a design gust speed. Gusts cause rapid fluctuations in the speed and direction of the wind, which recur with a frequency of from several a second to one every 5 or 10 seconds, or more. There is a corresponding variation in the sizes of gusts. It is considered that for the design of a window, a gust lasting about 3 seconds is the most significant.

It is therefore a happy accident that the instruments used in this country for the measurement of wind speed in normal meteorological practice – the Dines pressure-tube anemograph and the rotating-cup electrical anemograph – have a speed of response sufficient to measure gusts with a duration of about 2 to 3 seconds. Thus statistics of the magnitude and frequency of gust speeds measured by these instruments can be used for the design of window tests.

The returns from the meteorological stations in Britain which have an anemograph include a record of the highest gust measured on each day. There are now about 100 such stations, and Shellard (1965) has used the highest gust measured in each year at each of these stations in an analysis of the recordings. By means of Gumbel's extreme value theory the gust speed which is likely to be exceeded only once in 50 years has been calculated for each place.

Before the analysis, the recorded data were adjusted if the effective height of the anemograph was not the standard 10 metres above ground. It was assumed that the speed of gusts varied with height above ground according to the relation

$$V_{gh}/V_{g10} = (h/10)^{0.085} \qquad (27)$$

where V_{gh} is the gust speed h metres above ground, and V_{g10} is that at a height of 10 metres (Deacon 1955).

The calculated gust speeds have been plotted on a map and isopleths drawn, showing for each part of the country the gust speed which is likely to be exceeded not more than once in 50 years (Figure 86). The values apply to open and level sites, and to a height of 10 metres above ground. It must be emphasized that maps of this kind must be used with the greatest caution, and that they can only be used as a general guide to the wind speeds to be expected. In part this is because the local terrain may give rise to appreciable variations in the wind speed, and moreover that the nature of the surface can have a considerable effect on the local wind. Usually local roughnesses such as vegetation (which may vary in size from grass to large trees, either with or without leaves according to the season) or buildings and other structures, will cause a decrease in wind speed. However, in strong winds such local roughnesses cause turbulence, so that the wind becomes more gusty and in limited areas the wind speed may be greater than if there were no such obstructions or roughnesses present. The isopleths on the maps do not take into account any such local variations.

A further difficulty is that most of the anemograph stations used in preparing the map are at relatively low-level stations, and the map is therefore generally representative only of rather level, open country, at an elevation of not more than about 300 m above sea-level. If we add to this the fact that two stages in the preparation of the map involve a measure of subjective judgement (in converting the data to the standard height of 10 m, and in drawing the final isopleths), it becomes clear that, if figures read from the map are to be used in projects where safety is involved, it will be desirable to take expert advice, from the Meteorological Office or other suitable source.

There are two other points worthy of note. Firstly, some places may have greater exposure to wind than the generality of sites in the neighbourhood, because they are raised above the general level (and this of course applies also to buildings which are higher than the others around them). Secondly, it should be remembered that shelter from trees and existing buildings may not be the same through the whole life of a new building, for trees can be cut down and buildings demolished. It may often be wise to assume that the exposure to gusts of a new building is the same as if it were in open country without local shelter. The pattern of air-movement in a town is complex, and it is likely that in places the gust speeds may be little different from those outside the town, and particularly when the wind is strong.

It might be thought after reading all these provisos that it is hardly worth providing the map. This is not so, however, for although it needs using and interpreting with care, it would be worth having even if it only drew attention to the fact that there *are* real differences between the degree of exposure to strong winds at different places.

However, repeating the advice offered above, designers may prefer to take expert opinion on the appropriate wind pressure to be used in a particular circumstance, rather than attempt to estimate this from the maps in the way shown in this Section. The choice of method may depend on the factor of safety allowable for the construction which is in mind.

Table 37 Design gust speeds and pressure differences (at a height of 10 m above ground) for strength tests on windows

Exposure zone	Design gust speed m/s	Design pressure difference N/m²
A – normal	42	1620 (1600)
B – moderate	48	2120 (2100)
C – severe	52	2490 (2500)
D – extra severe	56	2880 (2900)

Design pressure differences based on a pressure coefficient of 1·5. Figures in brackets are the rounded-off values used on Figure 121.

The correction factors listed in Tables 23 and 24 have not been applied to the data used in this Section, so that all numerical values refer to airfield exposure.

Now, the pressure difference across a window when the wind is blowing depends not only on the wind speed, but also on the direction of the wind relative to the face of the window, on the position of the window in the building, and on the presence of openings in the various faces of the building. Thus to obtain the pressure difference, the calculated dynamic pressure due to the wind must be multiplied by a factor which is itself highly variable. Measurements by Newberry, Eaton and Mayne (1968) on buildings suggest that a suitable value for the factor is 1·5, so that for a gust speed of V_g m/s the pressure difference is

$$p = 0{\cdot}919\ V_g^2\ \mathrm{N/m^2} \qquad (28)$$

This value allows for the increased pressure and suction experienced near a corner, the basic force relation being $q = 0{\cdot}613\ V_g^2$.

The variation of these pressures across the country, based on the once in 50 years design gust speeds, is shown in Figure 121, this being in fact Figure 86 redrawn with isopleths of pressure difference instead of gust speed. The design gust speeds chosen to divide the country into four zones, and the corresponding pressure differences, are shown in Table 37. These apply to a height of 10 m above ground in an open situation. The zones delineated by these values are shaded in Figure 121.

The pressures across the windows in a building will be expected to increase with height above ground. If these increase as gust pressures in the free atmosphere do, the design pressures given in Table 37 will increase with height as shown in Figure 122. The data shown there imply, for example, that windows which just satisfy the strength

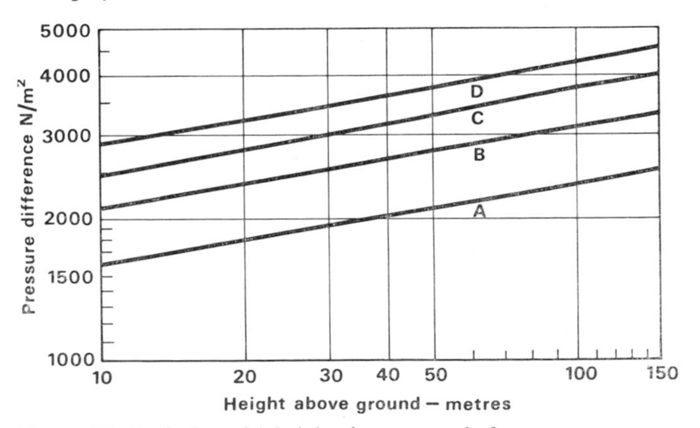

Figure 122 Variation with height above ground of pressure across a window due to gusts of wind.

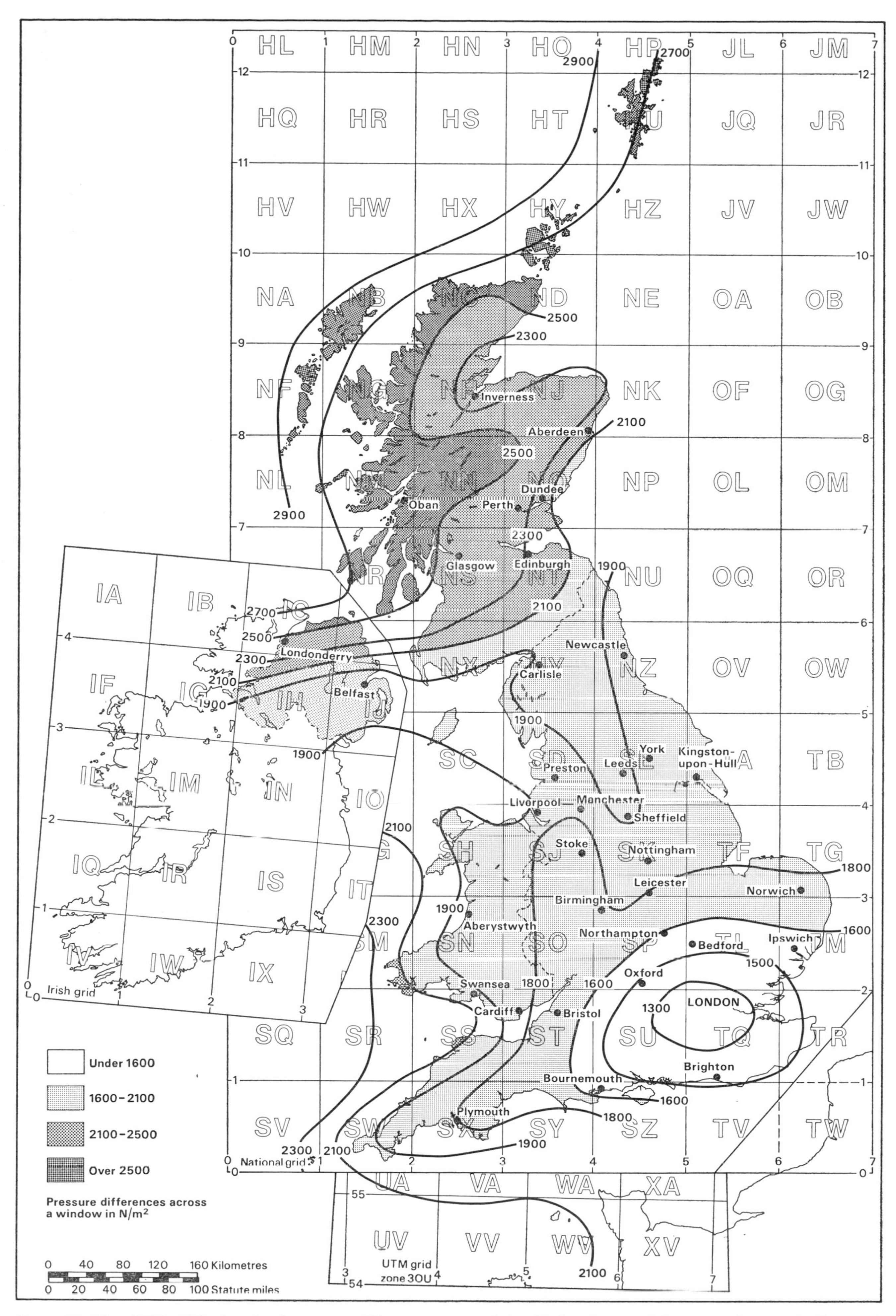

Figure 121 Map of United Kingdom showing pressure differences across a window, likely to be exceeded not more than once in 50 years at a height of 10 m above the ground in open level country. Based on the 3-second design wind speeds in Figure 86 and a pressure coefficient of 1·5. Suggested exposure grading zones indicated by shading

requirements for exposure zone A will not be suitable for buildings appreciably higher than 10 m, and it will be necessary to use windows which satisfy the requirements for zone B.

Design wind speeds for rain-penetration tests on windows

It was stated in the introductory remarks to this Section that 5-minute mean wind speeds would be used as the basis of the statistics of wind speeds for rain-penetration testing, since the tests by Agrément and other agencies commonly specified that the test pressure would be held for that period of time. Furthermore, on page 116 it was suggested that penetration of rainwater through joints was determined by wind pressure rather than by the amount of rainwater available, and it was shown that the stronger the wind, the more likely it was to be raining. It was therefore only necessary to consider statistics of wind speed.

The test pressures commonly used are shown in Table 38, together with the corresponding wind speeds based on the relation

$$p = 0{\cdot}674\ V^2 \quad \text{......................(29)}$$

the pressure coefficient being assumed to be 1·1 (and not 1·5 as for gust speeds). There are no published statistics of 5-minute mean wind speeds, but according to Shellard (1965) the extreme 5-minute mean wind speed is some 10 per cent greater than the corresponding hourly speed, the exact value depending on the roughness of the surroundings. Thus it should be possible to calculate with adequate accuracy for our purposes 5-minute mean pressure differences from statistics of hourly data.

Table 38 Rain-penetration test pressures used in Agrément tests

Test pressure difference across window		Corresponding wind speed	
mm H_2O	N/m²	5-min mean	Hourly mean
4	39	7·6	6·9
16	157	15·3	13·9
30	294	20·9	19·0
50	491	27·0	24·5
70	687	31·9	29·0

The 5-minute mean speeds correspond to a pressure coefficient of 1·1. It is assumed that the 5-minute mean speed is 10 per cent higher than the corresponding hourly mean speed.

The hourly statistics used were those derived by Shellard (1965) and more recently revised by N C Helliwell (unpublished). These can be presented in the form of a map of extreme once in 50 years hourly mean wind speeds (Figure 123) or in tabular form (Table 39). The map, which was produced in a manner similar to Figures 86 and 121, shows wind speeds to be expected at well-exposed places, such as airfields, of such a nature that the extreme gust speed (on a once in 50 years basis) is 50 per cent greater than the corresponding hourly mean speed. At places where the surroundings are rough, whether due to vegetation or to buildings, the hourly wind speeds may be much reduced.

However, the pressure experienced once in 50 years may well be thought too severe for tests of rain penetration, and the pressures to be expected on average not more than once in 10 years and once in 2 years have also been estimated. This calculation has involved assumptions which may not be fully justified. Thus, it is assumed that:

a. the reduction factor S_3 in Figure 88 can be applied to 5-minute mean wind speeds,
b. the hourly mean wind speed is two-thirds of the peak gust speed when the wind is strong,
c. the highest 5-minute mean wind speed is 10 per cent greater than the corresponding hourly mean speed when the wind is strong,
d. the reduction factors *b* and *c* can be applied to the calculated once in 10 years and once in 2 years design gust speeds,
e. in these conditions the appropriate value for the pressure coefficient is 1·1.

Table 39 Cumulative percentage frequencies of time with hourly mean wind speeds below the stated limits. Averages for 10 years 1957–66, with corresponding estimated 5-minute mean pressure differences ($C_p = 1{\cdot}1$)

Hourly wind speed m/s	5-min mean pressure N/m²	A London Heathrow	A–B Boscombe Down	B Manchester Ringway	B–C Cardiff Rhoose	B–C Belfast Aldergrove	C Edinburgh Turnhouse	C Glasgow Renfrew	D Tiree
2·1	4	20·75	23·94	20·26	11·78	18·51	23·30	26·92	11·72
4·1	14	52·30	52·11	48·07	40·87	42·31	47·59	52·83	24·44
6·2	31	77·52	75·57	74·27	67·12	67·99	71·01	76·73	42·81
8.2	54	91·66	90·07	89·50	83·82	86·15	85·59	91·01	63·34
10·3	85	97·64	96·34	96·37	92·73	94·74	93·65	96·66	77·18
12·4	124	99·40	98·86	98·82	97·00	98·03	97·49	98·74	89·15
14·4	168	99·86	99·70	99·72	98·97	99·39	99·05	99·62	94·59
16·5	220	99·98	99·96	99·94	99·66	99·79	99·67	99·90	97·88
18·5	277	100·00	100·00	99·99	99·91	99·93	99·90	99·97	99·28
20·6	343	—	—	100·00	99·99	99·97	99·97	99·99	99·77
25·7	534	—	—	—	100·00	100·00	100·00	100·00	99·99
30·8	767	—	—	—	—	—	—	—	100·00
5-min pressure exceeded in 1% of hours	—	110	130	130	170	150	170	130	260

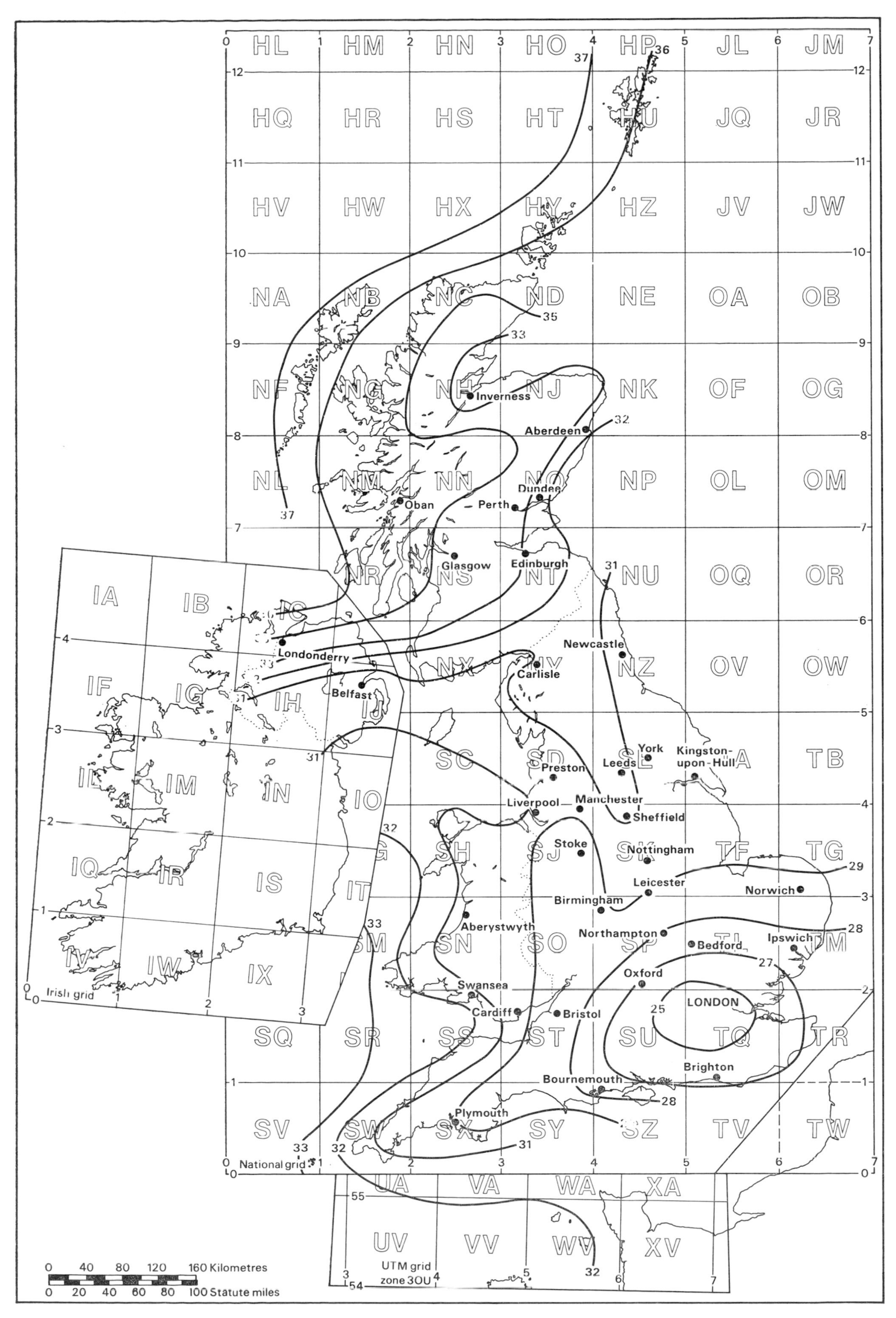

Figure 123 Map of United Kingdom showing hourly mean wind speed, in metres per second, likely to be exceeded on average only once in 50 years at a height of 10 metres in open level country. The values are exactly two-thirds of those of the basic 3-second gust speeds shown in Figure 86, rounded off to the nearest m/s

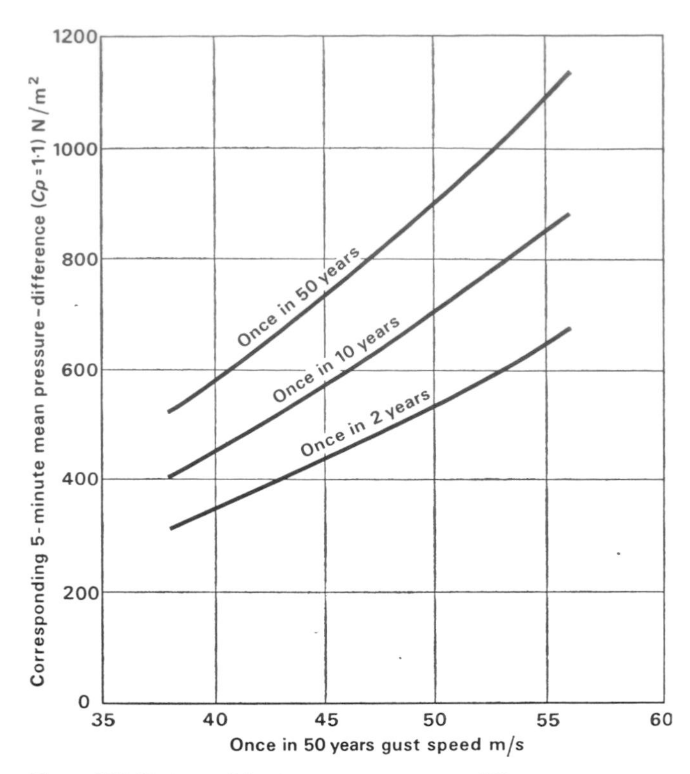

Figure 124 Estimated 5-minute mean pressure differences across a window, with pressure coefficient 1·1, in relation to the once-in-50-year design gust speed

Figure 124 shows the relation between the estimated 5-minute mean pressure differences across a window on the basis of these assumptions, and the corresponding design gust speeds which are shown in Figure 86. On the basis of this, the map in Figure 125 has been drawn, in which the isotachs of Figure 86 have been labelled with the pressure differences corresponding to the once in 10 years curve of Figure 124.

Typically in rough surroundings such as the centres of towns the extreme hourly mean speed at a height of 10 m may be only half the extreme gust speed, although this extreme gust speed may be little different from that at a more open exposure. However, it may not always be wise to use lower grades of windows in nominally sheltered places, because degree of shelter may be changed, moreover it would not be practicable to take account of all variations in local exposure when selecting windows or other components. This advice may be thought over-cautious, but should perhaps be borne in mind in some circumstances.

As in the case of the strength test, it was desired to divide the country into zones with different degrees of exposure. For the purpose the limits chosen for the exposure zones were:

	Extreme hourly mean wind speed m/s	Pressure N/m²
A – normal exposure	28	500
B – moderate exposure	32	650
C – severe exposure	35	760
D – extra severe exposure	37	880

The pressure differences are based on a pressure coefficient of 1·1. The limits were chosen to agree with those on the gust map, Figure 121, because as the gust factor is taken to be 1·5 at the stations providing the data for the maps, the 'once in 50 years' maximum gust is everywhere 1·5 times the 'once in 50 years' maximum hourly mean wind speed. Thus, for example, the 30 m/s isotach on the hourly map coincides with the 45 m/s one on the gust map. The hourly map is shown in Figure 123.

All the values are based on speeds at 10 m above ground. Figure 126 shows how the 5-minute mean wind pressure varies with height above ground in an open situation, on the assumption that its variation with height is the same as that for hourly mean speeds.

Table 40 gives more detailed information on how the pressure difference (*not* the wind speed) varies, both with height above ground and according to the local aerodynamic roughness. The information is in the form of correction factors, corresponding to the S_2 factors of Table 24 but in this case for correcting pressure differences directly, and based on data for hourly mean wind speeds. (There are no corresponding data for 5-minute mean wind speeds, but it is considered that the hourly data can be used for this purpose without sensible error.) Thus by multiplying pressure differences read from Figure 125 by the appropriate value from Table 40, a suitable

Table 40 S_2' **factors for correcting hourly mean** ***pressure differences.*** **These can be used for correcting the 5-minute pressures shown in Figure 125 to allow for local shelter and height of building.**

Height above ground m	Category 1	2	3	4
3	0·55	0·38	—	—
5	0·72	0·58	0·29	—
10	1·00	0·90	0·59	0·24
15	1·14	1·10	0·81	0·44
20	1·28	1·23	1·00	0·59
30	1·46	1·42	1·28	0·88
40	1·61	1·56	1·46	1·14
50	1·72	1·72	1·61	1·37
60	1·85	1·85	1·72	1·54
80	2·02	2·02	1·93	1·80
100	2·19	2·19	2·10	1·99
200	2·76	2·76	2·72	2·66

The definitions of the four categories are the same as those for Table 24 (page 88).

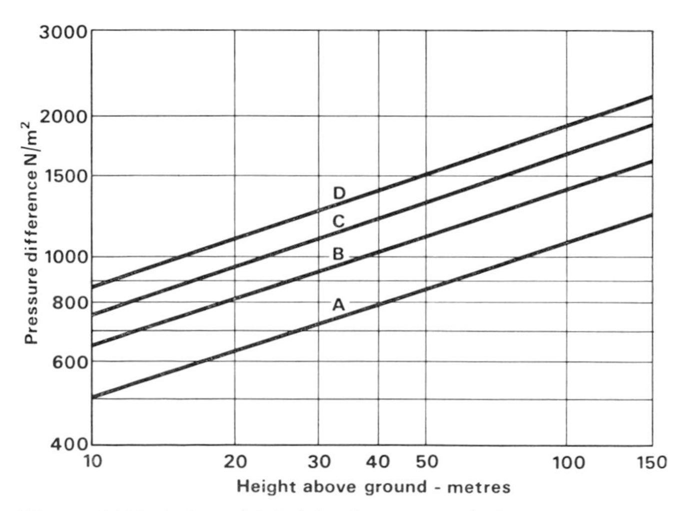

Figure 126 Variation with height above ground of pressure across a window corresponding to a 5-minute mean wind speed

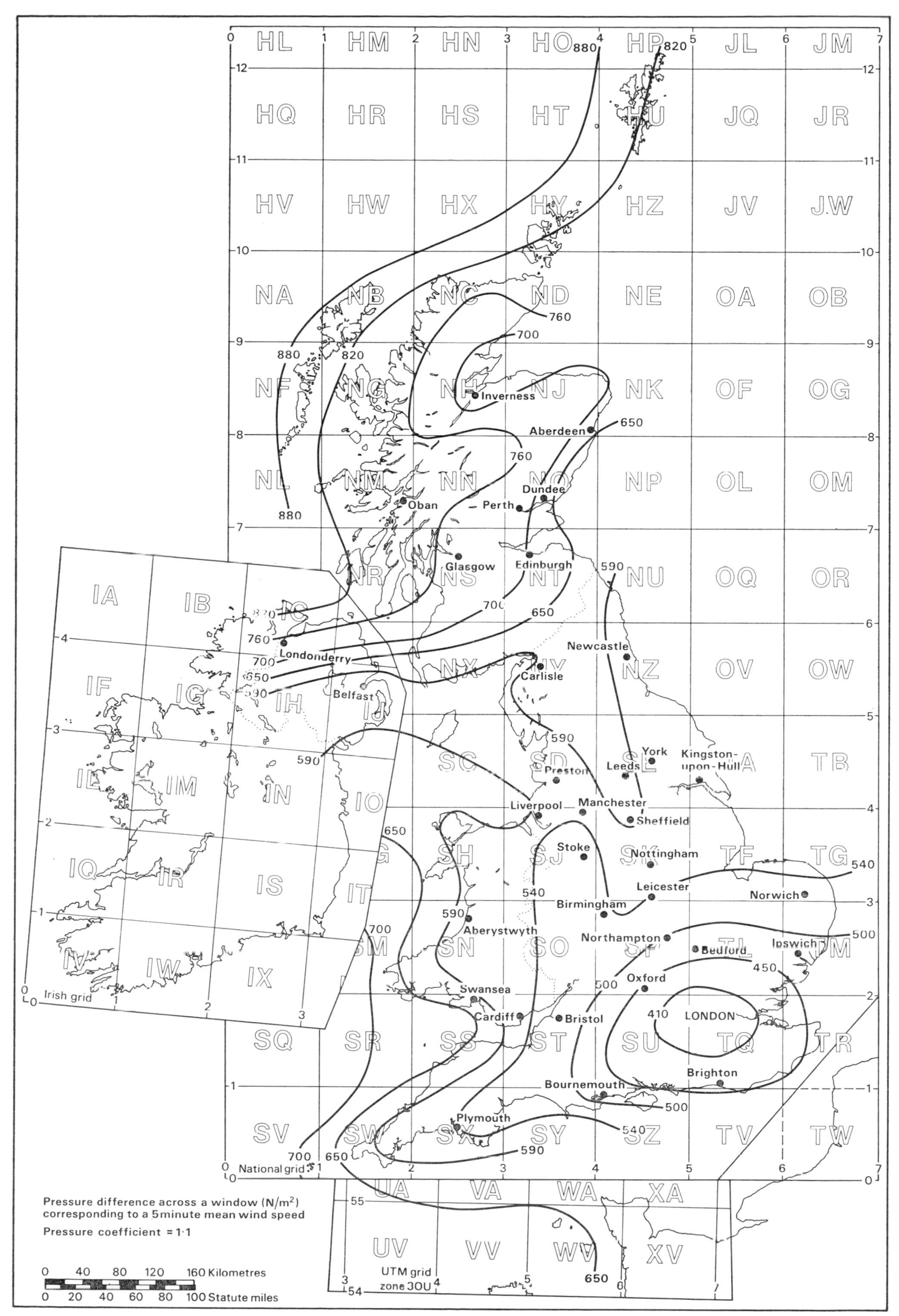

Figure 125 Pressure difference (in N/m²) which is developed across a window by that 5-minute mean wind speed which is likely to be exceeded on average only once in 10 years at a height of 10 m above the ground in open level country. Pressure coefficient = 1·1

Table 41 Pressure differences across windows in four exposure zones and four local exposure categories (using S_2' factors from Table 40). All are at a height of 10 m above ground

Exposure zones	Category 1	2	3	4
	N/m²			
Normal	500	450	290	120
Moderate	650	590	380	160
Severe	760	680	450	180
Extra severe	880	790	520	—

local value is found. Table 41 shows some examples, using as basic – Category 1 – values the limits chosen for the four exposure zones. No value is given for Category 4 in the extra severe zone, as there is no very large city in this exposure zone. Since in this study we are concerned with the pressure differences which develop during strong winds, the use of correction factors based on the vertical gradient of wind speed when the wind is strong would seem to be justified. These correction factors do not apply to situations when the winds are weak.

There is perhaps no completely satisfactory simple way of characterising areas of the country for exposure to wind for the present purpose. The most convenient way is to use statistics of extreme hourly wind speed as displayed in Figure 123. However, as can be seen from Table 39, the cumulative frequencies of speeds below the stated limits at the individual stations do not always show the same characteristic differences as those indicated by the extreme wind speeds. Thus, Renfrew, Glasgow, in zone C, has a lower annual mean wind speed than Ringway, Manchester, which is in zone B, indeed, its mean wind speed is little different from that at Heathrow, London. This is because Renfrew has an exceptionally high proportion of low wind speeds, although its extreme speeds are higher than anywhere in zones A or B. Even at the 99 per cent frequency level there is sometimes surprisingly little difference between places in zones A, B and C, except at especially exposed places such as Valley, Holyhead, which is just in zone C on the extreme speed criterion.

Table 42 Frequencies (per cent) of hours with 5-minute mean pressure differences exceeding the design values ($C_p = 1{\cdot}1$). Based on data for 10 years, 1957–66
Note: 0·01 per cent is equivalent to about one hour a year.

Place	London Heathrow	Boscombe Down	Cardiff Rhoose	Glasgow Renfrew	Tiree
Zone	A	A–B	B–C	C	D
Design pressure N/m²	Frequencies % of hours				
39	14	17	26	14	50
157	0·25	0·4	1·5	0·5	6·8
294	0	0	0·04	0·02	0·6
491	0	0	<0·01	<0·01	0·02
687	0	0	0	0	0·005

The proposed test pressure differences, based on the 'once in 10 years' 5-minute mean wind speeds, are appreciably greater than the currently used values.

It is of interest to see how often these currently used values are exceeded at various places (Table 42). It will be seen that 157 N/m² is exceeded (for an average time of 5 minutes in the hour) during about 22 hours a year at Heathrow, which is perhaps an unusually exposed place for the London area, and about 40 hours a year at Boscombe Down, on the outer boundary of the normally exposed zone A. At the most exposed place, Tiree, this pressure difference is exceeded during 7 per cent of the hours.

The 5-minute mean pressures are calculated from the relation $p = 0{\cdot}674V^2$, that is 1·1 times the dynamic pressure of the wind. While it is uncertain how much 5-minute mean pressures do in fact vary over the face of a building, it is observed that rain penetration *is* more severe at edges and corners, so that in the absence of more exact data it is considered reasonable to propose the use of the figures given above. However, as has been suggested above, it may be found desirable to reduce the pressures in the light of experience, by the use of reduction factors to take account of local shelter, the height of the building, and 'micro' factors such as the way the window is set into the wall.

Design air pressures for air-leakage tests

Present practice is to use a constant air-pressure difference across a window when testing it for rate of leakage of air, irrespective of the exposure conditions in which it is to be used. The value commonly used is 10 mm H_2O, or about 100 N/m². As can be seen from Table 39, this value is exceeded (for 5-minute periods) in some 2 to 3 per cent of the hours in zone A, but in upwards of 20 per cent of hours in the most severely exposed zone. It would be more logical therefore to specify that pressure which is exceeded for a certain proportion of the time.

The lowest line of Table 39 lists the 5-minute pressure differences which, it is estimated, are exceeded during 1 per cent of the hours at the places listed. On the basis of these and other data it is suggested that the test pressures for the four zones should be:

A	130 N/m²	C	210 N/m²
B	160 N/m²	D	260 N/m²

Table 43 Summary of suggested test pressure differences (N/m²) for use in testing windows and other lightweight wall panels. Based on wind speeds at a height of 10 m above ground

Zone (see Figure 121)	Strength test ($C_p = 1{\cdot}5$)	Water-leakage test ($C_p = 1{\cdot}1$)	Air-leakage test
A	1470	500	130
B	1945	650	160
C	2490	760	210
D	2880	880	260

Summary of suggested test pressures
Table 43 summarises the suggested pressure differences to be used for the three tests. All are based on wind speeds measured at a height of 10 metres above ground in an open situation. If it is required to make special allowance for height above ground, the test pressures for the strength and rain-leakage tests may be increased as indicated in Figures 122 and 126 respectively. The air-leakage test pressures may also be increased in the same ratio as those for the rain-leakage test. The data in the first column are based on 'once in 50 years' and in the second on 'once in 10 years' statistics and may therefore be considered too stringent. The frequency data given in Tables 39 and 42 will assist in choosing the most appropriate values for a particular purpose. Where safety is concerned, it may be thought desirable to use more stringent values than if a failure merely causes inconvenience or discomfort. For example, while the factor of 1·5 introduced in calculating pressure-drop may be thought desirable for the strength tests, it has been reduced to 1·1 for the other tests. Reductions may be applied to take account of local shelter.

Snow loading on roofs

The British Standard Code of Practice CP3, Chapter V, Part 1, gives the design snow load for flat or gently-sloping roofs as 718 N/m²*. No mention is made of the density of the snow, except indirectly when it gives the maximum depth as 61 cm. It is assumed that this load can be experienced in any part of Britain.

At present the statistics of the amount of snow lying at any one time at places in Britain are inadequate to improve on the assumption made in the Code of Practice. Snow cover can vary widely from one place to another, and great care must be taken in interpreting records. At heights of less than about 300 m, a total depth of 0·6 m of snow is very rare – Reynolds (1954) has calculated that on Merseyside such a depth might be expected once in 1000 years. At greater altitudes, the snow depth may be appreciably greater, and as much as 1·5 m has been reported (Bonacina 1957). These are, of course, the depth of undrifted snow. Drifts may be many times deeper.

Up to now, there has been little information on the density of the snow on the ground in Britain, a series of readings taken at Garston in the 1962–63 winter being the only continuous set showing how the density changed with time (Lacy 1964). However, since 1964 a network of stations taking observations of snow depth and density has been established in Britain.

*Given in the Code as 15 lb/ft².

The same observers will also be recording the depth of snow on selected roofs, and it is hoped that in time much useful information will be obtained. But it must be remembered that the snow on a roof may be in a different condition from that lying on the ground. A building is usually heated, so that the snow may be partly melted and compacted. Thus the depth and density of the snow on the ground may not be a good guide to that of the snow on the roof; moreover the snow on the roof is more subject to drifting, when fresh, than that on the ground.

Observations of the depth and weight of snow on roofs have been made in Canada and the USSR, but the results from these cold countries cannot be safely extrapolated to British conditions. However, an important point that has come out of recent studies in Sweden is that snow loads on pitched roofs may be appreciably higher than on flat roofs, rather than the opposite as has been generally supposed. This is because snow may be blown completely off a flat roof by wind, whereas on a pitched roof snow blown off the windward slope is likely to be re-deposited on the leeward slope, thus causing grossly eccentric loading. It is believed that some failures of roofs in upland Britain have occurred from this cause.

Earth temperatures

The temperature of the surface of the soil varies markedly between daytime and night-time. When the sun is shining on dry soil, the temperature in the uppermost millimetre or so may be 10 deg or more above that of the air. Conversely on a clear night the temperature of the surface of dry soil may be about 5 deg below that of the air above. Thus at night in clear weather it is possible for there to be a frost at ground-level, while at the standard measurement height of 1·2 m the air temperature is well above 0°C. The temperature of the surface of wet soil will not change nearly as much, while in cloudy weather the temperatures of soil surface and air will be nearly the same.

The diurnal variation of temperature in the ground decreases rapidly with depth. At a depth of 0·3 m the diurnal range of temperature in Britain rarely exceeds about 2·5 deg, and is less than 0·01 deg at a depth of 1·2 m. Seasonal changes in temperature can be detected at greater depths, to more than 10 m, while below about 30 m the temperature at any depth below open country is practically constant, the only change resulting from long-term changes in the climate. Below 30 m there is a gradual rise of temperature, because heat is being transported from great depths. The increase in temperature with depth in Britain averages about 1 deg C for each 30 m, below a depth of 30 m.

Table 44 Earth temperature variation (°C) with depth at Harestock at 09 00 GMT during the period 1896–1903.

Month	Depth								Mean air temperature 1880–1903 °C
	0·31 m (1 ft)	0·61 m (2 ft)	1·22 m (4 ft)	1·83 m (6 ft)	3·05 m (10 ft)	6·10 m (20 ft)	9·14 m (30 ft)	21·34 m (70 ft)	
January	4·61	5·56	7·28	8·50	10·16	10·69	10·25	9·95	3·0
February	4·22	5·00	6·50	7·67	9·44	10·59	10·30	9·96	3·8
March	5·50	5·83	6·56	7·28	8·82	10·41	10·32	9·96	5·3
April	8·11	7·61	7·33	7·56	8·56	10·19	10·34	9·96	7·9
May	11·39	10·44	9·06	8·50	8·68	9·96	10·30	9·96	10·9
June	15·17	13·72	11·39	10·17	9·21	9·83	10·25	9·96	14·3
July	17·44	16·11	13·50	12·17	10·03	9·79	10·19	9·96	16·0
August	16·83	16·11	14·33	13·11	11·03	9·87	10·14	9·97	15·7
September	15·06	14·67	14·11	13·33	11·72	10·08	10·12	9·97	13·6
October	11·11	11·94	12·72	12·72	11·96	10·33	10·14	9·98	9·2
November	7·78	9·06	10·72	11·44	11·74	10·55	10·19	9·97	6·4
December	5·28	6·56	8·67	9·83	11·07	10·74	10·26	9·97	3·9
Means	10·21	10·22	10·18	10·19	10·20	10·25	10·23	9·96	9·2

Table 44 (from Meteorological Office 1968b) shows the mean monthly temperatures, averaged over the 8 years 1896–1903, measured at Harestock, Hampshire, in a chalk soil in open country. The observations were made in a wooden tube, with precautions to avoid convection in the tube. Even at a depth of 9 m the annual variation in temperature can still be seen, although it is only about 0·2 deg. It will be seen that the annual wave of temperature lags more and more behind that at the surface as one moves downward, and at a depth of 9 m at this place the maximum temperature occurs in April (this being the peak from the previous summer) and the minimum in September. At Harestock during 1896–1903 the annual mean temperature was not quite the same at all depths. At 21 m it was about 0·25 deg lower than at a depth of 9 m. Moreover there was a perceptible upward trend of temperature through the year at 21 m. This variation probably resulted from the distinctly higher air temperatures experienced in the years after 1895, some 1 deg higher than in the 10 previous years.

Although the temperature in the ground at depths of about a metre is usually nearly constant, in porous material like sand or chalk it is possible for cold water from melting snow to percolate rapidly to an appreciable depth and cause a sudden fall of temperature. This is probably an unusual occurrence, and of little significance to building practice.

As can be seen from Table 44 the mean temperature at any depth down to some 20 m is somewhat above the annual mean air temperature at the same place. Observations in other types of soil confirm this, the difference averaging about 0·5 deg C; for example Building Research Station measurements at a depth of 25 m below Ashford Common, well away from any built-up area, show a temperature of 10·5°C in a clay soil. However, below buildings the soil temperature may be appreciably higher than the annual mean air temperature. For example, at a depth of 21 m in London Clay at Cloudesley Square, Islington, the temperature was 13·0° in 1962.

Another series of measurements, also in London Clay, was taken during the boring of a tunnel during October and November 1957, at a depth of 15 m. The measurements were taken daily at a time when the tunnel was being bored at a rate of about 4 m a day, a hole about 6 mm in diameter being bored at right angles to the tunnel face and a thermometer left in it until a steady reading was obtained. The reading was taken as near to the fresh face of the tunnel as possible so that the temperature of the clay will not have been affected by the air in the tunnel.

The readings obtained on successive days were (°C): 16·5, 15·0, 17·0, 16·5, 13·5, 16·5, 17·0, 17·0, 16·0, 16·0, 14·0, 16·0, 16·0, 16·0, 15·5.

There is a difference of 3·5 deg between the highest and lowest values, and these represent the difference between measuring points under buildings and those under streets or other open spaces. It appears therefore that under built-up areas of central London the temperature, at a depth of 15 m below the ground surface, is about 17°, some 6 or 7 deg above what would be expected if the town were not there. The differences of temperature between points immediately below heated ground floors and those just below the surface of streets will probably be greater still, but no figures are available.

Although the increase in temperature of the soil under London (and no doubt under other towns) is probably mainly due to the winter leakage of artificial heat downwards from buildings, as well as to heat from sewage and other effluents, and from buried electrical cables and underground railways, it is likely that a proportion is solar heat. A dark asphalt-covered surface absorbs more sensible heat than soil covered with vegetation, because of its higher absorptivity to solar radiation (85 to 90 per cent as against 75 per cent), and only a small proportion of the heat is used in evaporating moisture, whereas almost all the solar heat falling on vegetation is used in transpiring moisture. A proportion of the extra heat absorbed by the asphalt-covered surface will be transmitted downwards and will gradually warm up the ground below, but no figures are available to show the magnitude of this effect.

Depth of frost in the ground

In a prolonged cold spell the ground may freeze. In Britain, except in hilly country the depth to which the frost penetrates is rarely more than 10 or 20 mm, but in an exceptionally cold and prolonged spell, such as those which occurred in 1947 and in 1962–63, the ground may freeze to a depth of over 0·3 m, and even to about 0·6 m. In the latter winter the lowest temperature recorded under grass at a depth of 0·3 m was −3·2°C at Oxford on 24 January. Such a depth of frost will be reached only in well-drained sands or gravels not protected by a layer of deep snow. Pipes conveying very cold water from open reservoirs and buried within 0·6 or 0·9 m of the surface will increase the depth to which the soil is frozen. A rough guide to the depth of frozen ground may be obtained from the thickness of ice on small lakes and ponds.

The depth to which the ground freezes is determined by the environmental conditions and by the thermal properties of the soil, that is:
a. by the average air temperature during the spell of freezing weather
b. by the duration of sub-zero temperatures
c. by the thermal diffusivity of the soil and by the latent heat of any moisture held in the soil
d. by the insulating effects of any vegetation or snow cover.

A given site will freeze much less readily if it is water-logged than if the soil is well-drained. A cover of grass, or a layer of snow, will afford good insulation and much reduce the rate at which the ground can freeze.

There are differences in the way various soils behave when they freeze. Very fine sand, silts or chalk, when saturated with water, expand as they freeze to an extent which may be many times the volume expansion of water into ice. This is because at certain rates of penetration of the freezing-front into the ground, liquid water is drawn by capillary forces into the freezing zone from the unfrozen region. There is therefore an increase in the total amount of water in the frozen region and masses of ice known as ice-lenses may form. This process is responsible for most of the uplift and cracking of thin pavements and slabs, and is known as 'frost heave'. Heavy clays and coarse sands and gravels are much less likely to experience frost heave.

It is believed that serious damage from frost heave in this country has only occurred to roads, pavements and incomplete or unheated buildings, being especially widespread during the exceptional and prolonged cold winters of 1946–47 and 1962–63. Most of the trouble occurred in relatively open areas like suburban building estates, but it is not impossible that roads and paving in open shopping precincts and similar situations in towns could be affected. (Frost heave and other effects can also occur around cold-stores and storage tanks containing cold liquids such as methane, but this is outside the scope of the present book.)

The difficulty of excavating frozen ground also varies with the type of soil and with the degree of saturation at the time of freezing. Saturated silty soils are significantly more difficult to dig when frozen. Coarse sands and gravels are little altered and heavy clays affected to only a small extent.

A useful index of the depth to which a given sample of ground may be expected to freeze is given by the 'freezing index', which is the total number of degree-days below 0°C which have been experienced since the beginning of the spell. During the 1962–63 cold spell the total number of Celsius frost degree-days at Garston, Hertfordshire, amounted to 133. A relationship based on work in North America suggests that the depth of frost in three different soils at the end of this period would be as shown below. The figures apply to soils in open country:

Well-drained sandy gravel ....0·56 m
Saturated silty sand..............0·33 m
Saturated clay ....0·175 m

All other types of soil in open country, and with a low water-table, would be frozen to depths within the range indicated, except for organic materials such as peat. The depth of 0·56 m agrees well with the maximum depth of frost recorded under roads kept clear of snow. This depth is almost certainly the greatest to which the ground in lowland Britain has been frozen since 1890–91.

Warm nights in summer

Statistics of temperature and other parameters on hot days will be required in the design of air-conditioning plant for offices and factories which are normally only in use during daytime, but warm nights can also interfere with comfort and make sleep difficult. Although it is usually possible to ventilate freely and induce reasonable conditions in houses, this may not be possible in a noisy neighbourhood, close to an airfield or a busy road (an important point is that it must be possible to leave windows open at night without risk to security). Table 45 gives an indication of the frequency of warm summer nights in the London area. No information on the actual number of warm nights for a prolonged period was available, and the figures in the left-hand part of the Table (extracted from *London weather*, by Brazell 1968) show the number of months in a period of 80 years at Kew Observatory in each of which the minimum temperature shown occurred at least once. Thus the true number of warm nights may be somewhat higher.

Table 45 The frequency of warm nights in the London suburban area

Lowest temperature during the night	Kew observatory					London Weather Centre
	June	July	August	September	Total	Total
°C						
17·2	7	25	21	8	61	73
17·8	3	11	14	2	30	73
18·3	1	9	7	0	17	47
18·9	0	6	3	0	9	7
19·5	0	5	0	0	5	13
20·0	0	3	0	0	3	0
20·5	0	0	0	0	0	7
21·1	0	0	0	0	0	7
Highest minimum on record, °C	18·4	20·1	19·4	18·2		21·1
Total frequency					125	227

Kew Observatory. The figures are the number of months during a period of 80 years (1883–1962), in each of which the minimum temperature shown occurred at least once. From Brazell 1968.

London Weather Centre values are actual number of days in 12 years 1956–67, multiplied by (80/12) to give direct comparison with 80 years at Kew. Data from LWC manuscript records.

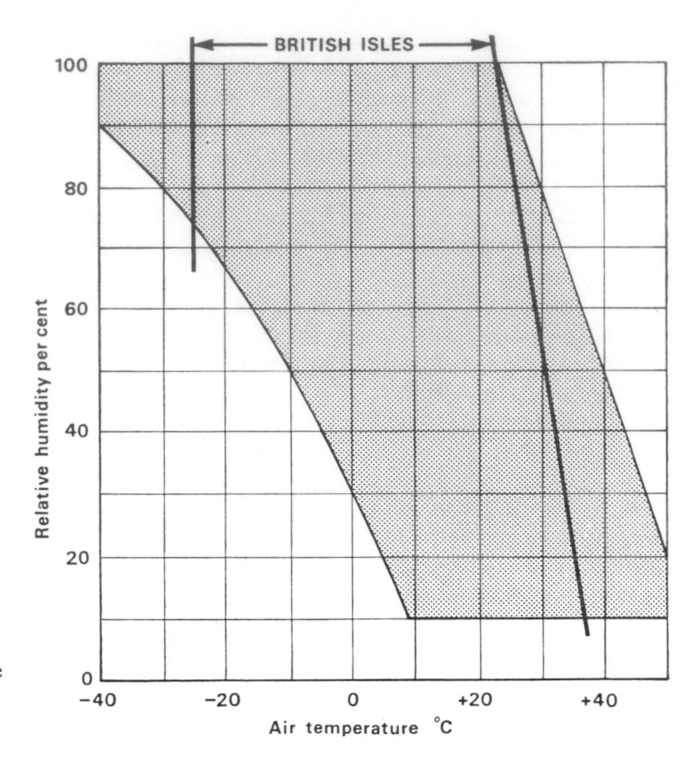

Figure 127 Range of temperature and relative humidity that can be experienced in the British Isles (bounded by lines through −25° and +38°C) and in continental Europe

The figures are representative of intermediate areas of suburban London, and it may be expected that warm nights will be more frequent in the closely built-up inner area. This is borne out by the figures in the right-hand column. These are based on temperatures on individual warm nights at London Weather Centre, Kingsway, in the 12 years 1956–67. For ready comparison the original frequencies were multiplied by (80/12) on the assumption that the 12 years examined were representative of the 80 years 1883–1962 at Kew. It appears that there were about twice as many warm nights (minimum air temperature 17·2°C or more) in central London as at Kew Observatory. Note that London Weather Centre temperatures were measured in a screen mounted over a flat roof some 30 m above street-level. It is believed that they were not dissimilar from temperatures which would have been obtained about 1 m above the street.

The Kew temperatures were also obtained in a non-standard screen, mounted 3 m above ground on the north wall of the Observatory. It is thought that the temperatures are broadly representative of conditions in the outer suburbs of London at a height of 1 m above the ground.

Not only were there more very warm nights in central London than at Kew, but on some of these occasions at least the minimum temperature was 4 deg C or more higher in London. However, as shown by Chandler (1965), there is usually a much bigger contrast between night-time temperatures in the open country (and extreme outer suburbs) and inner London, than there is between intermediate suburbs, such as Kew, and inner London.

The temperature excess of the town over open country at night depends on the weather conditions, being greatest on clear calm nights, and least in cloudy and windy weather. The temperature distribution over and around London on a number of occasions was described by Chandler (1965), although none of them were in exceptionally warm weather. Examination of temperatures on five exceptionally warm nights, when the minimum at London Weather Centre was in the range 23 to 22°C, showed that at Kew the night minima were from 2 to 5 deg C lower on these nights, and at Garston, a semi-rural station, were from 3 to 7·5 deg C lower than at Kingsway. Clearly more detailed information is required before a complete picture can be obtained.

Air temperatures and associated humidities

Figure 127 provides information on the range of air temperatures and associated relative humidities that can be experienced in Europe. The data used were extracted from published information (Meteorological Office 1958 a, b, and Tunnell 1958).

The temperature range shown in Figure 127 was obtained from the extreme values given for various places in Europe, except that some very low values, down to −50°C or below, from a few high-level places in Scandinavia and parts of extreme northern USSR were not included. The range in Britain is delineated by the lines through −25° and + 38°.

The humidity values are based in part on a study of the *Climatological Memoranda* and *Geophysical Memoir* mentioned above. The highest recorded dew-point for the British Isles is about 23°C – the right-hand boundary of the shaded area is based on the assumption that this dew-point is maintained with increasing air temperature. The lower boundary, a constant 10 per cent, is based on known occurrences of low humidity in Britain. The left-hand boundary is derived in part from some extra data supplied by the Meteorological Office, concerning humidities experienced at a Swedish station at temperatures down to −38°C.

It must be emphasised that the diagram does not purport to delineate all extreme combinations of temperature and humidity that are possible. The information for doing this is not to hand, and may not exist in usable form. It is believed, however, that conditions outside those represented by the shaded area are very rare. For example, a relative humidity as low as 10 per cent has been recorded only once at Kew in about seventy years, and this is probably representative of lowland Britain. Such low humidities occur more frequently, perhaps as often as once in five years, at altitudes of 300 m or more, when there is anticyclonic subsidence.

Detailed information on the simultaneous frequencies of dry-bulb and wet-bulb temperatures at a number of places in Britain is available in a series of *Climatological Memoranda*, which may be obtained from the Climatological Services Branch of the Meteorological Office, Bracknell.

Extremes of temperature and humidity

Over most of lowland England, the highest air temperature that is ever likely to be experienced with a relative humidity of 100 per cent is about 22°C, corresponding to a water-vapour pressure of 26·5 mb. The dew-point will be 22°C. For central Scotland the corresponding maximum temperature will be 19°C with a dew-point of 19° and water-vapour pressure of 22 mb.

The highest atmospheric water-vapour pressure on record in Britain is 27·1 mb, measured at Aberporth on 29 July 1948. The air temperature was 31·1°C, the wet-bulb temperature 24·7°, dew-point 22·5° and the relative humidity 60 per cent.

It is probably reasonable to take the values shown in Table 46 as occurring about once in 20 years at major inhabited places in England at altitudes up to about 200 metres above sea-level. The maximum values in Scotland will be about 3 deg C lower, or 3 mb vapour pressure lower. Lowest wintertime values in Scotland will be little different for the same range of altitude.

Table 46 Extreme values of temperature and humidity which may be expected about once in 20 years in lowland England

Maximum vapour pressure	25 mb
Maximum dew-point	21°C
Maximum dew-point averaged over about 6 hours	20°C
Maximum air temperature	36°C
Corresponding relative humidity	42 per cent
Corresponding wet-bulb temperature	25·7°C
Minimum water-vapour pressure	1·3 mb
Minimum dew-point	−20°C
Minimum air temperature	−20°C

Trouble with flues

Many of the heating appliances used in single-family houses nowadays are notably efficient. So much so that only small quantities of gases and heat find their way up the flue, especially in the case of 'slumbering' solid-fuel appliances (usually at night). In consequence, the flow-rate is so low that it is unduly sensitive to outside influences, to wind, to temperature, and even to temperature-gradients in the outside air. Disturbances of the flow have led to a number of tragedies when carbon monoxide in the exhaust gas has found its way into bedrooms and asphyxiated the occupants – several such instances have been reported in the last few years in both Britain and Belgium.

Although variations in the dynamic pressure acting on different parts of a house can cause changes and even reversals in the rate and direction of flow of the gases in a flue, these are normally only temporary because of the gusty nature of the wind.

However at least one case is on record of flue gases from a solid-fuel appliance being drawn into the house by wind pressure. The gas made its way into the bedroom, although fortunately the occupants were discovered and rescued before the effects became fatal. In this house the chimney was short, finishing below the level of the ridge of the roof, so that with the wind from certain directions a positive pressure could develop at the terminal. If the flue was, say, 4 m high, and the flue temperature near the base 15°C, with an outside air temperature of 5°C the stack pressure can be estimated from

$$\Delta p = 0{\cdot}043\, h\,(t_1 - t_2) \qquad (30)$$

so that $\Delta p = 0{\cdot}043 \times 4\,(15 - 5)$

$= 1{\cdot}7$ N/m^2 approximately (see, for example, Dick 1950).

If the wind speed was 3 m/s, the dynamic pressure would have been about 5 N/m^2. The total wind pressure force across a building is usually about the same as the dynamic pressure, so even at this low wind speed the flow up the chimney would have been reversed if, for example, an open bedroom window was in a low-pressure region. At such a wind speed and especially at night, the flow may have been quite steady, giving a constant flow of lethal gas through the bedroom.

There is at least one case on record in which it appears that flue gas from an appliance has found its way through cracks in the brick withes to an adjacent flue in the same stack and has been drawn into a bedroom with fatal results. The forces on the gas were weak and it is not possible to be quite certain how it was made to take this path. The wind was very light, almost calm, but the house was near the foot of a long slope and it is possible that in spite of the overcast sky a weak downslope flow of air developed during the night, strong enough to draw the flue gas through the house.

It is believed that all other cases reported have occurred when the wind was very light. It would seem that there are two possible conditions which might give trouble. The commonest is probably the inversion of the air-temperature gradient which occurs on clear still nights, and may even persist throughout the day in foggy weather (see page 77). If the air is calm, or if the wind is so light that there is no appreciable stirring of the air, the temperature inversion may have a depth of 100 m or more by morning. However the maximum gradient always occurs in the lowest metre or so of the air, so that the air close to the ground may be 5 deg colder than that 1 metre above the ground. Such a high gradient will only be experienced over a surface with a high resistance to heat flow from below, such as grass or fresh snow. Over ground with a lower thermal resistance, such as bare soil, concrete or an ordinary street, more heat can be conducted from the warm ground below, and the surface temperature remains higher, so that the temperature difference in the lowest metre of the air may be no more than 1 or 2 deg.

It is clear then that a heating appliance may be drawing in low-level air which is at a temperature lower than that of the air at the level of the chimney top. Because of the low rate of heat-input, especially when the appliance is 'slumbering', and the low rate of gas-flow, the gas-temperature has fallen to that of the chimney by the time the gas has risen about 2 m above the appliance. Very often the flue is on an outside wall, and its temperature is little different from that of the outside air, averaged over the preceding 24 hours or so. The buoyancy forces on the column of gas in the flue are thus small and quite small changes in the temperatures could reverse the direction of flow and force the gases into the house.

A second condition which might cause trouble is a sudden rise in air temperature. If the rise were sudden enough, being completed within perhaps 24 hours, and occurred with light enough winds, it would be quite possible to have external brick-built flues appreciably cooler than the outside air. Table 47 shows the average frequency of changes of temperature from one day to the next in the winter half-year at Garston. The statistics are based on 24-hour mean air temperatures (09 00–09 00) over a period of 20 winters, 1946–47 to 1965–66. It can be seen that a rise in mean temperature of nearly 10 deg has been observed, from one day to the next, and on another occasion a fall of over 10 deg.

In all these cases the forces acting on the gases were small. It is probable that no trouble would have arisen if the installations had been correct and well-maintained. In particular, it is important that flues from efficient appliances are small enough to ensure an adequate gas velocity at all times, and that the flues are lined with a smooth corrosion-resistant material so that cracks cannot develop. Further, the terminals should be in such a position that they cannot be subjected to positive pressures by wind flowing over a roof-ridge or other feature.

Table 47 Average seasonal frequency of changes in daily mean air temperature in 0·55 deg C steps (1 deg F). 20 winters, October to March 1946–47 to 1965–66

Change in mean temperature deg C	Number of rises	Number o f falls
0	17·85	
0·6	16·15	17·70
1·1	13·75	17·65
1·7	13·45	13·20
2·2	8·90	9·90
2·8	8·10	8·00
3·3	5·40	5·35
3·9	5·00	5·15
4·4	2·65	2·10
5·0	2·00	2·40
5·6	1·15	1·30
6·1	1·00	0·95
6·7	0·50	0·45
7·2	0·45	0·20
7·8	0·20	0·40
8·3	0·20	0·11
8·9	0·10	0
9·4	0·15	0·15
10·0	0	0·10
10·5	0	0·05
Mean total winter frequency	79·15	85·16
Mean change of temperature deg C	2·19	2·10

Condensation on windows

Statistics of outside air temperature can be used to calculate the frequency of condensation on the windows of a building, because the thermal capacity of the glass is so small that the steady-state equation can be used to compute the glass temperature without sensible loss of accuracy. Table 48 lists the conditions of temperature and humidity which have been assumed in the following calculations.

The conditions *a* and *b* were chosen to represent comfortably heated buildings, with two levels of humidity, 40 per cent being typical of offices and 53 per cent representing the rather more humid condition which may be expected in a dwelling house. Conditions *c* and *d* represent very poorly heated buildings. Obviously much more humid conditions may be experienced in kitchens and bathrooms, but these are normally transient and are not considered here.

When the glass temperature falls to the dew-point temperature, or below, condensation will occur. The critical outside air temperature which will cool the glass surface to the dew-point may be calculated as shown on page 131. This calculation assumes that the glass surface follows changes in the external air temperature without significant delay. It ignores the effects of variations in wind speed and radiation exchange on the outside surface heat-transfer coefficient. The last two columns in Table 48 show the computed values of the critical outside-air temperature necessary to induce condensation on the inside surface of the glass. One column gives the figures for single-glazed windows, the other for double-glazed windows.

Table 48 Assumed conditions of temperature and humidity in a building, and corresponding 'critical' outside air temperature which will induce condensation on window surface

	Temperature	Relative humidity	Dew-point	Vapour pressure	'Critical' air temperature	
					Single-glazed	Double-glazed
	°C	%	°C	mb	°C	°C
a	20	40	6·0	9·4	0·0	−19·9
b	20	53	10·0	12·3	5·7	− 8·5
c	10	76	6·0	9·4	4·3	− 1·4
d	13	82	10·0	12·3	8·7	4·4

The calculated value for condition *a* with double glazing, −19·9°C, is extremely rare in Britain, and is therefore not included in the subsequent tables.

The monthly average percentage frequency of hours for which the outside air temperature is at or below these critical values can be obtained from published data. Table 49 shows the values for five places in Britain typical of the most populated regions. Figure 128 is another way of presenting the data, and can be used to estimate the frequencies of other 'critical' temperatures calculated for different sets of internal conditions.

Calculation of the frequency of conditions which cause condensation on windows

The following notes describe how to calculate the frequency and duration of meteorological conditions which will give rise to condensation on windows. It is assumed that the conditions of temperature and humidity in the building are those given in Table 48, although the method is applicable to any other set of conditions.

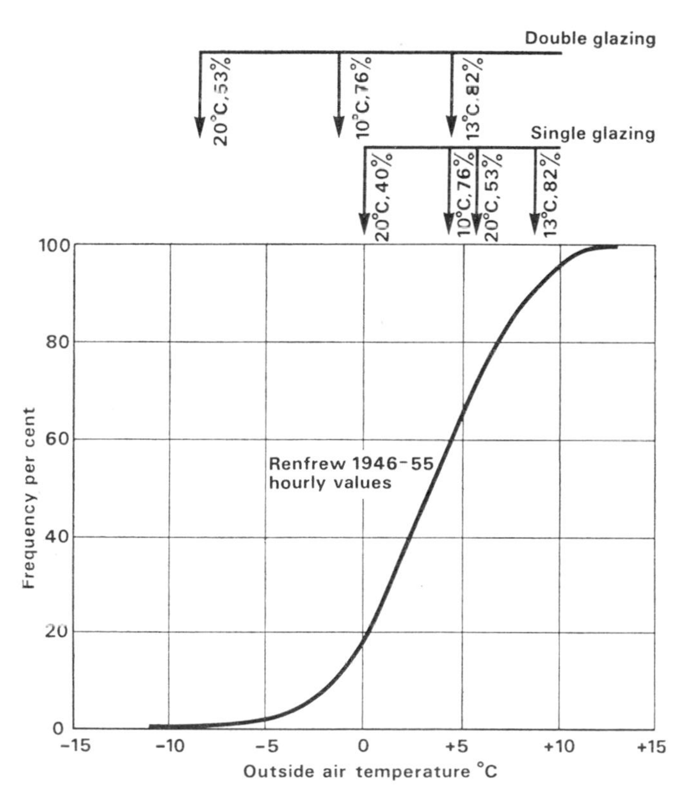

Figure 128 Frequency distribution of hourly values of outside air temperature during February at Renfrew. Average for 10 years 1946–55. The vertical arrows indicate outside air temperatures at which condensation will begin to form on windows under the specified inside conditions of temperature and humidity

(i) Single glazing

Considering first the case of single-glazed windows. The values for the inside and outside surface resistances to heat transfer are taken to be:

r_{so} = outside surface thermal resistance = 0·053 m²s deg/J (= 0·30 ft²h deg F/Btu)

r_{si} = inside surface thermal resistance = 0·123 m²s deg/J (= 0·70 ft²h deg F/Btu)

outside air temperature = t_o

inside air temperature = t_i

glass surface temperature = t_s

The thermal resistance of the glass sheet is assumed to be zero.

$$\text{Now } \frac{t_i - t_s}{t_s - t_o} = \frac{r_{si}}{r_{so}} = \frac{0{\cdot}123}{0{\cdot}053} = 2{\cdot}33$$

$$\text{so} \quad t_o = \frac{3{\cdot}33\, t_s - t_i}{2{\cdot}33} \quad \ldots\ldots\ldots\ldots\ldots(31)$$

At the critical temperature, when condensation is just forming, the glass surface is at dew-point temperature.

Example: take condition *a*

$$t_s = 6{\cdot}0°\text{C}$$

$$t_i = 20{\cdot}0°\text{C}$$

$$t_o = \frac{3{\cdot}33 \times 6{\cdot}0 - 20{\cdot}0}{2{\cdot}33} = 0{\cdot}0°\text{C}$$

that is, condensation will begin forming when outside air temperature falls to 0°C.

(ii) Double glazing

Take r_{so} and r_{si} as before.

Cavity resistance, r_c = 0·176 m²s deg/J (= 1·0 ft²h deg F/Btu).

Now, if t_s is temperature of glass surface in room,

$$\frac{t_i - t_s}{t_s - t_o} = \frac{r_{si}}{r_c + r_{so}} = \frac{0{\cdot}123}{0{\cdot}053 + 0{\cdot}176} = 0{\cdot}54$$

$$t_o = \frac{1{\cdot}54\, t_s - t_i}{0{\cdot}54} \quad \ldots\ldots\ldots\ldots\ldots(32)$$

and t_o may be calculated as above.

When these calculations are performed for the four sets of internal conditions specified in Table 48, the eight values of 'critical' air temperature t_o which will just give rise to condensation on the inner surface of the window are obtained, as shown in the example for single-glazing when the internal temperature is 20°C and the humidity 40 per cent. The frequency, or proportion of the time, for which the glass temperature is at or below this value can be obtained from statistics of air temperature. The most convenient way is to prepare a table of cumulative percentage frequency of occurrence of air temperature, which is then displayed as a curve like that shown in Figure 128. From this the proportion of the time for which the air temperature falls below the relevant critical air temperature can readily be read off.

The data used can be averages for a long period, like those used in Figure 128, which are 10-year means for the month of February, or can be those for a particular winter or other shorter period of interest, such as a particularly cold spell.

Table 49 **Percentage number of hours with dry-bulb temperatures at or below stated temperatures (°C) at five places**

(a) Aldergrove (N Ireland) 1946–55

Temperature	−8·5°	−1·4°	0°	4·4°	5·7°	9·4°
January	0·2	7·4	12·6	57·5	67·6	91·6
February	0·005	8·8	16·5	59·6	69·9	92·4
March	1·8	5·2	8·1	41·3	53·0	83·5
April	—	0·5	1·6	18·9	28·4	64·0
May	—	0·1	0·2	6·0	10·0	36·6
June	—	—	—	0·4	1·0	10·5
July	—	—	—	0·03	0·1	2·2
August	—	—	—	0·05	0·1	3·1
September	—	—	0·1	1·2	2·1	11·5
October	—	0·1	0·5	7·5	12·0	38·6
November	—	2·6	4·4	25·6	35·4	73·8
December	0·03	4·7	8·5	46·7	56·1	86·7

(c) Manchester 1946–55

Temperature	−8·5°	−1·4°	0°	4·4°	5·7°	9·4°
January	0·3	11·4	18·1	60·3	68·7	90·7
February	2·1	13·9	21·4	60·6	69·5	92·7
March	0·1	5·8	10·5	43·5	52·7	80·0
April	—	0·1	0·7	16·9	26·1	59·9
May	—	—	0·1	5·0	8·8	32·3
June	—	—	—	0·1	0·5	9·2
July	—	—	—	—	0·01	1·8
August	—	—	—	—	0·1	1·9
September	—	—	—	1·0	1·7	9·8
October	—	0·9	1·7	8·8	12·7	40·9
November	—	2·7	4·7	24·7	33·4	72·2
December	0·1	6·7	10·9	46·7	57·7	86·8

(b) Croydon (London) 1946–55

Temperature	−8·5°	−1·4°	0°	4·4°	5·7°	9·4°
January	0·3	10·1	16·8	53·9	63·4	87·9
February	0·4	12·9	19·3	54·7	63·6	87·2
March	—	3·9	8·7	39·5	48·1	75·3
April	—	0·2	0·7	12·8	20·7	50·6
May	—	0·01	0·05	2·3	4·9	25·3
June	—	—	—	—	0·1	4·5
July	—	—	—	—	—	0·6
August	—	—	—	—	—	0·6
September	—	—	—	0·4	0·9	7·1
October	—	0·5	1·5	8·5	11·7	31·9
November	—	1·8	3·8	21·1	28·5	62·1
December	0·01	5·4	9·2	39·8	49·4	79·0

(d) Renfrew (Glasgow) 1946–55

Temperature	−8·5°	−1·4°	0°	4·4°	5·7°	9·4°
January	3·6	12·6	17·9	60·0	69·2	91·3
February	4·4	11·2	17·7	62·8	71·8	91·6
March	0·6	5·5	9·7	43·1	54·7	84·7
April	—	0·5	1·6	17·7	26·8	64·4
May	—	0·01	0·2	5·2	9·3	33·7
June	—	—	—	0·2	0·7	9·0
July	—	—	—	0·04	0·1	2·4
August	—	—	—	0·1	0·3	3·5
September	—	—	0·1	1·2	2·2	11·0
October	—	0·9	1·9	11·1	15·5	42·0
November	0·1	5·0	7·7	27·6	37·0	73·1
December	0·1	8·5	13·5	48·3	59·6	86·3

The long-period data can be obtained from published sources, such as the *Climatological Memoranda* published by the Meteorological Office, which were used to produce Table 49, or from other analyses available from the same source. A useful source for data on specific periods of time is the *Daily Weather Report* of the Meteorological Office, which contains temperatures measured four times a day at about 50 stations in the British Isles. Obviously the statistics from this source will be less detailed, but they can be useful for making a quick survey of recent conditions when hourly observations of temperature are not available.

(e) **Turnhouse (Edinburgh) 1952–60**

Temperature	−8·5°	−1·4°	0°	4·4°	5·7°	9·4°
January	11·4	17·9	24·8	67·8	76·3	92·9
February	13·3	14·6	21·9	67·4	74·8	91·8
March	0·005	4·2	7·3	42·2	55·6	86·4
April	—	1·61	3·3	20·7	29·7	63·4
May	—	0·01	0·2	6·9	10·3	35·7
June	—	—	—	0·9	2·0	13·7
July	—	—	—	0·1	0·4	3·5
August	—	—	—	0·3	0·7	5·4
September	—	0·05	0·1	2·1	4·1	15·5
October	—	0·8	1·6	9·9	14·3	39·2
November	0·05	4·8	7·1	28·1	37·8	77·5
December	0·2	8·7	12·9	49·1	62·0	87·5

Sunshine probabilities

In Chapter 1 maps showing the variation of the duration of bright sunshine over the country were shown (Figures 43, 44, 45), these giving 30-year averages of actual durations in June, in December, and for the whole year. For some purposes it is more useful to have the information in the form of the proportion of the possible duration. Figures 129–132 show average means of percentage of possible bright sunshine in Britain during March, June, September and December. These are in effect the actual durations, expressed as a percentage of the time for which the sun is above the horizon. The values can be regarded as the proportion of the time for which the sun may be expected to be shining in an average month.

As described in Chapter 1, page 40, usually sunshine is not equally probable at all times of day. In winter the afternoons tend to be sunnier, after the clearance of early morning mist or fog. The same is generally true in summer at coastal places, but inland there is often less sunshine in the afternoon because clouds tend to build up as the day becomes warmer. However the exact pattern varies from place to place and from one year to another. Table 50, from Bilham (1938), shows the average diurnal variation of bright sunshine at Aberdeen, Kew and Falmouth, during March, June, September and December. During the sunniest hours there is on average about half the possible sunshine. In an exceptionally sunny month there may be, around the middle of the day, three-

Table 50 Average diurnal variation of sunshine. Hourly means (Local Apparent Time), 1881–1915

Month	4 h	5 h	6 h	7 h	8 h	9 h	10 h	11 h	12 h	13 h	14 h	15 h	16 h	17 h	18 h	19 h	20 h	Day
									Aberdeen (*Aberdeen*) – hours									hours
March	—	—	0·01	0·11	0·29	0·38	0·42	**0·43**	0·42	0·40	0·39	0·35	0·29	0·14	0·01	—	—	3·63
June	0·07	0·23	0·31	0·35	0·38	0·41	0·43	0·45	0·45	0·45	**0·46**	0·44	0·41	0·39	0·36	0·27	0·08	5·95
September	—	—	0·04	0·20	0·33	0·39	**0·42**	0·41	**0·42**	0·41	0·40	0·38	0·34	0·23	0·03	—	—	4·00
December	—	—	—	—	—	0·01	0·14	0·24	**0·27**	0·25	0·17	0·03	0·00	—	—	—	—	1·10
Year	0·01	0·06	0·11	0·17	0·23	0·30	0·36	0·38	**0·39**	**0·39**	0·37	0·31	0·24	0·18	0·11	0·06	0·01	3·67
									Kew (*Surrey*)									
March	—	—	0·00	0·09	0·23	0·33	0·37	**0·40**	**0·40**	**0·40**	0·37	0·35	0·28	0·13	0·01	—	—	3·36
June	—	0·16	0·34	0·41	0·44	0·48	0·50	0·50	**0·52**	0·51	0·51	0·50	0·48	0·45	0·42	0·28	0·01	6·46
September	—	—	0·02	0·18	0·33	0·43	0·50	**0·52**	**0·52**	**0·52**	0·51	0·51	0·44	0·31	0·05	—	—	4·84
December	—	—	—	—	—	0·05	0·17	0·21	0·22	**0·23**	0·20	0·09	—	—	—	—	—	1·17
Year	—	0·04	0·12	0·19	0·25	0·32	0·37	**0·40**	**0·40**	**0·40**	0·39	0·35	0·28	0·22	0·15	0·07	0·01	3·96
									Falmouth (*Cornwall*)									
March	—	—	0·01	0·16	0·37	0·43	0·46	**0·49**	0·48	0·48	0·47	0·44	0·39	0·21	0·01	—	—	4·40
June	0·01	0·23	0·38	0·40	0·44	0·47	0·49	0·53	0·53	0·54	0·57	**0·58**	**0·58**	0·55	0·49	0·30	0·01	7·28
September	—	—	0·06	0·27	0·43	0·49	0·52	0·54	0·55	**0·56**	0·54	0·53	0·49	0·36	0·07	—	—	5·41
December	—	—	—	—	0·01	0·14	0·27	0·31	**0·32**	0·29	0·25	0·13	0·01	—	—	—	—	1·73
Year	—	0·05	0·14	0·22	0·32	0·40	0·45	**0·48**	0·47	0·47	0·46	0·43	0·35	0·27	0·19	0·07	—	4·77

The values refer to periods of 60 minutes centred at exact hours of Local Apparent Time.
(From *Hourly Values from Autographic Records*, 1915; the values for Aberdeen have been amended).
After Bilham 1938, p 202.

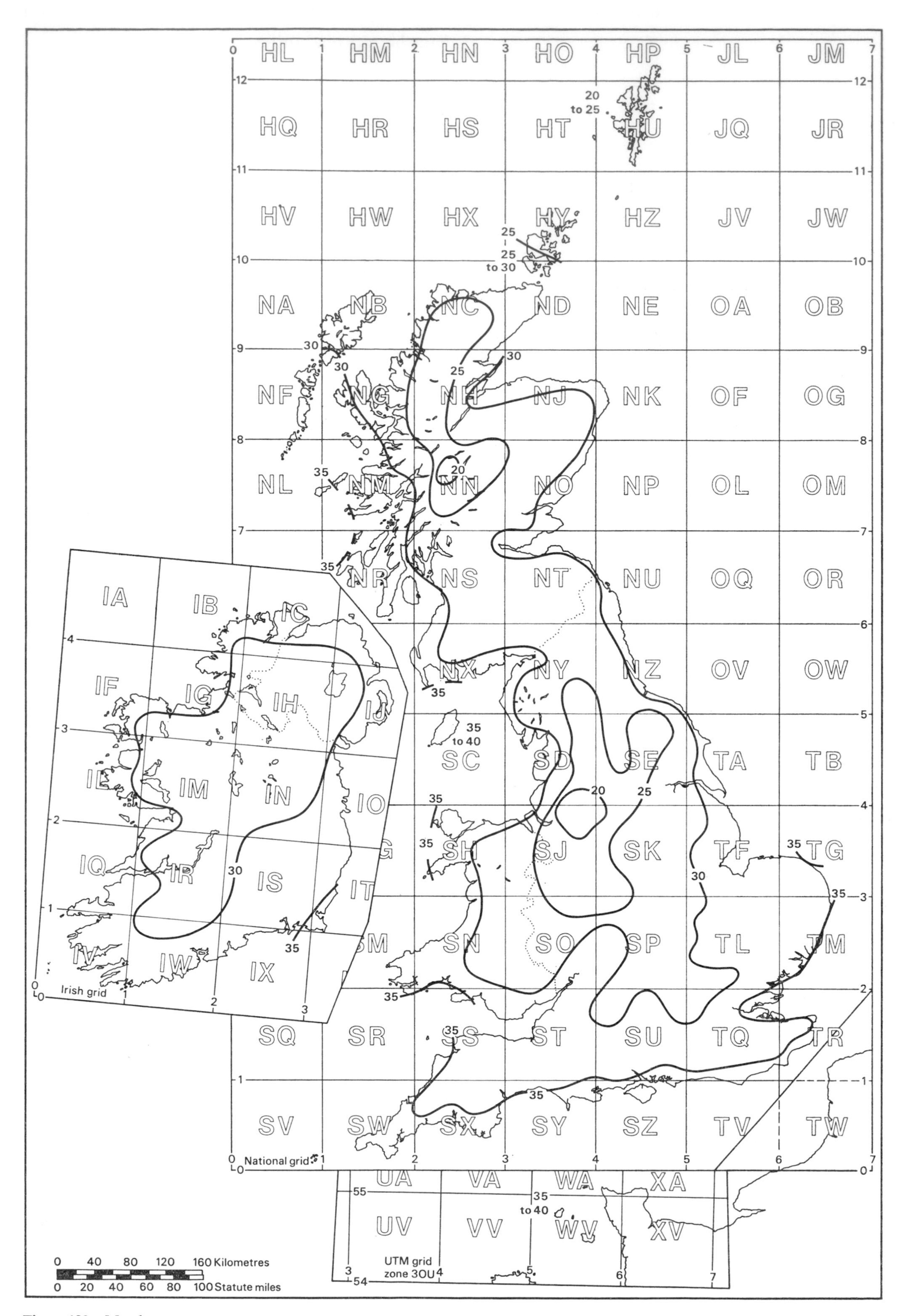

Figure 129 March

Figures 129-130 Maps of the British Isles showing average means of percentage of possible bright sunshine which occurred during the 30-year period 1901–30, in March and June respectively

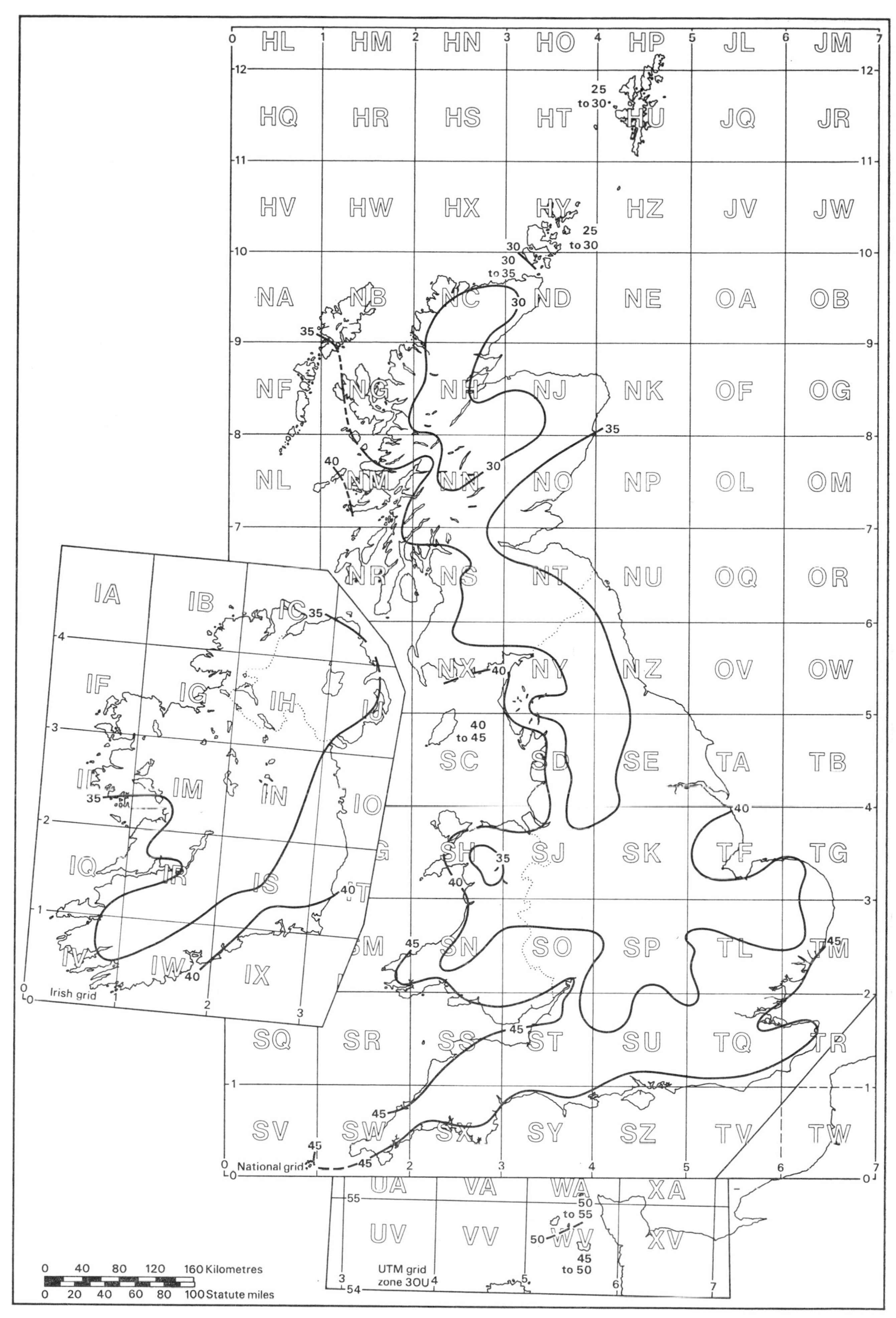

Figure 130 June

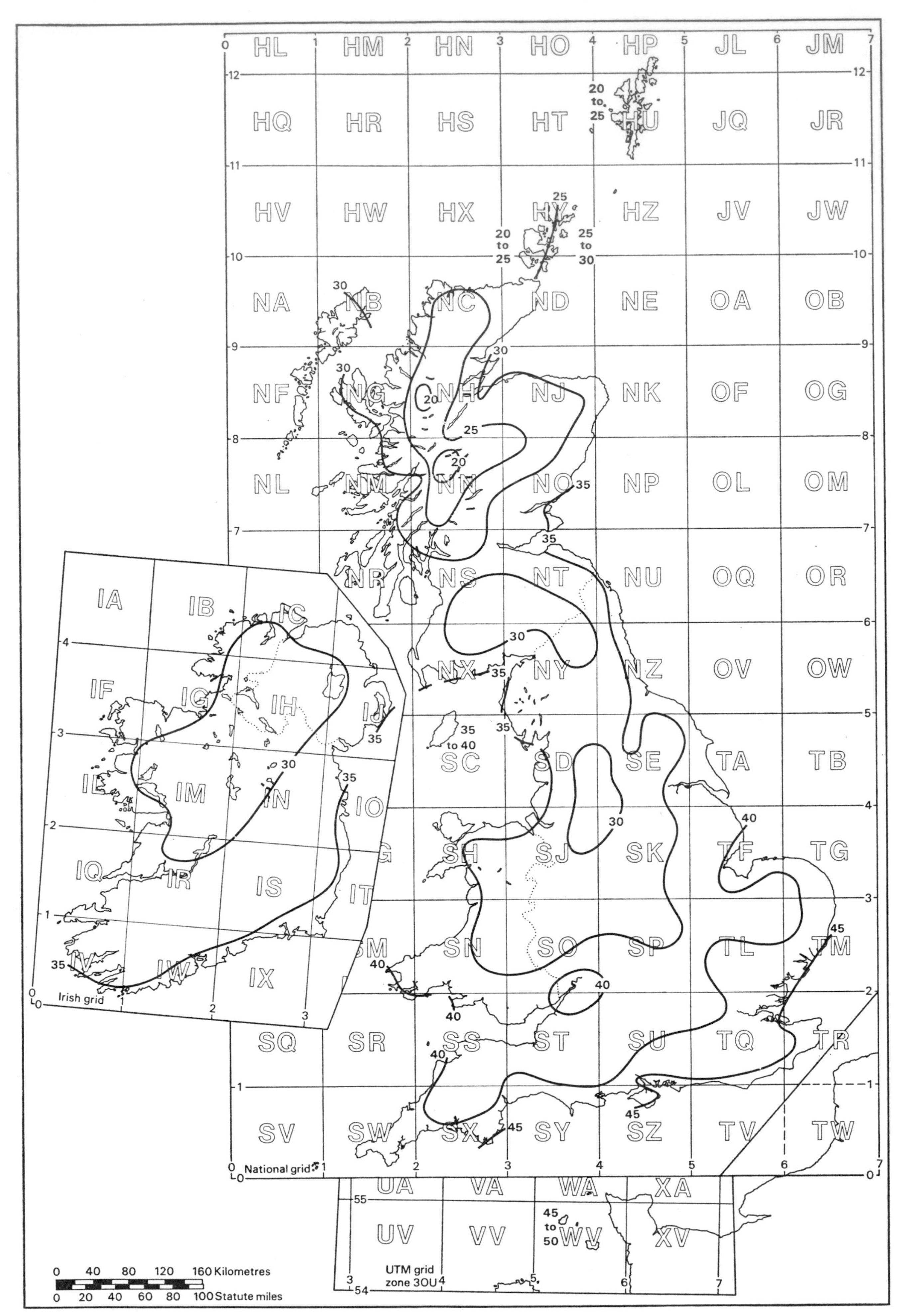

Figure 131 September

Figures 131-132 Maps of the British Isles showing average means of percentage of possible bright sunshine which occurred during the 30-year period 1901–30, in September and December respectively

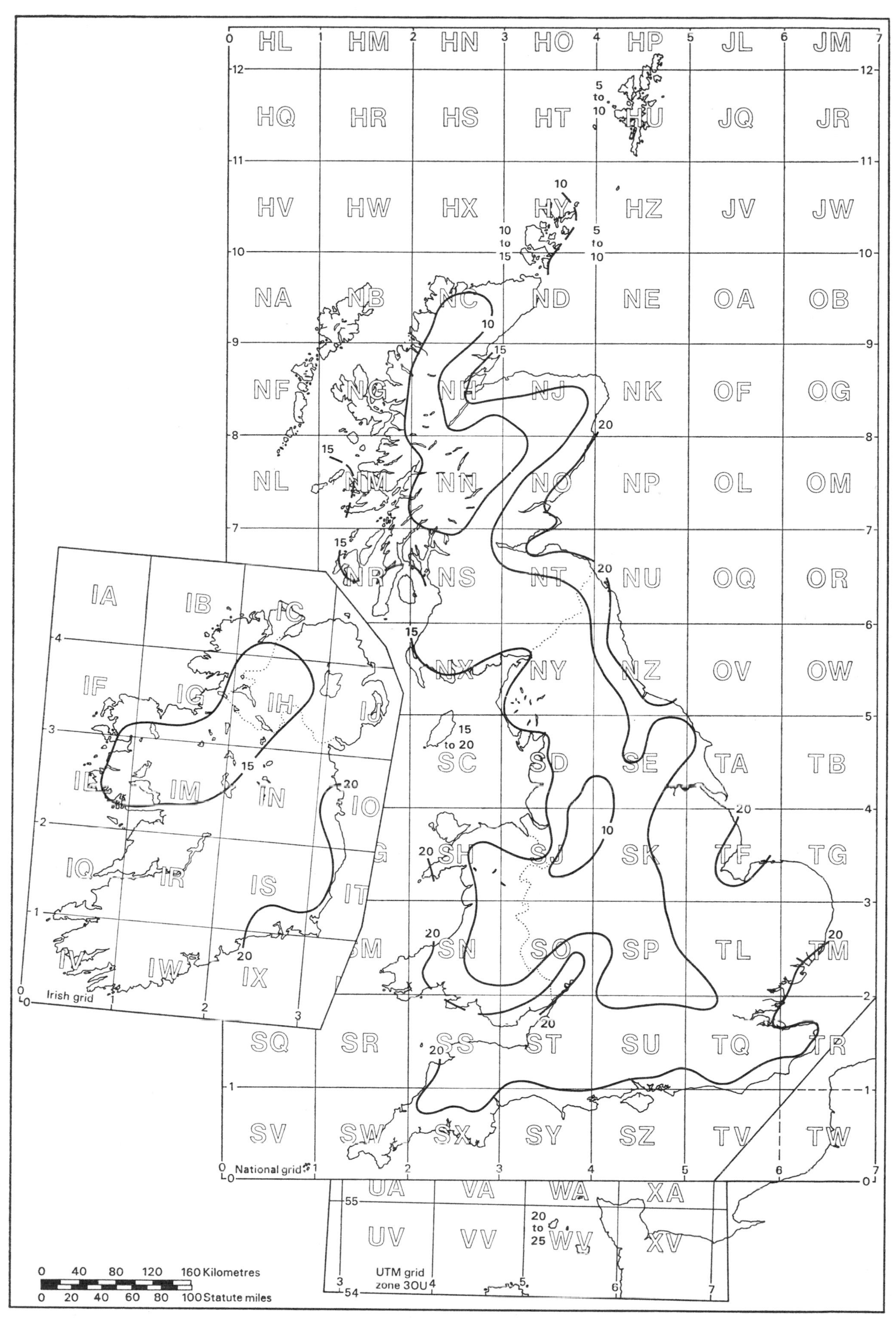

Figure 132 December

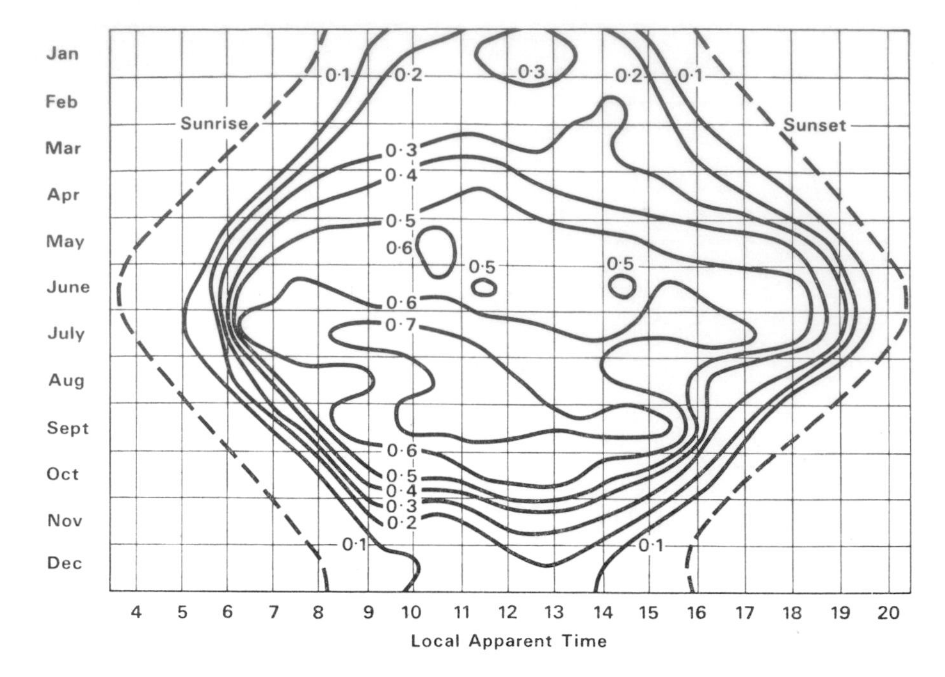

Figure 133 Average hourly totals of bright sunshine at Garston in 1959, a year with an unusually fine summer

quarters or more of the possible sunshine. Figure 133 shows the distribution of sunshine at Garston, Hertfordshire, during 1959, a year with an exceptionally sunny summer. September was particularly well-favoured, each hour between 10 00 and 15 00 LAT* having on average 0·7 h or more bright sunshine. A diagram such as this permits the probability of sunshine at any time of the day or year to be estimated.

Extremes of temperature on building materials

Table 1 in Chapter 1 gives the extremes of shade air temperature that have been recorded at various places in Britain during recent years. It would be reasonable to adopt a range of from −20°C to +38°C as the extremes likely to be experienced in open country in southern England during a period of about 100 years. In the Central Lowlands of Scotland the highest temperature likely is about 30 to 32°C. The temperatures of materials used in buildings may differ greatly from the air temperature, partly because of processes in the building, and partly because of the effects of the physical properties of exposed materials on the heat exchange at the surface exposed to the weather.

In cold weather the outside surface of exposed parts of the external cladding will normally be warmer than the air because of the transfer of heat from the inside of the building, unless the insulation of the components is good. If the component is well-insulated, and especially if the thermal capacity of the outer part is small, the temperature of the outside skin may fall below that of the air when the sky is clear and no solar radiation is falling on the skin. With thin sheet materials, backed by good insulation of low heat-capacity (for example, a metal sheet backed by fibrous insulation) it is reasonable to assume that on a clear night the outside surface temperature on a wall will be 3 deg C below air temperature, and on a roof surface the temperature will be 5 or 6 deg C below air temperature.

When sunshine is falling on a material, its temperature may be many degrees above that of the air. The difference in temperature depends on a number of factors, on the thermal conductivity and capacity of the material and of its backing, on the reflectivity of the surface to short-wave solar radiation, on its emissivity to long-wave radiation (that is, heat radiation), and on the local wind speed. A partly-clouded sky might produce a higher surface temperature than a perfectly clear one, for suitably-placed clouds can increase the intensity of the solar radiation by 20 per cent or more by reflecting radiation on to a surface which is already illuminated by the direct beam. At the same time, the partially clouded sky has a higher radiation-temperature than a clear sky and thus reduces the long-wave loss of radiation from the surface. It is estimated that the temperature rise of a thin black horizontal surface under such conditions might be as great as 7 deg C more than under a clear sky, although this may be partly offset by the lower air temperature normally experienced with a partly cloudy sky. Finally, the rise in temperature may also be affected by the moisture-content if the material can absorb water. The moist material will usually be darker than a dry one, and so will absorb more solar radiation. However a moist material has a higher thermal conductivity and capacity than a dry one, which will tend to reduce the rise of temperature at the surface. Furthermore evaporation of the moisture will absorb heat which would otherwise go to warming the material.

Thus it is a matter of some difficulty to calculate the temperature rise of a material, except in rather simple cases, when it is exposed to the climate.

The data given in Table 51 were obtained from measurements in south-east England and extrapolated to the assumed extreme air temperatures. They were obtained from a range of different constructions, and from them it will often be possible to make reasonable estimates for other materials and types of construction. Maximum temperatures on buildings in Scotland and northern England might be expected to be some 5 to 10 deg C lower than those in the Table.

Although the temperatures given for thin metal sheets can be taken to apply to the whole thickness of the sheets, there will be a large temperature-gradient in materials such as masonry and concrete, for which the figures apply only to the outer skin.

*LAT = Local Apparent Time, ie that given by a sundial.

Table 51 Estimated extreme temperatures on buildings (°C)

Constructions	Maximum	Minimum
Walls		
Light-coloured masonry wall, exposed concrete eaves, edges of floor slab	+ 50	−20
Similar construction, dark-coloured	+ 65	−20
Glass, ceramic tiles or metal, insulated behind: Black	+ 80	−25
White	+ 60	−25
Black metal tray, exposed behind clear-glass and insulated behind	+130	−10
Clear-glass in front of dark insulated background such as tray above	+ 80	−25
Aluminium mullion in a curtain wall, natural colour or white	+ 50	−15
Clear glass in a window	+ 40	−20
Roofs		
Thin black covering on: flat roof, insulated behind	+ 80	−25
pitched roof facing south	+100	−25
Top of 150 mm thick horizontal concrete roof, with 20 mm asphalt over	+ 45	−15
Exposed steelwork	+ 50	−20

Note: Data mainly from Building Research Station measurements.

The highest maxima in Scotland and northern England may be some 5–10 degrees lower than those shown in this Table. Lowest temperatures will be similar, except at higher altitudes, where they may be lower.

The temperature of unheated exposed objects at night

It is often supposed that unheated, freely exposed objects can be maintained, in cold weather, at a temperature appreciably above that of the ambient air, merely by surrounding them with insulating material.

In fact, unless heat is supplied to the object, the object will continue to lose heat, although at a lower rate than if uninsulated, until it is at the ambient temperature, providing this remains steady for a long enough period. Thus a small diameter water-pipe, containing still water, will freeze in a prolonged spell of sub-zero weather, even if wrapped in insulating material.

Drainage and soakaways

Soakaways are sometimes used to collect rainwater running off a roof or a paved area. The volume of the soakaway depends not only on the area of the drained impervious material, and the quantity of rainfall, but also on the kind of soil into which the soakaway is dug and on the level of the water-table at the time.

In a relatively impervious soil such as clay it is usually not possible to use a soakaway, because the water drains away so slowly that it would not be economic to dig a hole large enough to take the rain falling over a period of perhaps several days. Indeed, in winter the ground may be completely saturated for several months on end, so that it can take up no more water.

Even in pervious soils it may not be permissible to use a soakaway because of the risk of polluting sources of drinking water. Chalk, for example, is often highly fissured so that water may drain to a considerable depth without being filtered by the soil.

If the permeability of the soil is known, the required size of hole in relation to the area of paved surface which is to be drained, and to rainfall, can be calculated. Tables 52 and 53 give statistics of the frequencies of rainfall for periods of up to 24 hours. However, it will usually be sufficient to use the empirical rules for the sizing of soakaway pits which are given in Building Research Establishment Digest 151. The rules are based on a mean rainfall rate of 15 mm/h lasting for 2 hours, which event is estimated to recur on average not more than once in 10 years (Table 53).

The frequency of intense falls of rain

It has been customary to use, in calculations for the sizing of guttering, drainpipes and so on, a formula obtained by Bilham (1936) which relates the amount of rainfall, the time during which the rain fell, and the likely frequency of recurrence of such a fall. Other relationships of a similar nature have also been used. Later, after additional data had been accumulated (Holland 1964) the tables, which were formerly based on the Bilham formula, were revised. For full details of the method of revision the article by Holland should should be referred to, but Table 52 gives in summary form his recommended relationship between rainfall duration (from 2-minute duration up to 24 hours), return period (from 1 to 100 years) and rainfall amount. Table 53 gives similar information, but with rate of rainfall as one variable instead of amount of rain. Figure 23 in Chapter 1 displays some of the data graphically. We may interpret the Tables and Figure by saying that the rates or amounts quoted are those which it is likely will occur on average not more than once in the return period selected.

More recently still the Meteorological Office has made an exhaustive study of the peak daily rainfalls measured each year at about 7000 stations in the UK, as well as many other data, including those from continuous records on charts of rainfall at about 200 places. As a result it is now possible to prepare Tables similar to 52 and 53 for individual places. The Meteorological Office Rainfall Enquiry Bureau can compute and supply such Tables by return of post plus 1 day for computation. It is desirable to use these more precise data for designing valley gutters and in other situations where overflowing could have disastrous and expensive results. Although in most places the data of Tables 52 and 53 apply to within about plus or minus 25 per cent, in certain parts of the country there are significant differences from them. However, for ordinary domestic guttering it may not be worth while taking the differences into account.

Some information is also available about the area of ground to which an extremely high rate of rainfall applies. This shows that an area of 1 km^2 will experience an amount of rainfall, averaged over the whole area, equivalent to about 90 per cent of the peak intensity measured at a point in the area. An area of 10 km^2 will experience some 70 per cent of the peak rainfall.

When using the Tables for the design of guttering it should be remembered that there will be a 'time of concentration'. That is to say, it will be some time, perhaps one or two minutes, after intense rain has begun to fall, before water is flowing in the gutters and downpipes at the full rate corresponding to the rate of rainfall. If one assumes a time of concentration of 2 minutes, Table 53 shows that a rate of 75 mm/h lasting 2 minutes can be expected on average nearly once a year, and a rate of 150 mm/h lasting 2 minutes about once in 30 years. The procedure for calculating sizes of gutters and downpipes will be found in BRE Digests 188 and 189.

The drainage of roads is not covered by this book and the reader is referred to Road Note No 35 *Guide to engineers for the design of surface water systems,* published by HMSO, London. The background to this work is described in *Urban drainage in the United Kingdom* by C P Young (CIRIA Research Colloquium on rainfall run-off and surface water drainage, Bristol, April 1973).

Information on heavy falls of rain in periods of an hour or more can be found in the *Flood studies report* (Meteorological Office 1975), Volumes II and V. This report is particularly concerned with those heavy falls which are liable to cause flooding. For the first time it has been possible to produce reliable statistics of very heavy falls, such as 2-day amounts which recur as rarely as once in 1000 years, or even once in 10 000 years.

Lightning strikes

The frequency of thunderstorms varies over the country, being generally greatest over southern and eastern England. Figure 134 shows the distribution of days on which thunder was heard during the period 1955–64 – the figures representing the number of days per 10-mile square. Chalmers (1957) stated that at a place with 30 thunderstorm days per year there are about 4 ground strikes, that is to say flashes to ground, per square kilometre. If 'thunderstorm days' are the same as 'days with thunder heard', Figure 134 suggests that in the southern half of England there are from 1 to 2 strikes per square kilometre each year, with a rather unexpected peak in the Manchester area. There would seem to be less than one strike per square kilometre over most of south-west and northern England, Wales and Scotland. There are considerable differences in detail between Figure 134 and a map for the period 1901–30 (*Climatological atlas of the British Isles,* Meteorological Office 1952, page 99), possibly because of the relatively short period used for Figure 134.

The figures quoted above may be underestimates, for a recent study by Atkinson (1966) showed in the years 1951–60 an average of about 67 thunder outbreaks a year in south-east England. This figure is larger than 'days with thunder' for two reasons. Firstly on some days there was more than one thunder outbreak. Secondly, the value applies to the whole of the area considered, and each individual outbreak will only have been heard over part of the area. However it does suggest that there may be more thunder days a year, and hence lightning strikes, than is indicated by the statistics used in Figure 134. But it is worth remarking that because of increasing noise from road and air traffic it is becoming more and more difficult to obtain reliable statistics of 'thunder heard'. Furthermore, flashes from electric railways and other equipment can make suspect reports of distant lightning at night.

There is some suggestion that certain places may be unduly prone to being struck by lightning. No reliable statistics have been obtained, and the events that have been reported, of one house struck twice or more, or of three strikes within a hundred metres or so in a period of a couple of years, may be no more than chance.

British Standard Code of Practice CP 326: 1965, *The protection of structures against lightning*, shows how to evaluate the need to protect different types of structure, and how to design and install protection devices if they should be needed.

Table 52 The variation of quantity of rainfall (millimetres) with duration and return period

Duration	Return period (years) 1	2	5	10	20	50	100
minutes	mm	mm	mm	mm	mm	mm	mm
2	2·3	2·9	3·6	4·2	4·7	5·5	6·1
2½	2·7	3·4	4·2	4·9	5·6	6·5	7·2
3	3·1	3·8	4·8	5·6	6·4	7·5	8·3
3½	3·4	4·2	5·4	6·2	7·1	8·3	9·3
4	3·7	4·6	5·9	6·8	7·8	9·2	10·2
4½	4·0	5·0	6·3	7·4	8·5	10·0	11·1
5	4·2	5·3	6·8	7·9	9·1	10·7	12·0
5½	4·5	5·6	7·2	8·4	9·7	11·4	12·8
6	4·7	5·9	7·5	8·9	10·2	12·1	13·6
7	5·1	6·4	8·2	9·7	11·3	13·4	15·1
8	5·4	6·8	8·9	10·5	12·2	14·6	16·5
9	5·7	7·2	9·4	11·2	13·1	15·7	17·8
10	6·0	7·6	10·0	11·9	13·9	16·8	19·0
11	6·2	8·0	10·5	12·5	14·7	17·7	20·2
12	6·4	8·3	10·9	13·1	15·4	18·7	21·3
13	6·7	8·6	11·3	13·6	16·1	19·6	22·3
14	6·8	8·8	11·7	14·1	16·7	20·4	23·3
15	7·0	9·1	12·1	14·6	17·3	21·2	24·3
16	7·2	9·3	12·4	15·1	17·9	22·0	25·2
17	7·4	9·5	12·8	15·5	18·4	22·7	26·1
18	7·5	9·7	13·1	15·9	19·0	23·4	27·0
19	7·7	9·9	13·4	16·3	19·5	24·1	27·8
20	7·8	10·1	13·6	16·6	19·9	24·7	28·6
25	8·5	10·9	14·8	18·3	22·0	27·6	32·2
30	9·1	11·6	15·8	19·6	23·8	30·0	35·2
35	9·6	12·2	16·6	20·7	25·3	32·2	37·9
40	10·1	12·8	17·3	21·6	26·6	34·0	40·3
45	10·5	13·3	18·0	22·4	27·7	35·7	42·5
50	10·9	13·8	18·6	23·2	28·7	37·2	44·4
55	11·3	14·2	19·2	23·9	29·5	38·5	46·2
60	11·6	14·6	19·7	24·5	30·3	39·7	47·8
70	12·2	15·4	20·7	25·7	31·8	41·8	50·7
80	12·8	16·1	21·6	26·8	33·1	43·6	53·1
90	13·3	16·7	22·4	27·8	34·3	45·2	55·3
100	13·8	17·3	23·2	28·7	35·4	46·6	57·2
110	14·2	17·8	23·9	29·5	36·5	47·9	58·8
120	14·6	18·4	24·5	30·3	37·4	49·2	60·4

Table 53 The variation of rate of rainfall (millimetres per hour) with duration and return period

Duration	Return period (years) 1	2	5	10	20	50	100
minutes	mm/h	mm/h	mm/h	mm/h	mm/h	mm/h	mm/h
2	69·2	85·8	107·8	124·6	141·5	164·2	181·5
2½	65·2	80·9	101·8	117·8	134·1	156·0	172·9
3	61·6	76·5	96·5	112·0	127·8	149·1	165·5
3½	58·5	72·7	91·9	106·9	122·3	143·0	159·1
4	55·6	69·3	87·9	102·4	117·4	137·7	153·5
4½	53·1	66·2	84·3	98·4	113·1	133·0	148·5
5	50·8	63·5	81·0	94·9	109·1	128·7	143·9
5½	48·7	61·0	78·1	91·6	105·6	124·8	139·8
6	46·8	58·7	75·4	88·6	102·4	121·3	136·1
7	43·4	54·7	70·6	83·3	96·6	115·0	129·4
8	40·5	51·3	66·5	78·8	91·7	109·6	123·7
9	38·0	48·3	63·0	74·9	87·4	104·8	118·6
10	35·9	45·7	59·8	71·4	83·5	100·6	114·1
11	33·9	43·4	57·1	68·2	80·1	96·8	110·1
12	32·2	41·3	54·6	65·4	77·0	93·4	106·4
13	30·7	39·5	52·3	62·9	74·2	90·3	103·1
14	29·3	37·8	50·2	60·6	71·7	87·4	100·0
15	28·1	36·2	48·4	58·4	69·3	84·8	97·2
16	27·0	34·8	46·6	56·5	67·1	82·3	94·6
17	26·0	33·5	45·0	54·7	65·1	80·1	92·2
18	25·1	32·3	43·5	53·0	63·2	77·9	89·9
19	24·3	31·2	42·2	51·4	61·5	76·0	87·7
20	23·5	30·2	40·9	49·9	59·8	74·1	85·7
25	20·4	26·1	35·6	43·8	52·9	66·2	77·2
30	18·2	23·2	31·5	39·1	47·6	60·1	70·5
35	16·5	21·0	28·4	35·4	43·3	55·2	65·1
40	15·1	19·2	26·0	32·4	39·8	51·0	60·5
45	14·0	17·7	24·0	29·9	36·9	47·6	56·7
50	13·1	16·5	22·3	27·8	34·4	44·6	53·3
55	12·3	15·5	20·9	26·0	32·2	42·0	50·4
60	11·6	14·6	19·7	24·5	30·3	39·7	47·8
70	10·5	13·2	17·7	22·0	27·3	35·8	43·4
80	9·6	12·1	16·2	20·1	24·8	32·7	39·8
90	8·9	11·2	14·9	18·5	22·9	30·1	36·8
100	8·3	10·4	13·9	17·2	21·3	28·0	34·3
110	7·8	9·7	13·0	16·1	19·9	26·2	32·1
120	7·3	9·2	12·3	15·2	18·7	24·6	30·2

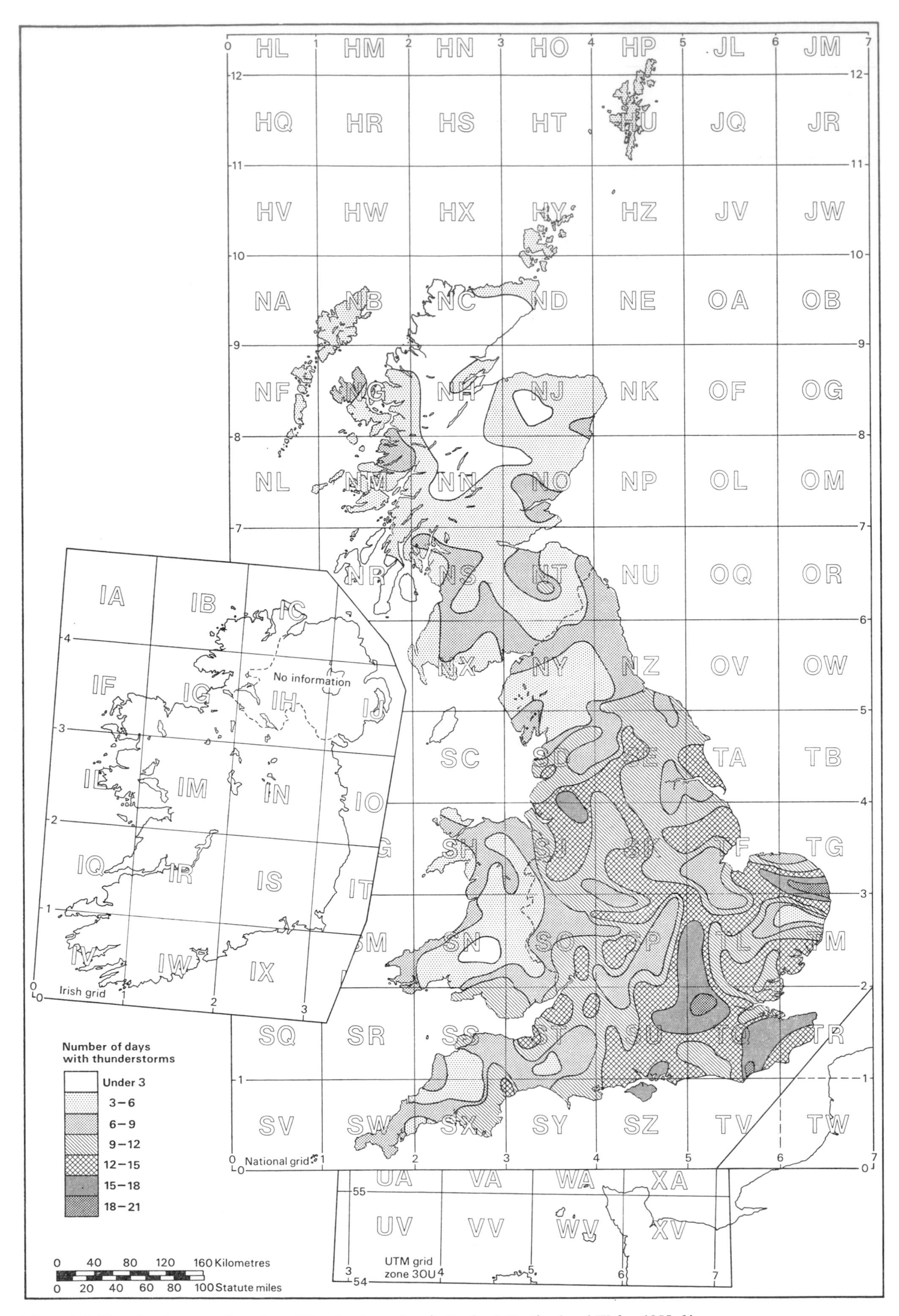

Figure 134 Map showing annual number of thunderstorm days in England, Scotland and Wales, 1955–64. (Electrical Research Association)

Air infiltration

Section A4 of the *IHVE Guide to Current Practice* (1971 edition) deals with the calculation of air infiltration into buildings. A design mean wind speed of 9 m/s at a height of 10 metres above ground in open country (that is to say, airfield exposure) is assumed in the calculations.

However no information is given on the expected frequency of such a wind speed at any place, nor how it may be expected to vary from one part of the country to another. The data given below are intended to give some guidance in this respect.

The average annual number of hours during which a speed of 9 m/s is exceeded, expressed as a percentage of all hours, is shown for a number of places in Table 54. This is derived from the Meteorological Office analysis of data for the ten years 1957–66. The wind exposure zones are those delineated in Figure 121, based on the 'once in 50 years' extreme hourly mean wind speeds. As was remarked on page 124, the frequencies of intermediate wind speeds do not always agree with this grading. For example, Cardiff, Rhoose, has more hours with speeds of 9 m/s than might be expected, while Glasgow, Renfrew, has only about half as many as might be expected. All the values may be taken to be for open country as specified in the *IHVE Guide* procedure, at a height of 10 m above ground.

Although the average speed of a strong wind may be taken to increase with height above ground, approximately according to equation (2):

$$V_{10} = V_h \times \left(\frac{10}{h}\right)^{0.16}$$

the rate of increase of the frequency of occurrence of a given speed with height has not been determined. In the absence of better information, it is probably not unreasonable to assume that in strong winds the frequency of occurrence of a mean hourly speed varies with height according to a similar law.

$$\frac{f_h}{f_{10}} = \left(\frac{h}{10}\right)^{0.16} \quad \ldots\ldots\ldots\ldots\ldots\ldots (33)$$

However this relation is certainly not true at moderate or low speeds. The exponent 0·16 only applies to open country with short vegetation. In towns the frequency of an hourly mean wind speed of 9 m/s will be perhaps half of that in the open country nearby, although the frequency of this speed around exposed towers or other high buildings may be little different from that at the same height over open country.

Winter basic design air temperature

A winter basic design temperature is required for calculating the correct size of heating plant for a building. In the study made by the Brunt Committee (1955) and published as *Post-War Building Study No 33*, an analysis was carried out on daily mean air temperatures at a number of places in Britain during 25 winters. The period of observations was mostly 1925–50. Although the first part of this period was one of generally mild winters, the last ten years especially included some cold winters, including the exceptional one in 1946–47. The present tendency towards more frequent cold winters, and cooler weather generally, does not yet invalidate the findings published in *Study No 33*, although if it continues there may be a case for considering the adoption of rather lower basic design temperatures for the sizing of heating appliances. This is because the design temperature was chosen to be such that the average number of days in a heating season with mean temperature below the design temperature

Table 54 Percentage of hours during which an hourly mean wind speed of 9 m/s is exceeded. Averages for 10 years 1957–66

Place	Effective height of anemograph m	Map reference	Altitude m	Wind exposure zone	Time %
Lerwick	10	HU 453397	82	D	29
Kinloss	13	NJ 069625	5	C	13
Edinburgh, Turnhouse	10	NT 159739	35	C	12
Tiree	12	NL 999446	9	D	32
Glasgow, Renfrew	12	NS 480667	5	C	6
Prestwick	10	NS 369261	16	C	10
Dishforth/Leeming	10	SE 305890	32	C	7
Waddington	10	SK 988653	68	B	7
London, Heathrow	10	TQ 077769	25	A	6
Boscombe Down	16	SU 172403	126	A–B	8
Manchester, Ringway	10	SJ 818850	76	B	8
Holyhead, Valley	12	SH 310758	10	B–C	25
Aberporth	10	SN 242521	133	B–C	21
Cardiff, Rhoose	10	ST 060670	62	B–C	13
Plymouth, Mt Batten	13	SX 492529	27	B	15
Belfast, Aldergrove	10	IJ 147798	69	B–C	11

Table 55 Typical degree-day report issued by the British Gas Corporation during the winter 1974–75. The numbers represent Fahrenheit degree-days

AREA	YEAR	Sept	Oct	Nov	Dec	Jan	Feb	March	April	May	TOTAL
1. THAMES VALLEY	1954/74 (a) 1973/74 1974/75	108 123 175	236 309 427	451 515 426	577 575	622 504	563 494	533 533	359 373	212 260	3,661 3,686
2. SOUTH EASTERN	1954/74 (a) 1973/74 1974/75	137 127 179	278 336 438	497 520 440	630 573	680 514	610 497	553 522	409 379	258 270	4,052 3,738
3. SOUTHERN	1954/74 (a) 1973/74 1974/75	96 82 139	211 291 371	422 420 370	548 520	599 463	538 446	485 469	335 306	204 221	3,438 3,218
4. WESTERN	1954/74 (a) 1973/74 1974/75	88 81 194	195 245 361	384 356 361	484 488	536 404	508 430	471 474	340 312	226 250	3,232 3,040
5. SEVERN VALLEY	1954/74 (a) 1973/74 1974/75	120 111 188	252 342 397	465 455 440	586 545	634 502	580 494	523 548	379 391	237 253	3,776 3,641
6. MIDLAND	1954/74 (a) 1973/74 1974/75	133 123 214	276 349 443	497 498 492	622 588	670 562	614 515	561 568	407 423	256 266	4,036 3,892
7. LANCASHIRE	1954/74 (a) 1973/74 1974/75	138 128 210	276 354 443	500 512 490	623 593	663 536	602 506	549 533	402 369	253 251	4,006 3,782
8. NORTH WESTERN	1954/74 (a) 1973/74 1974/75	170 167 246	309 390 480	521 575 515	634 649	676 524	618 509	572 562	426 432	286 272	4,212 4,080
9. NORTH EASTERN	1954/74 (a) 1973/74 1974/75	179 186 246	320 397 466	544 582 532	666 659	709 606	641 543	599 608	460 511	319 300	4,437 4,392
10. YORKSHIRE	1954/74 (a) 1973/74 1974/75	144 141 208	285 377 416	514 553 518	641 629	684 580	612 518	566 561	410 421	264 278	4,120 4, 058
11. EASTERN	1954/74 (a) 1973/74 1974/75	130 121 182	267 359 437	494 540 450	631 595	677 557	599 500	543 531	387 412	238 268	3,966 3,883
12. S E SCOTLAND	1954/74 (a) 1973/74 1974/75	170 174 266	314 393 472	530 584 549	637 654	681 582	631 545	584 618	444 504	316 295	4,307 4,349
13. W SCOTLAND	1954/74 (a) 1973/74 1974/75	194 196 283	343 411 474	562 582 585	668 674	715 577	649 544	595 616	440 408	310 292	4,476 4,300
14. E SCOTLAND	1954/74 (a) 1973/74 1974/75	177 177 249	327 394 441	584 574 536	661 645	695 578	628 537	579 603	432 462	306 305	4,389 4,275

NOTE: (a) Average for 20 year period
Monthly values are expressed in Degree Days below 60°F. These are based on maximum and minimum temperature recorded in a screen at Climatological Stations reporting to the Meteorological Office

was equal to the assumed thermal time lag of the building under consideration (one or two days as the case may be). Clearly if cold winters become more frequent, the average number of days with temperatures below given limits will increase, so that basic design temperatures will have to be lowered to compensate. If this is not done there will be an increasing number of occasions on which the heating plant will be unable to supply enough heat to the building.

The Committee suggested (page 13 of their study) that for Great Britain and Northern Ireland it was appropriate to adopt a value of 29°F (say −2°C) for structures with a 2-day thermal time-lag, and of 26°F (say −3·5°C) for structures with a 1-day thermal time-lag. These correspond respectively to buildings with solid concrete floors and brickwork walls about 220 mm thick, and to lightweight, single-storey buildings (usually of sheeted construction). It might be necessary to lower these temperatures for places at higher altitude.

The Committee remarked that if comfort was to be achieved during the whole of the coldest winters it would be necessary to lower the basic design temperatures by 5 or even 8 deg F (ie to −4·5°C or −6°C for '2-day' lag buildings) and to −6°C or −8°C for '1-day' lag buildings. In practice, some types of heating plant are conventionally designed to have an inherent overload capacity which may well cope with extreme conditions, but other types do not have this flexibility.

If a 2-day spell has a mean temperature of −2°C, it is reasonable to assume that for about half this time the temperature is lower than −2°C. Thus the temperature is below the design value for about 24 hours, or 3 per cent of the month in which the spell occurs. At London, Heathrow, the air temperature in the 10 years 1957–66 was below −2°C for about 3 per cent of the time during the six months October to March, and 4·7 per cent of the time during the four coldest months (December to March). In this period January had the greatest proportion of cold hours, with 8·4 per cent below −2°C.

In other countries the basic design temperature chosen is often based on the temperature which is exceeded for a given proportion of the time. In Canada, for example, it is that temperature which is exceeded for 97·5 per cent of the time in the coldest month, in USA that temperature which is exceeded for 97·5 per cent of the time in the four months December to March. Both of these methods applied to British data would give a lower design temperature than that obtained by the Brunt Committee's method for buildings with a 2-day thermal lag. For example, at London Airport (Heathrow) in the period 1957–66, during the ten Januaries the hourly temperatures were below −5°C for about 2·5 per cent of the time, and during the four months December to March they were below −3·5°C for about 2·5 per cent of the time. The latter temperature is the same as that found for buildings with a 1-day thermal lag using the method proposed by the Brunt Committee. If, as at Kew in 1925–50, there was on average during 1957–66 at Heathrow one day a year with a mean temperature of −3·5°C or below, it seems that about five-sixths of hours with a temperature of −3·5°C or below occurred on days when the *mean* temperature was greater than −3·5°C.

Heating degree-days

Degree-days, or accumulated temperatures below a given base-temperature, are often used as a guide to average seasonal heating requirements. They are the product of a mean temperature difference and of time measured in days. It is assumed that the heat losses from a building are proportional to the difference between a base-temperature and the mean air temperature during the period being considered. The base-temperature is usually taken to be some 2 or 3 deg C below the mean air temperature in the building, this difference being accounted for by stray heat-gains, including solar radiation, whose cost is not chargeable to heating. For example, if the base temperature of a building is taken to be 18°C and the mean air temperature outside on a given day is 5°C, the temperature difference is 13 deg C. The heat required to maintain the building at 20 or 21°C is assumed to be proportional to the difference of temperature (13 deg C), and the total degree-days for that day is 13 × 1, ie 13; similarly for every other day in the month or heating season. Adding the individual daily values gives the total degree-days for the period. Alternatively, the total may be obtained by multiplying the mean temperature-difference for the period by the number of days, for example if the mean January temperature is 3°C, the mean temperature-difference between this and the base temperature is 15 deg, so that the mean January degree-days will be 15 × 31, or 465. It must be remembered, when comparing Figure 135 with Table 55, that a degree-day total expressed in Fahrenheit units will be 1·8 times larger than if it were expressed in Celsius units.

Average heat losses will be taken to be proportional to average degree-days, calculated from long-period mean temperatures (for example 1921–50). Heat losses during a particular month or winter can be compared with the degree-days for that period, using for example the data published regularly by the British Gas Corporation. Table 55 is an example from one of the recent reports of this body.

The map shown in Figure 135 was produced by H C Shellard of the Meteoreological Office from statistics of air temperature for the period 1921–50. It is calculated on the basis of Fahrenheit degree-days and a base-temperature of 60°F (15·6°C), but has been rescaled for Celsius degree-days.

The map is based on temperatures reduced to sea-level, and it is necessary to make an auxiliary computation for stations whose altitude exceeds about 30m. The sea-level map is preferable to one based on actual temperatures, because in hilly country it is not possible to draw a map on this small scale which will show every local variation. In his paper Shellard (1956b) gives an exact method of calculating average monthly and annual degree-days for any place from the sea-level map. In addition to the data given in that publication, the monthly mean air temperatures and the standard deviation of these means need to be known. They can be found in two further publications, Meteorological Office (1956a) and Shellard (1956a). A less accurate method in which the correction to the value read off the map may be in error by as much as 10 per cent, is to assume a constant increase of 2·0 degree-day for every metre of altitude above sea-level. The error in the total degree-days should not then exceed about 2 per cent.

Local variations in topography do not affect average or total seasonal conduction heat-loss to any marked degree. A place in a valley bottom may well be colder at night than another place at the same altitude on level ground, but equally it will usually be warmer during the day. This will compensate for

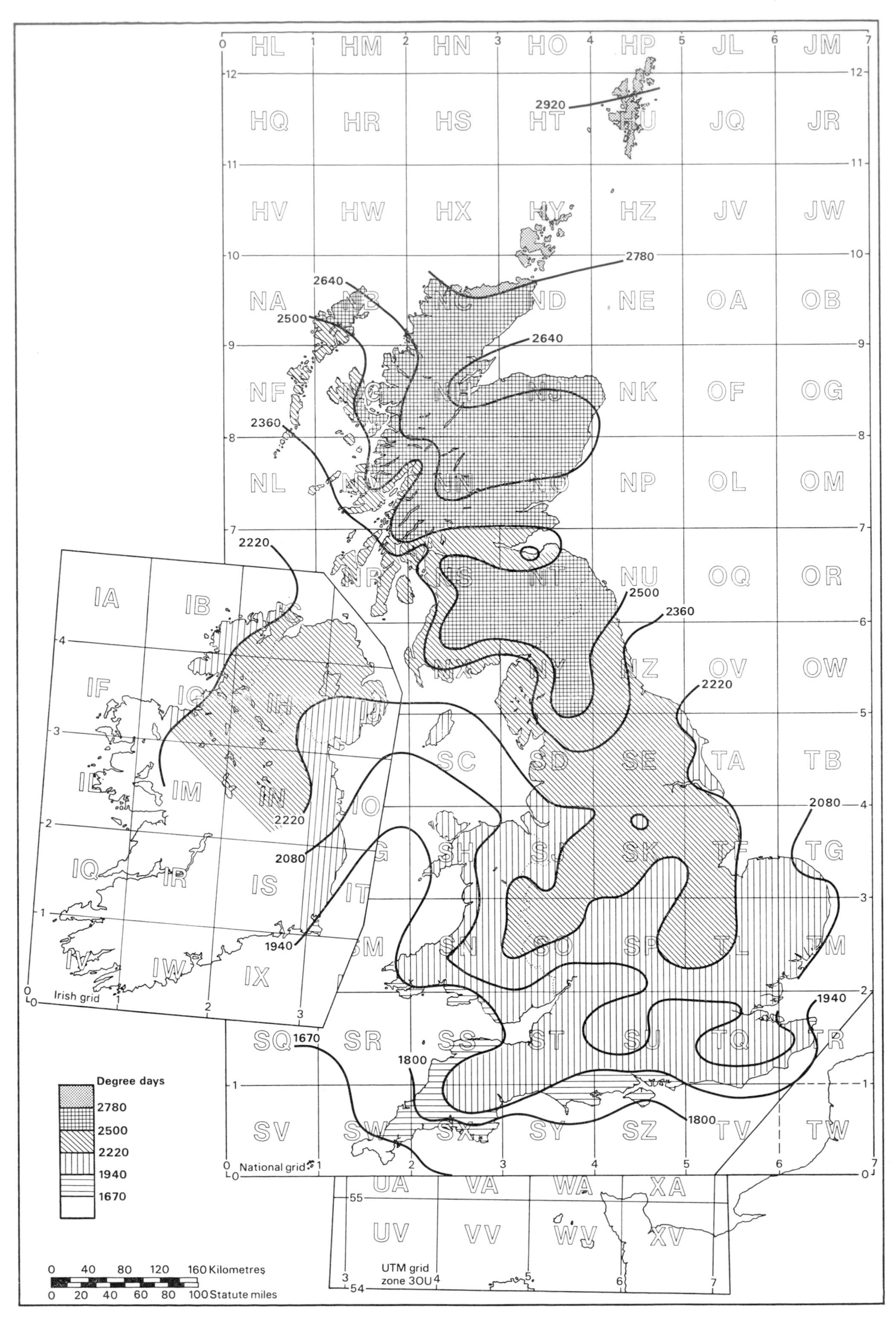

Figure 135 Map of annual average degree-days for the period 1921–50, below a base temperature of 15·6°C. (Shellard 1956)

the colder nights, at any rate with continuously heated buildings of relatively heavy construction.

However, the map does not allow for the rather warmer conditions which prevail in towns. Figure 59 (page 61) shows how the annual degree-days vary over London (Chandler 1965), there being a reduction of some 10 per cent in the middle as compared with places of similar altitude outside. The reduction is probably about the same for other large towns in Britain, but there is no information at present on how far up above the surface this effect is apparent. It is likely that buildings 50 to 100 m high which are well within the town will get the benefit of the 'heat island' effect, but those within, say, a kilometre or so of the edge may tend to be chilled at night by colder air percolating inwards from the open countryside.

The variation of heat requirements from one year to another

The degree-day method gives an approximate estimate of heat-loss variations, but there is an appreciable variation in the other climatic elements that affect heat consumption, from one heating season to another. It would be useful to be able to estimate the likely variations in the fuel consumption of a building, or of a complex of buildings, using the data for all significant variables.

This can be done, if an equation connecting the heat losses from the building and the relevant climatic elements can be formed. Such an equation might take the form

$$H = a + b\Delta t + cV\Delta t + dS \quad \ldots\ldots\ldots (34)$$

where:

H = fuel requirement
Δt = difference between building and outside air temperatures
V = mean wind speed
S = duration of bright sunshine

and a, b, c, and d are constants.

Data of the duration of bright sunshine were used because they are more readily available than values of solar radiation intensity. The values of the constants can be obtained experimentally (Lacy 1951), but usually it will be necessary to estimate them from the known or assumed characteristics of the building.

The variance in the fuel requirements of the building from winter to winter may be determined without calculating the requirements for each winter. It can be derived from the variances of the individual climatic elements by the following relationship;

$$\mathrm{var}(H) = b^2\mathrm{var}(\Delta t) + c^2V^2\mathrm{var}(\Delta t) + c^2(\Delta t)^2\mathrm{var}(V) + d^2\mathrm{var}(S) \quad \ldots\ldots\ldots (35)$$

It is assumed that the mean temperature of the building is constant, so that the variance of Δt is the same as the variance in the outside air temperature.

Strictly, when making statistical calculations of this kind, all the variables should be truly independent of one another. In meteorological studies the various elements are nearly always correlated to some degree, and this must be borne in mind.

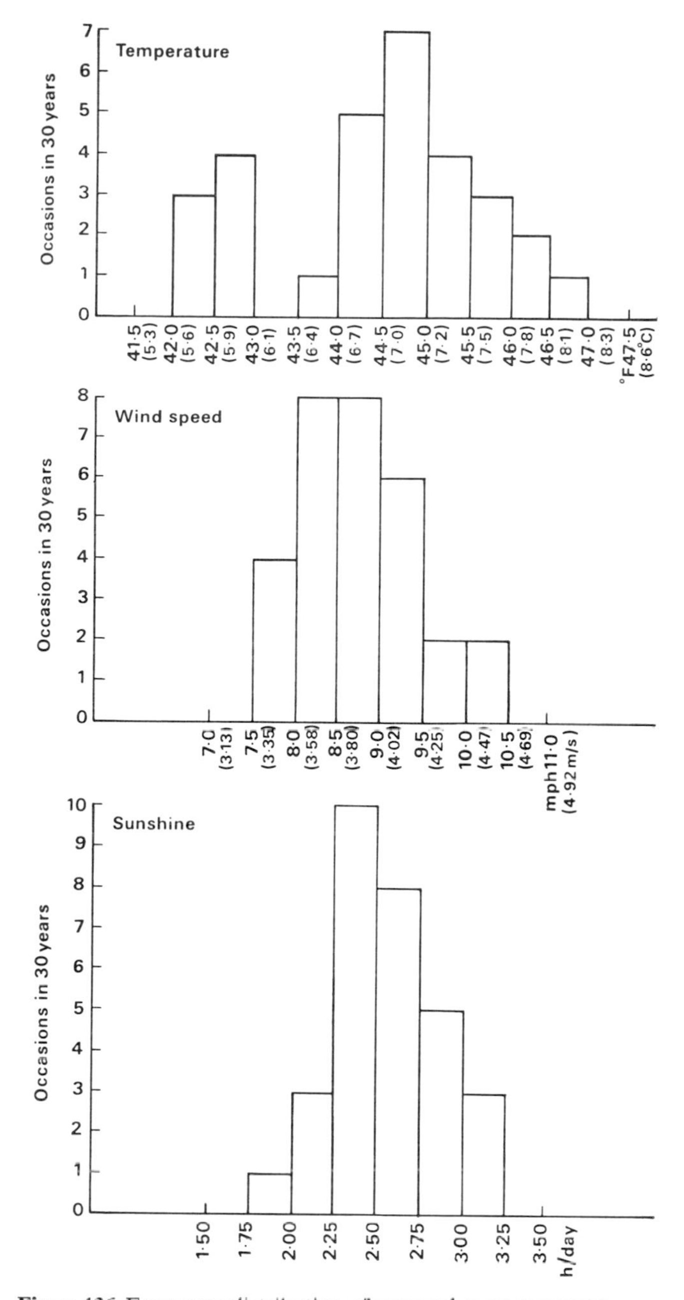

Figure 136 Frequency distribution of seasonal mean temperature, wind speed and daily sunshine at Kew Observatory in 30 heating seasons, 1919–49 (October to April)

In the study referred to above, the equation determined for a particular house was:

$$H = -13{\cdot}87 + 3{\cdot}02\Delta t + 0{\cdot}098v\Delta t - 1{\cdot}01S \quad \ldots (36)$$

where H was the daily heat consumption in kilowatt-hours, Δt was the mean temperature-difference between house and air in degrees Fahrenheit, V was the mean wind speed in miles per hour and S was the duration of bright sunshine in hours.

The variance in the seasonal heat requirements was estimated from the climatic data obtained at Kew Observatory during the 30 winters (October to April inclusive) 1919–20 to 1948–49. Figure 136 shows bar charts of the frequencies of the individual elements, seasonal mean temperatures, wind speeds and daily mean sunshine amounts. It will be seen that the distributions of wind speed and of sunshine may be regarded as approaching the statistical normal, but that of air temperature clearly may not. This, too, may introduce some error into the calculations, but probably not enough to alter the conclusions significantly.

Table 56 **Analysis of the climatological data from 30 heating seasons, 1919–20, to 1948–49 at Kew Observatory**

	Temperature °F	Wind speed miles/hour	Sunshine hours/day
Highest value	46·7 (1947–48)	10·3 (1940–41)	3·18 (1948–49)
Lowest value	42·1 (1946–47)	7·5 (1943–44)	1·99 (1946–47)
Overall mean	44·4	8·7	2·55
Standard deviation	1·26	0·72	0·29
Variance	1·59	0·515	0·086

A brief analysis of the data is given in Table 56. The dates of the highest and lowest values are given in brackets alongside the appropriate figures.

Applying the variances for the three elements given in the Table to equation (35) it was found that for the house to which equation (36) applied, the coefficient of variation* in the seasonal heat requirement was 8·1 per cent. Bearing in mind that some error may be introduced by the two factors mentioned above, it can be concluded that in two seasons out of three the house would have required an amount of fuel within about ± 8 per cent of that used in an average season. The heat requirement in extreme seasons might vary from the mean by about ± 15 per cent.

It is worth commenting on two of the constants in equation (36). The negative value of the constant a, −13·87, arises because the house gains some heat from solar radiation, equivalent in fact to 13·87 kW h per day in this case. It is because of this and other stray gains that it is usual to assume a base temperature for buildings a few degrees below the controlled temperature (which was 65°F in this case). It will also be noticed that the value obtained for the sunshine constant, d, was also negative. This is because the clear days that produce much sunshine in winter are usually associated with clear cold nights. On such nights the long-wave radiation loss from the house to the clear sky is greater than average, so that the 24-hour heat requirement is greater than average in spite of the greater amount of sunshine in the short winter day.

Most of the variance in the heat requirements is clearly due to the variance in the air temperature. The relative effect of the other two elements will depend on the design of the building.

Solar radiation data for building design

Some basic information on sunshine and solar radiation is given in Chapter 1, pages 39–49. As stated in the Introduction, it was decided that it was unnecessary to reprint in the present book all the considerable mass of information that is included in the *IHVE Guide 1970*. The tables of solar radiation printed therein are based on work done at the Building Research Establishment, which is reported in a number of papers, in particular those by Loudon (1965, 1968), Loudon and Danter (1965), and Petherbridge (1969). The radiation tables in the *Guide* are based on measurements of global solar radiation on the 5 per cent of days with the highest radiation totals at Garston in the six years 1957–62, so that the values quoted would on average be exceeded on 2½ per cent of days.

*Coefficient of variation is standard deviation expressed as a percentage of arithmetic mean.

Evaporation from bathing pools

Enquiries about bathing pools are becoming more frequent. The loss of water by evaporation is a common source of concern. Table 57 has been produced to give a guide to the rate at which water may be lost from a bathing pool in the open, according to the time of the year, the temperature of the pool, and the wind speed in the free air at 10 metres above ground.

If there is enough heat to supply the latent heat of evaporation, the rate of evaporation of water from a pool may be assumed to follow the relation

$$\varphi = a.V^{b}(e_w - e) \quad \ldots\ldots (37)$$

where V is the wind speed at a specified height above ground, e_w is the saturation vapour pressure at the temperature of the pool, e is the actual water-vapour pressure in the air, and a and b are constants which must be found by experiment.

A number of measurements have been made, the precise conditions under which the experiments were made not always being specified. Possibly the most appropriate results for the present purpose were those of Illig (1952) and Penman (1940), and combining results from these two sources it appears that a suitable relation for fairly rough terrain will be

$$\varphi = 1{\cdot}02 \times 10^{-2}\, V_{10}^{0.94}(e_w - e)\ \text{kg/m}^2\text{h} \quad ..(38)$$

where the wind speed at a height of 10 metres is in metres per second and the vapour pressures in millibars.

The actual vapour pressures used are mean values for the month, and the corresponding values of $(e_w - e)$ are given in Table 58 for a range of pool temperatures. The mean monthly atmospheric vapour pressures used in the above Tables are given in Table 59 and apply broadly to the London area. In the winter time, these values can be used without appreciable error for the whole country, but in the summer there is more variation in vapour pressure. Nearer the coasts of southern England, vapour pressure in summer is about 1 to 2 mb higher than in the London region, and in northern England and Scotland 1 to 2 mb lower.

The water temperatures used are those of a heated pool. The temperature of an unheated pool will on average be about 2 deg C above that of the air in midsummer, and at about air temperature in the winter. It will tend therefore to lose less water by evaporation than a heated pool.

It is likely that loss by evaporation from a well-used pool is of no practical significance because so much water is carried out by bathers as they climb out, but no statistics are available.

Table 57 Evaporation rates from outdoor bathing pools

$e_w - e$ (mb)	5	10	15	20	25	30
Wind speed m/s	Evaporation rate kg/m²h					
0·5	0·03	0·06	0·09	0·12	0·15	0·18
1·0	0·06	0·12	0·18	0·24	0·30	0·36
2·0	0·12	0·24	0·36	0·48	0·60	0·72
3·0	0·18	0·36	0·54	0·72	0·90	1·10
4·0	0·24	0·48	0·72	0·96	1·20	1·45
5·0	0·30	0·60	0·90	1·20	1·50	1·80
6·0	0·36	0·72	1·10	1·45	1·80	2·20

The wind speeds used are those measured at a height of 10 m above ground.

Table 58 Outdoor pools – average monthly water-vapour pressure-differences $e_w - e$ (mb). For outside vapour pressures e see Table 59

Water temperature °C	Month							
	February January	December March	November April	May	October	June	September	August July
16	11	11	10	8	7	6	5	4
18	14	13	12	11	10	8	7	7
20	17	16	15	13	12	11	10	9
22	20	19	18	16	15	14	13	12
24	23	22	21	20	19	17	16	16

Table 59 Average outside water-vapour pressure e (mb) in London region

mb	Month								
	February January	December March	November April	May	October	June	September	August July	Annual average
e	7	8	9	10	11	13	14	14	10

Solar heaters for bathing pools
Is it worth while building a solar heater for an outdoor pool? This is a question that is frequently posed. A solar heater is a device which absorbs solar radiation, the heat therefrom being used to warm water. The short answer is that a pool is itself an effective absorber of solar radiation. However, most of the absorbed heat goes to evaporate the water of the pool, only a relatively small proportion being lost by direct conduction to the air and to the ground. It follows, therefore, that the most economical way of heating the pool is to seal the surface of the water while the pool is not being used, and thus eliminate the loss of the latent heat of vaporization. There are obvious mechanical difficulties, but it is worth trying to use a transparent plastic sheet with high impermeability to water vapour, such as polythene. This will transmit much of the incoming solar radiation while greatly reducing the rate of evaporation.

Weathering problems

Although weathering of materials is perhaps the most obvious way in which the climate affects a building, it is by no means the best understood. It is obvious that certain effects are caused by weather, but the exact way in which this occurs is rarely understood. Until there is this understanding, the climatic elements which are significant cannot be clearly defined, and it is therefore impossible to quote suitable statistics.

The following paragraphs are in consequence little more than introductory notes to the subject.

Deterioration of plastics
The plastics materials used in buildings deteriorate in two ways. Those containing colouring pigments fade, following absorption of solar radiation in the visible band (400 to 700 nm wavelength) and the ultra-violet band (295 to 400 nm), the significant wavelength depending on the material and pigment.

Others, including the translucent and transparent materials which are mainly used in sheet form for roof lights and the like, degrade by absorption of ultra-violet radiation. The exact mode of breakdown is uncertain, but it is thought that radiation in the wavelength bands from about 300 to 350 nm breaks bonds in the long-chain molecules. The visible result is usually a yellowing of the material.

At present the climatology of the radiation responsible for the attack on plastics can only be described in general terms. It can be said, for example, that the intensity of the ultra-violet radiation will be greatest where the atmosphere is relatively free from artificial pollution. Thus we will expect the degradation to proceed fastest in the open country, and especially near the sea and at higher altitudes. Regular measurements of ultra-violet radiation in several narrow wavelength bands are now being made at a number of places, both with recently developed instruments (Harris 1973) and with a material which degrades in a predictable manner (Davis et al 1973). Before long therefore it should be possible to say with more precision how ultra-violet radiation varies from place to place, and its distribution with season of the year.

There is a further complication with composite materials such as glass-fibre reinforced plastics sheeting, for the deterioration initiated by solar radiation opens up cracks that permit the ingress of water, which then attacks the material at the interface between the glass fibres and the plastics matrix.

It follows that the time of occurrence of rain and of dew in relation to the attack by radiation may be of significance, as well as the duration of the wetting and, probably, the temperature of the material. The time for which the material remains wet in turn depends on other meteorological factors, temperature, humidity, wind speed and on the availability of energy to be converted to latent heat.

It might be expected that in the early stages the attack would proceed most swiftly in bright showery weather, with alternations of high radiation from a sun shining through a clean atmosphere, and short showers of rain. Another danger time might be in the early morning, after a clear night, when the sun shines from a clear sky on material which is covered with dew. But at this time the reaction might proceed more slowly since the atmospheric temperature will be low. At other times the material may remain wet for long periods, especially in misty or foggy weather when evaporation rates are low. Thus if the material is already partly broken down, such weather will permit attack to continue without interruption.

Figure 137 Mould growth on the outside wall of a cold room. Behind the right-hand part of the wall is a continuously warm laboratory; behind the left-hand portion, which is darkened by fungus, is a laboratory kept below freezing point

Mould growth on materials

No statistics of the conditions necessary for the growth of moulds or lichens on building materials are known, but observation indicates that in the British climate an unfavourable year may encourage their growth, while in a normal year little or none may be observed. For example, it was observed that after an unusually dull and wet summer there was a thick growth of mosses on a 5-year-old north-facing roof of sand-faced concrete tiles, while roofs facing in other directions had none.

The unfavourable conditions may also be created by factors other than the weather, by leaking pipes or gutters, or by high humidity created in the wall by some process in the building. Figure 137 shows a north-facing rendered wall of a laboratory, part of which is darkened by a growth of mould. Behind the right-hand, clean section of the wall is a laboratory which is continuously maintained at a temperature of about 30°C, so that the outer surface of the wall is normally a degree or two above air temperature. Behind the dark portion of the wall, on the other hand, is a cold laboratory, maintained at about −2°C, so that for the greater part of the year the temperature of the outside surface is below air temperature. Often, in fact, its temperature is below the dew-point of the outside air. Thus there will often be condensation of atmospheric water vapour onto the wall surface, and during much of the remainder of the time the rate of evaporation of moisture from the wall be much reduced because of its low temperature.

No figures of moisture content for this wall are available, but it is clear from the presence of the mould that it must be considerably wetter than a normally heated wall.

This picture also shows many cracks in the rendering, some of which appear to be random ones caused by moisture movements; other more regular ones may follow the joints in the underlying structure and be caused by moisture or thermal movements. Others near the top of the wall which run at 45 degrees to the horizontal are probably shear-failures caused by thermal expansion and contraction of the roof-slab with diurnal changes of temperature and radiation.

Thermal control of a growth of crystals on a wall

An example of the growth of crystals on the outside of a curtain wall is shown in Figure 138. The panel was formed of two sheets, 25 mm apart, of rigid plastic material about 3 mm thick, joined together by ribs of the same material about 50 mm apart, the cavities being filled with a fibrous insulating material. It appears that the crystalline substance (which was not identified) was washed off the window above the panel,

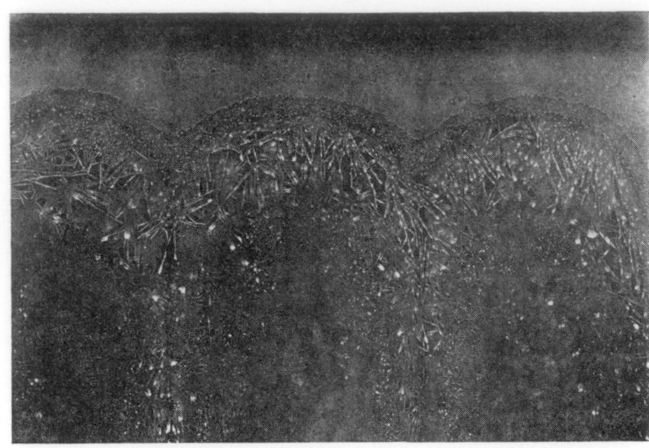

Figure 138 A pattern of crystallisation on a wall, controlled by temperature distribution. A solution washed down from the window or sill above was evaporated by heat coming through the panel

or possibly from the aluminium sill between the window and the panel, and the solution evaporated by heat coming through the panel. The warmest parts of the outside surface of the panel would be the area nearest the sill (because the metal conducted heat to the outer leaf and some of this heat would be transmitted downwards along the plastic material) and the parts opposite the ribs (because more heat would be transmitted through the ribs than through the insulating fill of the cavities). There will have been more evaporation at the top of the panel and over the ribs, so giving the greatest concentration of crystals at these points. This phenomenon occurred on a north-facing wall, soon after it was erected, and the crystals were subsequently washed off by the winter rains. It does however serve to show that quite minute differences in temperature can have a significant effect.

Frost patterns on buildings

A deposit of rime on the corner of the wall of a building is illustrated in Figure 139. Rime is a deposit of ice, which occurs when supercooled water droplets at a temperature of below 0°C touch a surface which is also below freezing point. The droplets freeze on impact, and if the process continues for a long enough period, the deposit may build up to a great

Figure 139 A deposit of rime on the corner of a wall

depth, up to 0·1 metre or more on free-standing unheated objects such as fences, towers or transmission lines. In the present case the deposit is quite thin, probably because the fog was not very dense and the wind speed was low.

On a building, the rime deposit occurs on a corner primarily because this part of the wall is usually the coldest. Figure 140 shows a horizontal section through the corner of the building which appears in Figure 139, with the probable course of the 0° and 5° isotherms. Because the thermal resistance of a corner is less than that of the adjacent parts of the wall, the heat-flow lines diverge and the temperature is less than that of the adjacent, uniform parts. Even if the whole surface was below 0°, there would probably be more rime at the corners, because there is a greater tendency for droplets to impact at these points as the air flows past the building. However, rime will build up on a wall surface if its temperature is low enough, as is shown in Figure 92 (page 91). Here, the main part of the wall consisted of a sheet of thin hardboard with no insulation, so that its temperature was above freezing point, probably about 5°. The hardboard was fixed to a wooden framework, and the insulation afforded by the wood was sufficient to allow the temperature of the outer surface of the board where it was fixed to the framing to fall below 0°.

Small objects attached to the building may acquire a thick deposit of rime (Figure 141). The object shown here is a thermocouple used to measure the temperature of the air near the test wall. The loop at the end is of wire 0·19 mm in diameter, bent into a loop 25 mm in diameter, the centre being 155 mm from the wall. Rime has built up on the wire, mainly on the side from which the wind has been blowing. The amount is proportional to the number of water droplets which have passed, and thus is an integration of the total airflow past the point, assuming constant concentration of the drops in the air. In fact, study of the appearance of the rime on a number of thermocouples enabled a picture of the airflow around the building to be built up. The thickness of the deposit decreases near the wall, mainly because the rate of airflow decreases as the surface is approached, but very near the wall there will be a rise of temperature. However, inspection of the picture shows that there was ice on the wire to within a millimetre or so of the wall surface.

It may also be expected that irregularly shaped pieces of a building, such as carved stonework, will experience differences in temperature, and in susceptibility to impact of super-cooled water droplets. Projecting parts, and edges, will probably experience rather lower temperatures in winter, and will tend to collect more rime.

In lowland Britain, rime does not occur often, but on rather rare occasions, possibly about once in 10 years, suitable conditions may persist for a few days (as in December 1962) and significant deposits can accumulate. At higher altitudes rime can occur quite often, and with strong winds blowing the supercooled cloud, large deposits of rime may occur. Figure 142 shows the frost pattern at the foot of a wall of a terrace of houses in Manchester. The walls are of brickwork about 230 mm thick, with a raised wooden ground floor, the space under the floor being ventilated by air bricks. The terrace runs due north-south (Lacy 1966).

Figure 141 The accumulation of rime on a thermocouple wire, demonstrating the mean speed and direction of the air flow. The amount of rime is proportional to the number of droplets which have passed

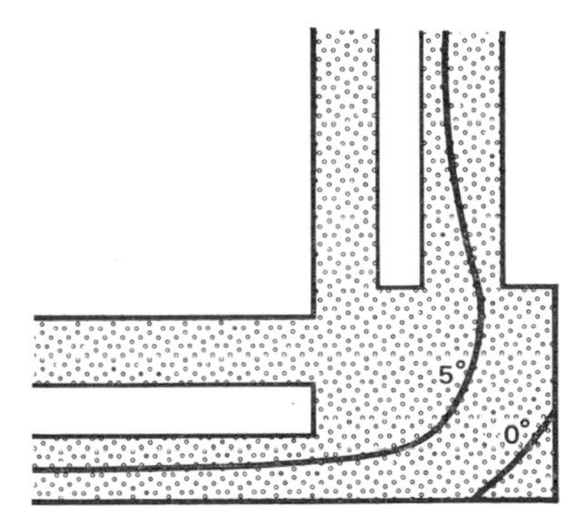

Figure 140 Cross section of the wall in Figure 139 with the probable run of the 0°C isotherm

Although little snow remains in the street, the outside of the east-facing wall is still white with frost, up to an irregular line which rises in places to as much as 700 mm above the pavement. A dark patch at intervals, some 200 mm up, shows where relatively warm air emerging from the crawl-space under the floor has melted the ice (or has prevented it from forming). It is not certain why the lower part of the wall should have been so cold. It is known that the front room of the house nearest the camera was unheated throughout the winter (and the same was probably true of the other houses), the occupants using only the rooms at the back of the house. The street may well have been frozen to a depth of some 300 mm, as happened in many places during this spell, while this area is said to be frequently subject to temperature inversions. In the early morning the wall is shielded from direct sunlight by adjacent houses, but their shadowing effect ceases by about 10 00 GMT. These factors together presumably account for the low temperature of the base of the wall.

Figure 142 The pattern of ice on a wall during prolonged cold spell

Figure 143 Rime deposited by fog-laden air flowing through a cast iron grille

The ice on the surface may have resulted from condensation of water vapour from the air, or may be partly rime as in Figure 139. Less probably, because the bricks are very dense and have low porosity, the ice may have been extruded from the pores of the bricks. The very irregular boundary to the ice is rather puzzling. It may be partly explained by variations in the thermal properties of the brickwork, but possibly some of the ice is derived from water splashed up from the footpath, which would give a random pattern. (It looks as if the footpath has been salted to melt the snow.) The end wall of the house faces south and apparently received sufficient sunshine to keep its temperature above 0°C.

Figure 143 shows a cast-iron grille covering a ventilating hole for the underfloor space of a building at Garston. The wall faces west and the hole is about 500 mm above the ground. The morning of 14 January 1959 was foggy, with a maximum temperature of −4°C, and air entering the ventilator has deposited rime on the grille. The rime is thickest on those parts of the grille farthest from the sides, there being little on the parts in good thermal contact with the wall, except at the extreme tip of some of the triangular projections. In this case cold air was entering the ventilator, whereas the ventilators seen in the wall in Figure 142 were sources of relatively warm air.

Fires caused by sunshine

Fires caused in buildings by sunshine may be regarded as a rather extreme form of weathering. They seem to be quite rare, unless it is that the evidence is usually destroyed by the fire. An unequivocal case was described by Reuben (1968), as follows:

'Returning home after a week's absence during which the house had been locked up, a lady had found two lines of rather severe burns on the side of an expensive well-type dressing table in the bedroom and no sign of any entry, forced or otherwise. Completely mystified, she had called in the Fire Brigade and after expert investigation the telephone call to the Manchester Weather Centre provided confirmation for their theory.

'A magnifying mirror had been left in the well of the dressing table in what, unfortunately, turned out to be just the correct position for it to reflect the sun's rays, entering through a window facing south-south-west, on to the woodwork enclosing the well (Figure 144). The focal length of the mirror was 18 in. and its distance from the burns was just 18¾ in! The two sets of parallel burns which were puzzling the Fire Brigade experts, were explained when the Weather Centre records showed 3·2 hr of sunshine on the 26th and 5·6 hr on 29 October, the difference in elevation of the sun on those 2 days being sufficient to cause the registration of two well-defined traces. The breaks in the traces, coinciding on both days, were seemingly due to the sun's rays being obstructed by a clump of poplar trees in the garden in front of the window.

Figure 144 Burns on a dressing table caused by the sun's rays reflected from a magnifying mirror. (Photograph by Manchester Fire Brigade)

'The Fire Brigade investigating officer remarked that they had often suspected the sun as the cause of fires, but this was the first direct proof they had come across. The one fortunate circumstance was that a full container of hair lacquer was just out of the line of reflection, otherwise, in the words of the Fire Officer, there would probably have been an explosion, a gaping hole in the ceiling and once again no real evidence left.'

Damage to roofing by hailstones

Occasionally, as has been described on page 29, severe thunderstorms may occur, with a fall of giant hailstones. In ordinary hailstorms in Britain the stones are usually fairly small, no more than about 10 mm in diameter. The severe storms, on the other hand, can produce hailstones which may be 75 mm or more in diameter, with a mass of 100 grams or more. These will of course smash any glass that they strike, and sometimes extensive damage to rooflights and glasshouses has been reported. In an account (Meteorological Office 1936) of a storm which occurred on 22 September 1935 in Northamptonshire, it was reported that large quantities of 21-oz horticultural glass were afterwards supplied to places in an area some 60 km long by 15 km wide, including about 1000 m^2 to one place alone. The damage to one hospital at Northampton amounted to over £600.

These large hailstones can also crack roofing tiles, and even other forms of roofing. In this 1935 storm, holes of up to 100 mm in diameter were punched in a newly-erected garage roof made of asbestos-cement sheet ⅛-in thick (3·2 mm). In another storm, on 1 July 1968, corrugated plastic roof sheet was punctured.

Chapter 4 The effects of weather on construction processes

Interference with construction by the weather

Although it seems to be generally agreed that bad weather can cause great delays in construction work, there are few reliable statistics of the amount of delay caused. One firm may claim that the weather does not delay them – they just 'press on regardless'. Another will admit that they expect that a job undertaken during the winter months will take twice as long as if it was done during the summer, and their estimates of time and cost take this into account. As the annual expenditure in Britain on building construction and civil engineering is now over £9000 million, even a small proportional saving could represent a great deal of money.

In an examination of the time lost at five housing sites, which was logged as time lost when men stopped work during the day because of inclement weather, Clapp showed (1966) that this amounted to from 0·7 to 1·7 per cent of the total site man-hours. As might be expected, the losses depended not only on the weather, but on the stage of the work, for some jobs are more likely to be affected by the weather than others. Because this type of analysis was not contemplated when the study was undertaken, no records of weather on the site were made. There is, therefore, no record of the details of the weather that caused the interruptions. The analysis was based on monthly data from the nearest available climatological stations.

However, the investigation showed the complexity of the ways in which the inclement weather affected the rate and cost of the building work. It was not merely a case of particular tasks being stopped. If one job was rained off, the men might be put onto another, indoor job, which was not their normal one, so that it would be done less efficiently than by operatives used to it.

The extent of the interference clearly varied from one contract to another, and appeared to affect a sub-contractor to a different extent from a main contractor.

Some damage was done to the incompleted buildings, but the man-hours required to make this good amounted at the worst site to less than 10 per cent of the other losses from bad weather. At this site most of the damage was to brickwork, about 10 per cent to plaster, and 5 per cent to roof timbers. About half the damage was caused by frost, the rest by wind. At other sites the damage was quite negligible.

Clearly it is not yet possible to formulate any general rules about the extent of the hindrance caused by inclement weather.

Although the occasional exceptionally cold winter, with frost lasting for some weeks, can cause great delay to constructional work, on average more delay and loss of working time is caused by rain. Some statistics of this are given on page 154, although it must be admitted that there is no exact correlation between the data and the time actually lost. The data relate primarily to time lost because the rain causes discomfort to operatives in the open, but there may also be direct interference with processes by the rain, especially to painting and rendering operations.

Probably the second most important climatic factor is wind. Strong winds may hinder the operation of cranes, especially when lightweight components such as panels and shuttering are being handled (see page 159). The wind also increases discomfort when it is raining, and especially at low temperatures. The wind-chill index may be relevant here (see page 166).

The frequency of high and of low temperatures may be important in the casting of concrete and the laying of masonry. When there is appreciable wind in association with low atmospheric humidity, the phenomenon of the 'plastic cracking' of freshly laid concrete slabs can be troublesome (page 168).

The transport of materials to and on the site may be hindered by snow and fog – these factors have already been discussed in the chapter on town planning (page 74).

Attempts have been made to assess the combined effects of two or more climatic elements on construction work. The first of these was by Cottis (1960); he produced statistics of the average frequency of occurrence of rainfall with an intensity of 0·5 mm/h or more and of air temperatures at or below 0·8°C. Tables were given of the average percentage frequency of each element, and of both together, for each month, during the hours 06 00 to 18 00 GMT, the data being from London, Heathrow Airport, and averaged for 10 years. Similar statistics have been computed for the Manchester, Glasgow and Newcastle areas. The intention was that these figures would give an assessment of the likely interference with outside work by rain, and with frost-sensitive work such as concreting and masonry.

More recently, Boobyer and Brookes (1967) have devised a formula for calculating the proportion of 'inclement weather' affecting building work which can be expected on average, and how much this differs from average in any month or series of months. It is based on the number of days when rainfall was 0·5 mm or more, and the number of days when the air temperature fell to 0°C or below. Multiplying factors are included which are intended to correlate with the degree of disruption caused to different phases of a building contract. So far, no record of the use of these approaches has been published. Although the correlations between the climatic elements chosen and the building operations are probably not high, it would be interesting to know if either or both of the methods are useful in practice.

In the paper mentioned above (Clapp 1966) a diagram was reproduced, Figure 145, showing how the percentage of bad-weather time varies with air temperature, rainfall and the stage of the work. This offers a guide to what may be expected on a housing site on which conventional methods of construction are used. It will be expected that non-traditional, prefabricated types of construction will be affected to a different degree, and multi-storey, high-rise building in a different way again.

In the following paragraphs, some of the problems of the construction processes are discussed in more detail. There is much to be learnt yet about the various problems and for the most part only a rather general guide can be offered.

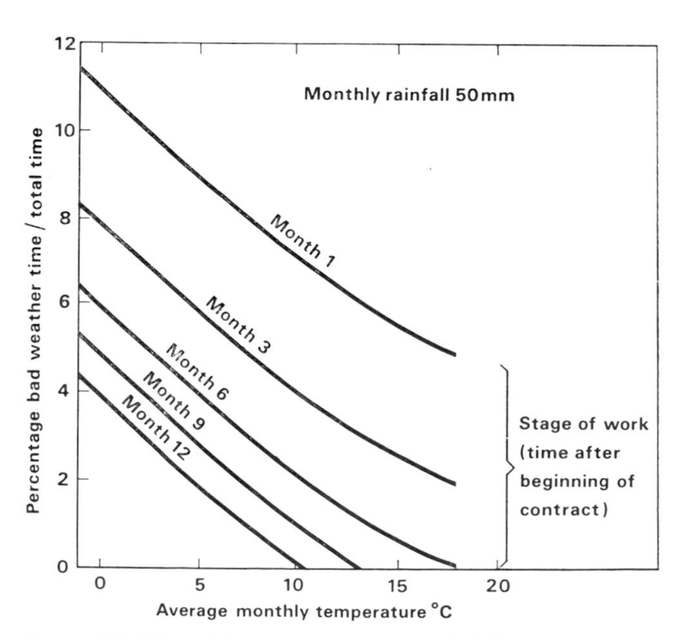

Figure 145 Effect of time, temperature and rainfall on percentage bad-weather time. The figure shows average losses for a monthly rainfall of 50 mm. For each change of rainfall by 25 mm, the percentage bad-weather time increases or decreases by 0·7 per cent. (From Clapp 1966)

It may be mentioned here that the Meteorological Office has started a service for constructors which is called CLIMEST (see Appendix 1, page 171). The aim is to provide, quickly, climatological information on average rain, temperature and wind in forms suitable for long-term planning of work. This is in addition to the usual forecasts of the weather to be expected in the next few hours or days, which are provided on demand (or by prior arrangement) by forecast offices. The various forecasting services are listed in DOE Advisory Leaflet 40, the text of which is reprinted as Appendix 2 (page 173).

Delays to work caused by wet weather

Wet weather may interfere with many different construction processes, indeed it is thought that in Britain more delay is caused by rain than by any other climatic element. For example, painting may have to be stopped whenever there is rain, whatever its intensity. The proportion of the time for which it is raining at a rate of 0·1 mm/h or more varies from under 5 per cent in the driest parts of south-east England to over 15 per cent in the wettest parts of Scotland (Figure 146). However, surfaces may remain wet for much longer times than are indicated by this map (see page 165).

Many outdoor jobs may be stopped when it is raining, merely because it is too unpleasant to continue working, whether or not there is any actual physical interference with the particular process. It is not possible to specify an exact rate of rainfall at which workers will stop working. The rate may depend on the kind of work being done and perhaps on the state of the ground at the site, but may also vary with other meteorological factors, especially with temperature and wind (Figure 145). It will be expected that low temperatures and high winds will increase the sensitivity to rain, so that work will stop at a lower rate of rainfall. There may also be non-meteorological parameters which have a significant effect, financial ones such as the methods of payment, and psychological ones which may be denoted by the general morale on the site.

A rate of rainfall of 0·5 mm in an hour has been suggested as a value which may be significant, perhaps because it is a statistic which can readily be obtained from existing rainfall records. There is no other logical justification for this value. Its artificiality is emphasized by the fact that the 0·5 mm of rain can fall continuously throughout the hour, in which case it has a true rate of 0·5 mm/h, or it may fall in one or more showers of short duration, giving at least 0·5 mm of rain altogether in the hour. The 0·5 mm might fall in 5 minutes or less, so that the rate of rainfall is in fact 5 mm/h or more. Clearly these different kinds of rain-spell could be expected to have different effects on the behaviour of workpeople.

However, as a first step in giving guidance on this problem, hours in which at least 0·5 mm of rain has fallen (wet hours) have been taken to be those in which work is likely to be stopped. Statistics of the average number of such hours each month, between the hours of 07 00 and 17 00 GMT, have been prepared for some 30 stations in Britain. These numbers have been converted to percentages of the 10-hour working day and are shown in Table 60. Thus each figure in the Table is the percentage of wet hours by day, between 07 00 and 17 00 GMT.

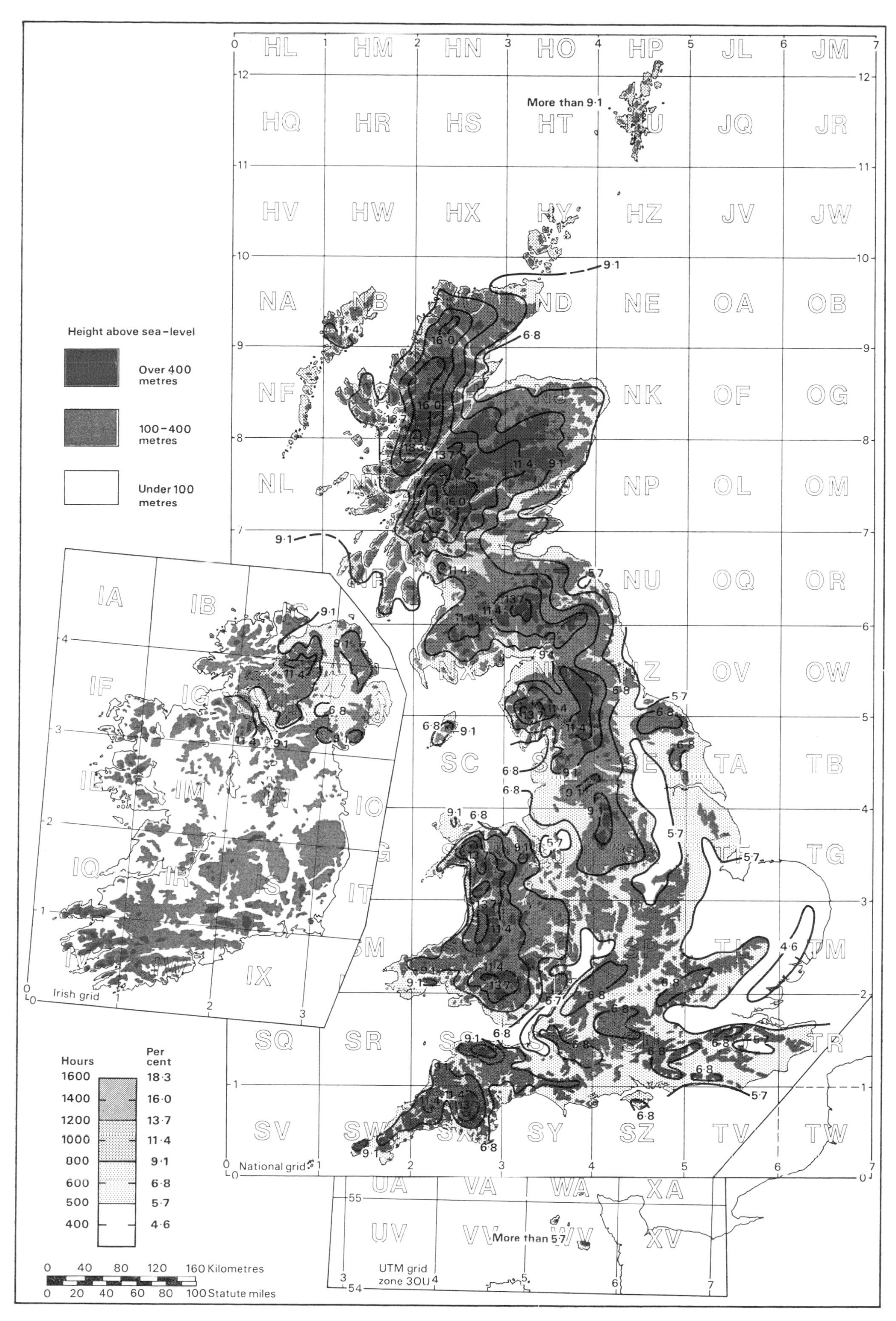

Figure 146 Average annual duration of rainfall in the British Isles as a percentage of total time (1941–70)

Table 60 Monthly mean of wet hours by day at 28 places, expressed as a percentage of the working day (defined as 07 00 to 17 00 GMT), 1931–60

Wet hours are those with 0.5 mm or more of rainfall.

Place	Annual rainfall mm	Annual wet hours %	Monthly percentages – wet hours											
			Jan	Feb	Mar	April	May	June	July	Aug	Sep	Oct	Nov	Dec
Aberdeen	854	5·8	6·5	5·0	3·9	3·3	4·5	4·3	7·8	7·4	5·3	7·1	5·7	8·1
Aldergrove	846	5·8	7·1	5·0	3·5	4·0	5·2	4·7	6·8	6·8	6·3	4·5	5·0	8·7
Birmingham	764	4·7	5·5	5·3	3·9	4·3	3·9	3·0	5·2	5·5	5·0	4·5	5·3	5·2
Bradford	840	5·5	6·1	5·3	4·2	4·0	4·5	4·3	6·5	7·7	5·7	5·5	5·7	6·1
Cambridge	558	3·8	4·2	4·2	2·3	2·0	3·2	3·3	4·8	4·2	3·7	4·5	3·7	5·8
Cardiff	1065	5·8	7·1	5·3	4·2	5·3	3·2	3·0	5·8	6·8	5·0	8·4	7·0	8·1
Cranwell	600	3·9	5·5	5·0	2·9	3·0	3·2	3·0	4·5	4·8	3·0	3·5	4·0	4·8
Douglas, IOM	1139	5·5	7·1	6·7	3·5	4·7	3·2	3·7	3·9	5·5	6·3	6·1	7·7	8·1
Durham	658	3·9	4·8	4·6	1·9	3·0	2·6	3·7	4·8	5·2	3·7	4·2	4·3	4·5
Edinburgh	677	4·7	4·8	4·6	2·9	3·7	3·6	3·7	6·5	6·1	4·7	4·2	5·0	6·5
Exmouth	796	5·0	6·1	6·0	3·9	4·3	3·5	3·7	3·2	5·2	4·0	5·5	7·3	7·7
Glasgow (Renfrew)	995	6·6	8·7	7·4	4·2	4·3	5·2	4·3	6·5	6·5	7·7	7·4	7·0	10·3
Gt Yarmouth	604	4·7	5·5	5·3	2·9	3·0	3·9	4·0	4·8	4·8	4·3	5·8	4·7	6·8
Holyhead	863	5·3	7·1	6·4	2·9	4·0	4·2	4·0	3·2	4·8	6·0	6·1	7·3	7·1
Ilfracombe	1025	5·6	6·8	6·0	2·6	5·3	4·2	3·3	5·2	4·8	6·3	7·1	7·7	7·7
Lerwick	1012	7·1	10·0	7·9	5·5	5·3	4·5	4·3	4·2	5·2	8·3	9·7	11·0	9·7
Liverpool	723	5·2	6·1	5·3	3·2	5·0	4·2	4·3	5·8	6·5	5·7	4·8	5·0	6·8
London (Kew)	594	4·0	3·9	3·9	2·6	2·7	2·9	3·3	4·8	4·2	3·7	5·2	5·0	5·8
Nairn	618	4·5	4·8	4·6	2·9	2·0	4·5	4·3	6·9	4·8	4·3	5·5	4·0	4·8
Oxford	661	4·3	5·8	3·5	2·3	4·0	2·9	3·7	4·8	4·8	3·7	5·2	4·3	6·8
Pembroke	1160	6·3	9·0	5·3	4·8	5·3	4·5	5·0	4·8	5·8	7·0	8·1	8·7	8·4
Plymouth	951	5·8	7·4	6·0	4·5	5·7	4·2	3·3	4·8	5·2	4·7	6·1	8·7	9·4
Scarborough	648	4·5	6·1	5·3	2·6	3·7	2·9	4·0	5·2	6·1	3·3	4·2	5·7	5·5
Scilly	845	6·1	8·7	6·7	5·2	5·3	3·9	3·7	4·5	5·2	5·3	7·1	7·7	10·0
Southampton	804	5·2	7·4	5·0	2·9	3·7	3·5	3·7	3·2	5·2	5·0	6·1	8·3	8·4
Stornoway	1060	7·6	11·3	8·9	5·2	5·7	4·8	4·3	6·8	7·1	5·0	10·0	10·3	12·0
Tiree	1128	7·7	11·3	7·8	6·9	4·7	4·2	4·3	6·5	6·9	9·7	8·7	9·7	12·0
Wick	775	5·8	8·4	6·0	4·5	4·0	4·5	3·7	5·2	7·1	5·0	6·1	6·7	8·1

(Meteorological Office data (*Monthly Weather Survey*) 1931–60)

Table 61 Relation between annual average rainfall and annual average number of wet hours (0·5 mm of rain or more in an hour) during the working day (defined as 07 00 to 17 00 GMT)

Annual average rainfall mm	Annual average number of wet hours % of time
500	3·6
625	4·3
750	4·9
875	5·6
1000	6·3
1125	7·0
1250	7·6
1500	8·9
1750	10·2
2000	11·5
2500	14·1

A comparison of these figures with those of mean rainfall at the same places shows that the percentage of wet hours by day at a place in any given month, or over the whole year, is approximately proportional to the corresponding average monthly or annual rainfall. Although the values at individual places may differ by some 15 per cent or more from the average value, the figures given in Table 61 will probably be sufficiently accurate for operational planning, especially in view of the difficulty of specifying any definite criteria. The values in the Table for rainfalls of 1250 mm a year and more are extrapolated, and likely to be less trustworthy than the others, but will serve to indicate likely amounts of wet hours.

Figure 147 shows how the annual average percentage of wet hours varies over the country. This map is in fact one of average annual rainfall, for the period 1916–50, suitably rescaled. Clearly a map to this scale cannot show all local variations, and it may be necessary to estimate figures for a particular place from local rainfall data, with the aid of Tables 60 and 61. Table 60 shows that the percentage of wet

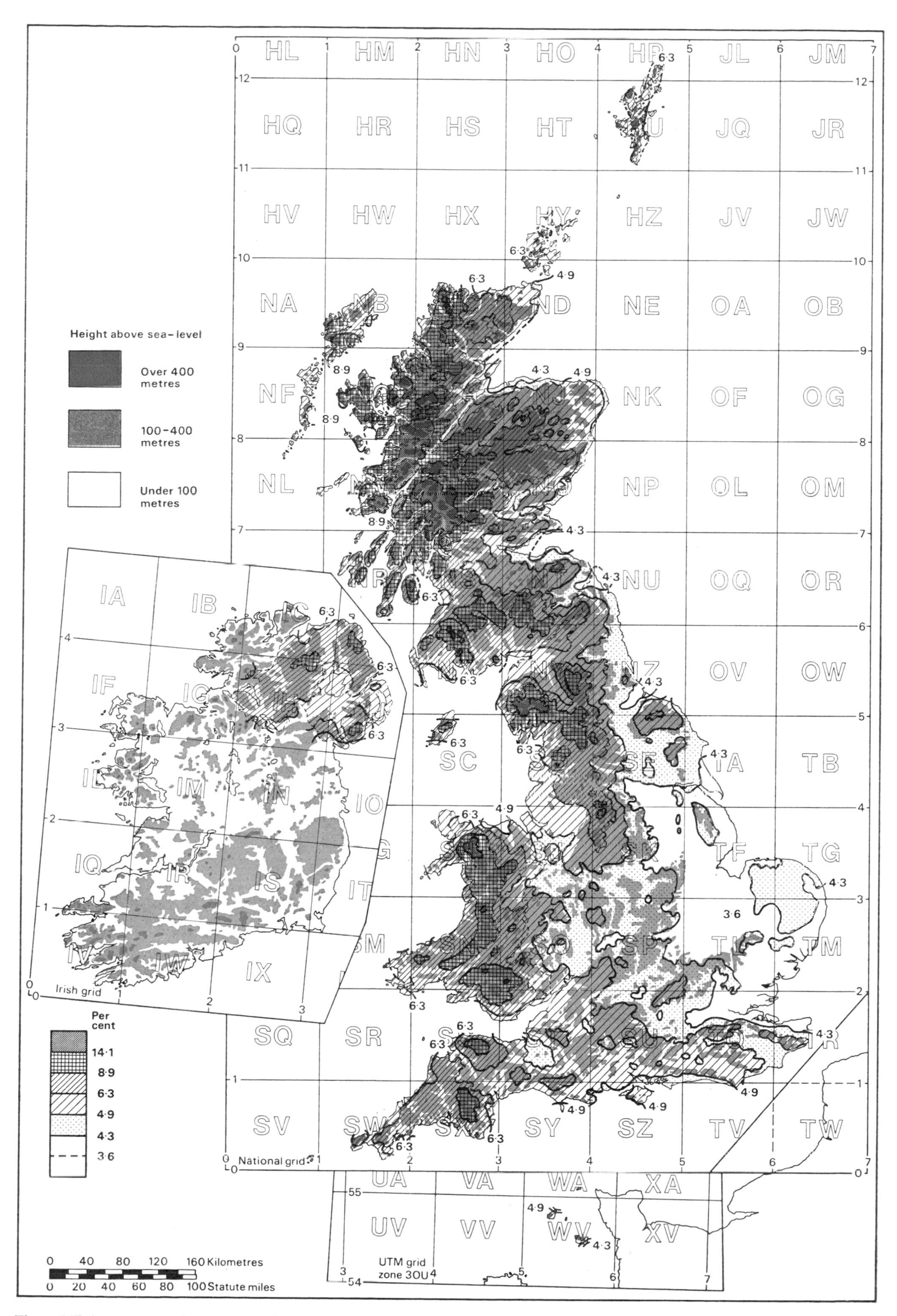

Figure 147 Average annual percentage of wet hours by day in the United Kingdom, 07 00–17 00 GMT

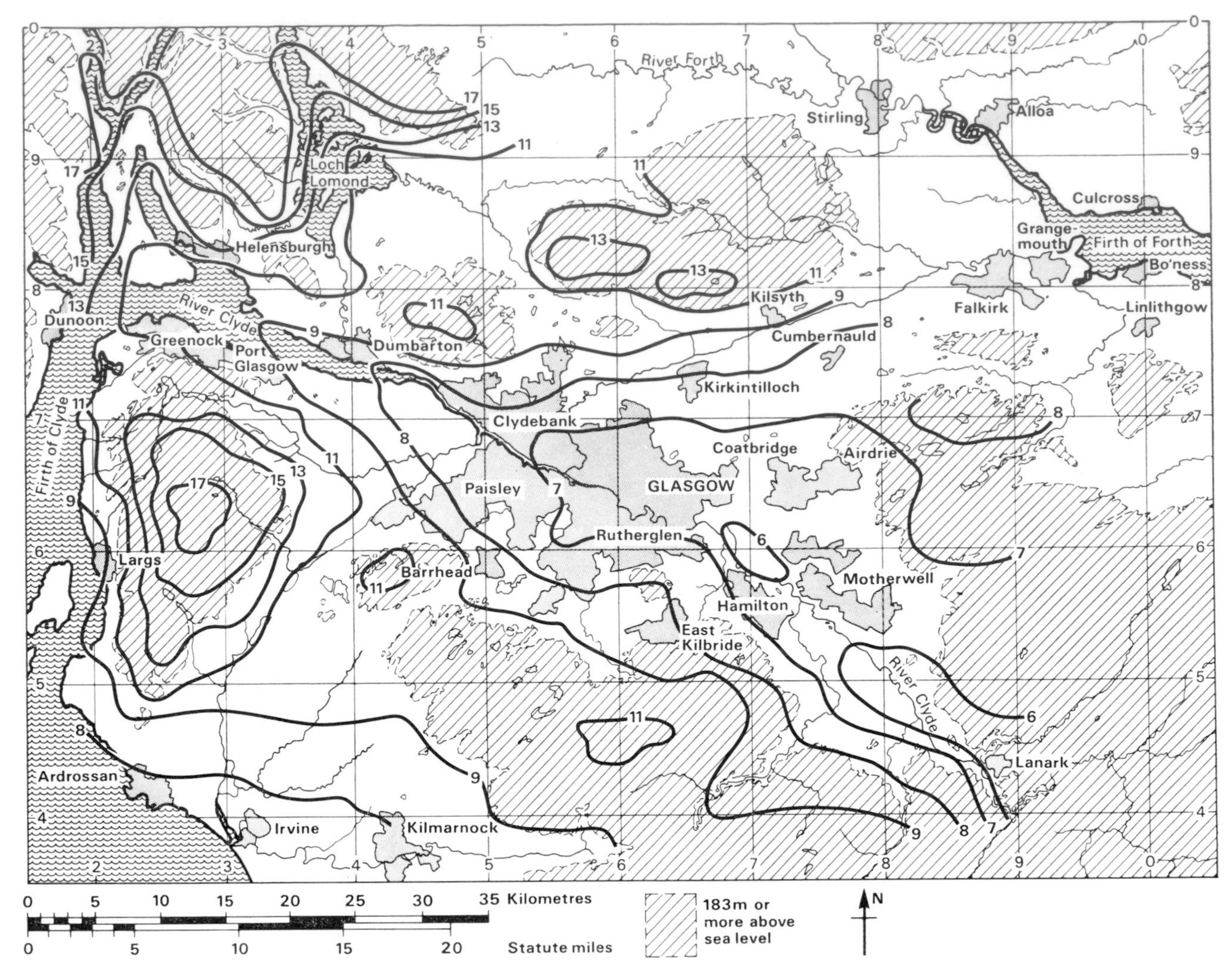

Figure 148 Average annual percentage of wet hours by day in the Glasgow region. Percentage of hours between 07 00 and 17 00 GMT during which 0·5 mm or more rain falls

hours by day is generally greatest in early winter (November–December) and least in spring and early summer (March to June), but there are appreciable differences in the pattern of the variation between one place and another. Assuming that the patterns are typical of regions, the monthly percentages of wet hours can be estimated from Table 60 for other places, after making suitable adjustments for differences in rainfall.

Figure 148 is an example of differences within a region and shows how the annual average percentage of wet hours varies over the Glasgow area. In the inhabited areas there is a range of some two to one, while there are greater values still on the higher hills. Like Figure 147, this is based on rainfall maps, with a relationship similar to that in Table 61, but based on rainfall data from Renfrew Airport alone, so giving rather longer durations of wet time than those shown in Table 61.

There are no records to show whether in fact the interruptions to work vary in the same ratio, or whether workpeople accustomed to more severe weather will go on working in conditions that will cause those used to a more genial climate to stop. However, it is probably reasonable to assume that interruptions to a given type of job will be proportional to the percentages of wet hours shown in these Tables and Figures. Some confirmation seems to be afforded by the fact that the 'wet hours' and temperature data for southern England agree approximately with the curves in Figure 145.

Effect of snow on building operations

Snow may hinder building operations in a number of ways. Falling snow causes discomfort to workmen in the open, although usually not to quite the same extent as rain. Moreover snow does not fall at as high an intensity as rain. Of course it is more of a nuisance in that it covers materials and equipment, and may partly melt and then refreeze, so cementing together individual items. These effects are probably minor compared with the effects of snow on transport. On the site itself the going may be made difficult, while the transport of men and materials to and from the site may become difficult or even impossible.

The frequency of snowfalls sufficient to hinder transport has already been considered in Chapter 2 (page 74). Figure 149 is a map of the Glasgow region, with superimposed on it lines showing the frequency, in days per year, of mornings with snow lying to such an extent that half or more of the ground was covered, during the years 1912–38. Some individual spot readings are average values for the period 1954–63 (taken from the *Monthly Weather Report*). Most of these are higher than the older values and presumably reflect the higher frequency of cold winters in that period. However snowfall is so variable from year to year, and from place to place, that no detailed comparison can be made. In the period 1954–63, on about one morning in three of those when snow was lying, the depth of the snow was 7·5 cm or more at these Clyde Valley stations (Figure 150).

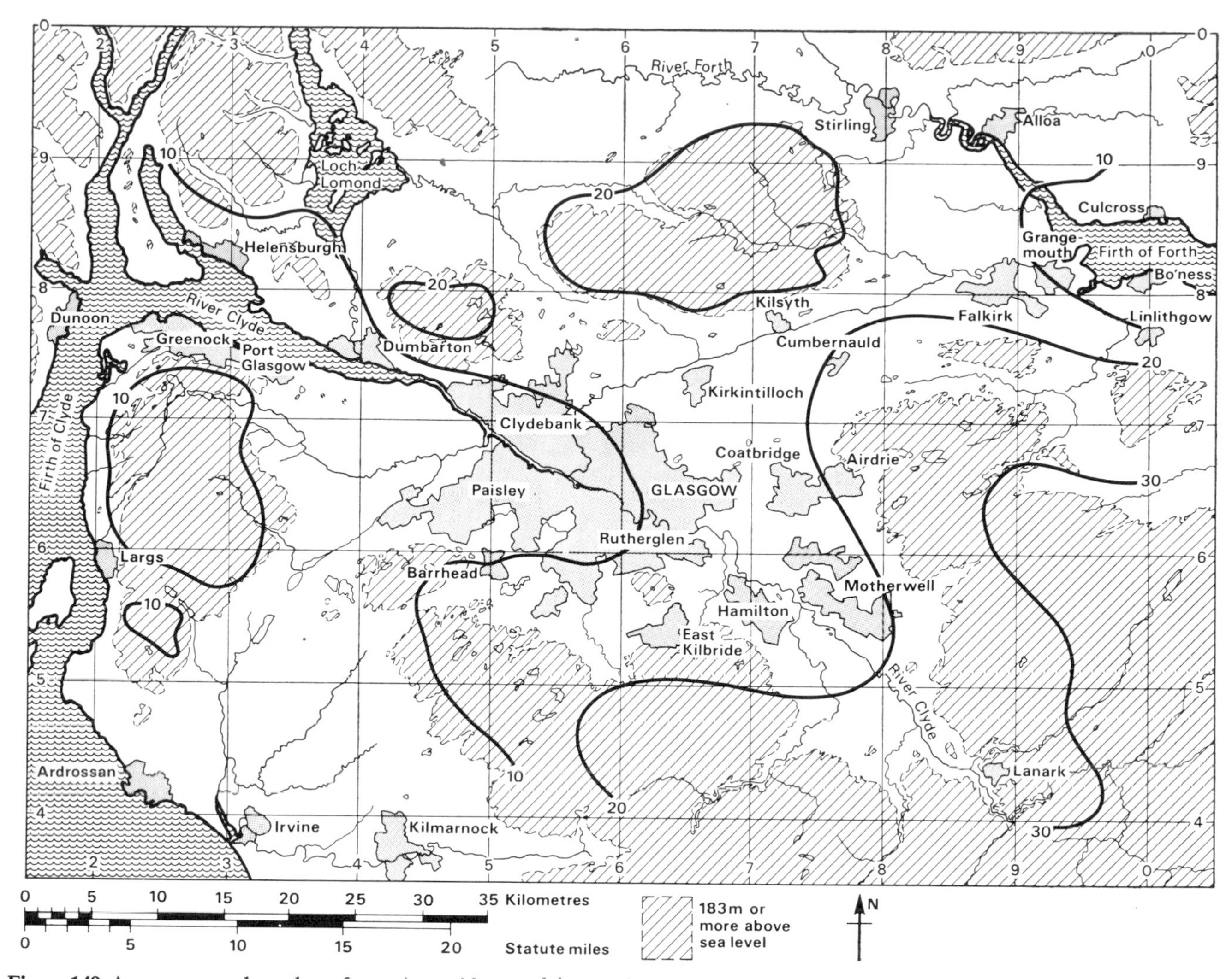

Figure 149 Average annual number of mornings with snow lying at 09 00 GMT in the Glasgow region. Isopleths 1912–38 mean.

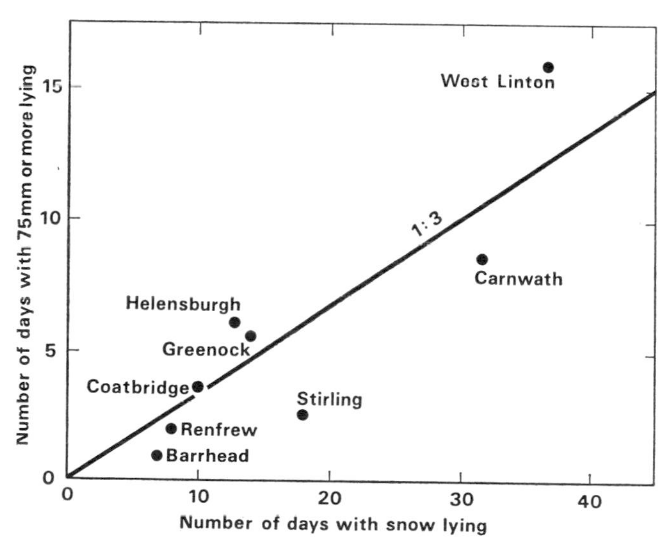

Figure 150 Relation between annual number of mornings when half or more of the surrounding country was snow-covered, and the number of days when the depth was 75 mm or more, Central Scotland, 1954–63

Effect of wind on operation of cranes

According to BS 2573, cranes are designed to be usable as long as the wind pressure on them does not exceed about 240 N/m². This is said to correspond to a wind speed of 17·2 m/s (note that the coefficient in the relation $p = cV^2$ is not the same as that used for calculating pressures on windows). When the wind speed exceeds 17·2 m/s the crane must be taken out of service and the jib allowed to swing freely. The British Standard does not make it clear that a gust speed is meant, but this must be so, for a mean speed (averaged over, say, 1 minute or more) is meaningless unless a gustiness factor is also given. A crane will be responsive to gusts of quite short duration, possibly of 10 seconds or less. According to the roughness of the surrounding terrain, gusts may be from 30 to 100 per cent higher than the mean speed of the wind.

The Standard goes on to say that with the jib free, the crane is designed to withstand a load of up to 1630 N/m², corresponding to a wind of 44·3 m/s. If this load is exceeded for 10 seconds the crane will overturn.

With some kinds of load, for example lightweight panels or shuttering, the crane, or at least the load, may well become unmanageable at lower speeds, probably at 12 m/s or less (see page 161).

Statistics of the number of hours in each month during which gusts of over 17·2 m/s are measured, are published annually

Table 62 Average number of hours in which there are gusts exceeding 17 m/s during each month at a number of places in Britain. Period 1950–59

Station	Jan	Feb	Mar	April	May	June	July	Aug	Sept	Oct	Nov	Dec	Year
Lerwick	225	167	148	84	57	54	29	36	87	157	172	233	1449
Stornoway	197	132	117	106	56	46	24	47	99	159	140	199	1322
Dyce	53	34	21	15	15	8	4	5	19	25	28	42	269
Bell Rock	163	116	95	70	53	39	14	39	97	130	140	184	1140
Leuchars	40	33	16	18	15	12	6	17	19	26	33	64	299
Edinburgh	65	53	34	35	15	19	7	18	42	48	51	93	480
Tiree	198	127	119	69	35	29	22	46	100	169	149	259	1322
Paisley	49	30	22	24	17	12	6	15	24	31	36	72	335
Renfrew	55	36	30	28	24	15	8	17	29	36	36	81	395
Prestwick	77	37	23	20	16	8	7	15	41	42	55	105	446
Eskdalemuir	86	54	43	48	42	23	15	27	54	57	52	102	603
Point of Ayre	101	53	37	33	18	14	14	23	65	66	75	127	626
Durham	65	43	31	35	22	14	11	11	33	40	37	78	420
South Shields	58	48	34	25	15	14	9	12	25	34	30	55	359
Cranwell	59	40	40	32	21	11	17	17	41	36	35	69	418
Gorleston	68	36	40	14	17	4	14	9	12	34	40	48	340
Mildenhall	38	28	26	17	15	4	9	8	21	20	22	45	253
Felixstowe	52	34	36	31	24	13	28	17	37	35	38	58	403
Dunstable	15	9	11	8	6	1	3	2	5	7	10	23	100
Cardington	46	32	34	25	23	7	16	11	20	27	24	63	328
Shoeburyness	63	53	48	33	29	13	25	24	35	34	52	67	476
Edgbaston	29	17	20	15	9	3	3	5	15	10	20	35	181
Keele	39	24	29	11	16	7	9	8	21	11	22	46	243
Kingsway*	34	31	20	25	8	10	7	10	11	8	20	21	205
Hampton	37	19	23	12	16	5	9	7	12	13	18	33	204
Kew	35	27	25	19	18	6	14	10	13	14	18	33	232
Croydon	38	29	26	15	13	4	9	8	13	15	21	35	226
Thorney Island	33	33	22	20	13	7	14	14	22	22	28	41	269
South Farnborough	26	21	19	12	9	1	5	3	9	9	15	28	157
Abingdon	26	27	27	20	13	4	11	8	18	12	13	31	210
Larkhill	52	46	36	32	25	9	15	12	28	17	35	58	365
Sellafield	67	32	27	24	16	19	18	30	69	59	62	83	506
Fleetwood	94	54	40	38	32	30	33	33	83	74	69	107	687
Southport	75	42	32	50	34	16	32	26	89	56	62	82	596
Speke	69	42	38	41	35	19	17	27	64	47	47	85	531
Bidston	107	56	50	60	41	32	44	38	90	78	74	127	797
Ringway	55	43	38	25	24	8	11	16	39	24	38	63	384
Valley	101	60	37	28	34	26	25	31	73	87	73	133	708
Aberporth	149	88	92	46	42	24	27	27	66	90	82	162	895
Plymouth Hoe	57	46	56	25	23	13	10	8	27	26	50	60	401
Lizard	167	115	102	60	42	24	25	26	59	67	123	199	1009
Scilly	192	160	130	91	50	32	26	33	70	81	149	201	1215
Aldergrove	63	35	35	34	21	16	10	21	44	36	45	77	437

*Period 1945–54

Meteorological Office 1968a

by the Meteorological Office, in the Annual Summary to the *Monthly Weather Report*. Table 62 is based on a 10-year average summary of these data, and shows how the average annual and monthly number of hours with gusts of over 17·2 m/s varies over the country. It is convenient for planning to have the data expressed as the percentage of hours, but this could be misleading, for it is likely that there are rather more hours with high gusts in the daytime than at night, since near the ground the wind speed is usually highest in the middle of the day. A further complication is that cranes close to a partly-completed building may be subjected to higher gust speeds than if they were at some distance from the building. The building can induce small, but violent eddies. For example a gust of over 31 m/s has been measured 1·5 m above the top of a building in Glasgow, some 30 m high, when the highest gust 12 m above ground at Renfrew Airport (just outside the city) was about 22 m/s.

The Meteorological Office statistics of gusts are taken from anemographs, the standard height for which is 10 m above ground. However, local conditions can and do require that instruments be placed at greater heights. Care should be taken to ensure that any data used include knowledge of the effective height of the anemograph. In open country the speed of the gusts increases with height, the variation with height in strong winds following approximately equation (27), that is:

$$V_{g1}/V_{g2} = (h_1/h_2)^{0.085}$$

where V_{g1} and V_{g2} are the speeds of the gusts at heights h_1 and h_2 (see also Figure 29). The increase of gust speed with height may be somewhat greater *over* the general roof-top level in towns, but as far as is known it is not likely to be very different from that over open country.

Because the speed of gusts increases with height above the ground, it is obvious that the number of gusts with a speed of 17·2 m/s or more must also increase with height. At present there are no data on this subject, but it is perhaps not unreasonable to suppose that the number of gusts of a given speed increases in about the same proportion as the speed of the gusts increases. The same argument may be expected to apply to gusts of lower speed. Thus if a particular load can be handled in a wind gusting up to 12 m/s, it may be that on certain days the load can be dealt with satisfactorily at ground-level, but that at a height of 30 m or more it becomes unmanageable, because at the greater height there are many more gusts of and even exceeding 12 m/s. Figure 29 also shows the variation with height of the hourly mean wind speed over unobstructed, smooth country.

As would be expected, strong gusts are more frequent in winter than in the middle of the year. Table 62 shows the number of hours with gusts of 17·2 m/s or more at a number of places, during each month. In some places there may be on average ten times as many hours with strong gusts in a winter month as in a summer month.

Generally speaking, the frequency of strong gusts is greatest in western and northern Britain, and is greater on and near coasts than inland. But the frequency also increases with altitude above sea-level in a given region. Figure 151 shows that in the Glasgow-Edinburgh region there is a general increase in the frequency of days with gale as the altitude increases. A day with gale is one with a ten-minute period in which the mean wind speed exceeds 17·2 m/s. There is a considerable variation in the frequency from one place to another, because of local differences of shelter, but the general trend is clear. For places without wind measuring equipment, visual estimates of gales are the only measures of strong

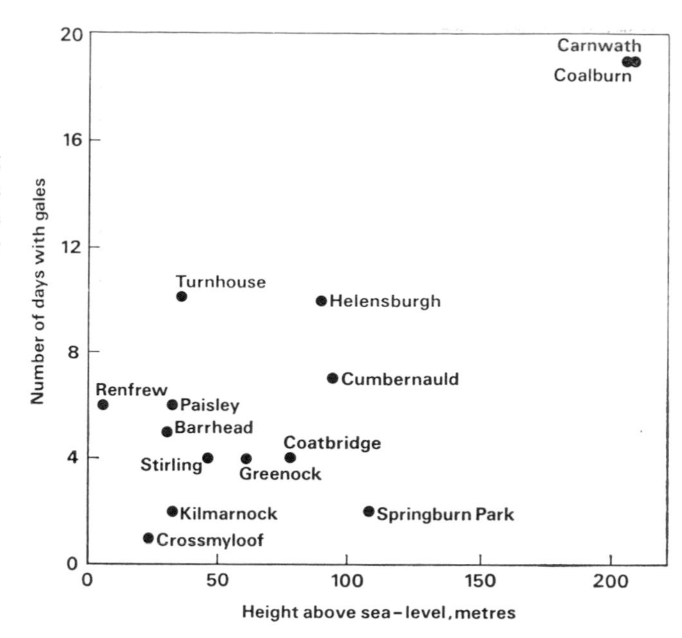

Figure 151 Annual number of days in the Glasgow region with gale-force winds, ie, 17·2 m/s for at least 10 minutes continuously. Means for period 1954–67

winds that are available, but it is thought that the estimated numbers of days with gales will be approximately proportional to the number of hours with gusts of 17·2 m/s or more.

It is not to be expected that there will be a simple correlation between the data given in this Section and the actual loss of time on a job. Most of the wind measurements are made on fairly open level sites, while most building is carried out on sites with rather rough surroundings, so that the wind is likely to be more turbulent. Usually the mean wind speed will be rather less than at an airfield outside a town, but the gustiness caused by buildings may mean that maximum gust speeds are little lower than those at the airfield. In a proportion of the hours included in the statistics there will be only one gust of more than 17·2 m/s, but it is likely that most of these include many gusts approaching the limiting value. Thus even a recording of a single strong gust is usually a warning that the wind conditions are nearing the critical state. An exception to this statement is afforded by the gusty winds occurring in the vicinity of thunderstorms or, sometimes, cold fronts. In these conditions a strong, squally wind can occur for a few minutes, with quite moderate winds before and after.

Providing that due account is taken of local degree of shelter, it is probable that the statistics will give a useful indication of the likely frequency of interruption of work when cranes are being used. Taken in conjunction with past experience at sites whose gust frequency can be estimated, it should be possible to estimate the likely frequency at a new site quite accurately from the data given here.

Reliable statistics of time lost because of wind (or of any other climatic element) are hard to obtain, but it has been reported that at one site in Liverpool, lost time amounted to 33 per cent during one winter. It is believed that most of this was due to wind, because of difficulties in handling rather light shuttering for concrete. At another site in south Wales it was necessary to cut down the size of the shuttering to half the normal in order to reduce the loss of time on windy days. Even at ground-level there may be problems at exposed sites (Figure 152).

Figure 152
Difficulties in erecting a structure during a strong wind. Men struggling to hold partitions upright until they can be permanently secured. The strength of the wind is shown by the antics of the tarpaulin cover, and the local variation in direction by the flow of the smoke and of the men's hair

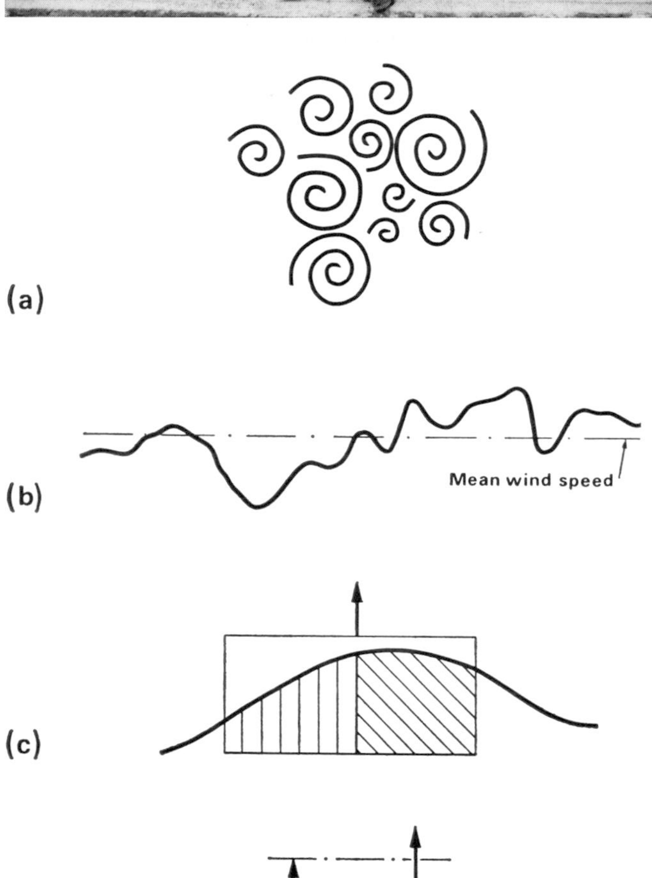

Figure 153 Wind forces on a suspended panel. In a gusty wind, the pressures on the two halves of a vertical panel will often differ, so that there is a resultant turning moment on the panel about the middle
(*a*) Schematic sketch of gusts existing at any instant
(*b*) Instantaneous wind speeds on a line perpendicular to the mean wind direction
(*c*) Wind pressure on a rectangular panel, held perpendicular to the mean wind direction
(*d*) Wind forces on such a panel (viewed from above)
(*e*) Forces on a panel nearly aligned with the wind direction

Wind effects on freely suspended panels

Wind speed at any point is continually varying, because eddies are formed by mechanical stirring of the air or by thermal 'bubbling'. These eddies are of a variety of sizes and strengths (Figure 153). Correspondingly, the speeds and directions measured simultaneously at a number of points may all be different.

Figure 153(b) shows a typical cross-section along a line perpendicular to the mean wind direction, with the wind speed at each point at a given instant. The pressure at any point on a flat surface normal to the wind is proportional to the square of the wind speed, and the distribution of pressure on the horizontal centre-line of a flat rectangular panel is sketched in Figure 153(c). The mean pressure on each half of the panel is proportional to the shaded areas under the curve. Since in this example the total force on the right-hand half is greater than that on the left-hand half [Figure 153(d)], there is a resultant turning moment about the middle, and if freely suspended the panel will tend to twist.

In the following calculations the effect of the wind on a panel 5 m long by 2·5 m high, suspended freely from the middle of the top edge, is considered (Figure 154). If the wind were perfectly uniform and steady, with no gusting, the force on the panel when perpendicular to the wind may be assumed to be that due to the dynamic pressure head of the wind (see page 89), and is thus given by

$$F = 5 \times 2{\cdot}5 \times 0{\cdot}613\, \bar{u}^2 \text{ newton}$$

where $\bar{u}$ is the wind speed (m/s) normal to the panel. The force at each end of the panel necessary to restrain it (assumed to be half the wind force acting on the panel) is shown in Table 63 and by curve A in Figure 154.

A man can exert a horizontal force equivalent to about two-thirds of his weight. Thus an average man weighing 80 kg can exert a force of about 500 N, so that in a perfectly steady wind two men could just resist the force due to a wind of about 11·5 m/s on this particular panel.

In fact the natural wind is usually turbulent, and consists of a succession of gusts and lulls with associated fluctuations in wind direction. These changes in speed and direction are very irregular and are made up of oscillations with periods ranging from a fraction of a second to many minutes, with eddies with which they are associated ranging in size from very small to very large. Because of these non-uniformities, a turning or

Table 63 **Restraining forces (in Newtons) required at ends of a panel 5 m long by 2·5 m wide when it is exposed to the wind**

Wind speed m/s	90° to uniform wind	90° to non-uniform wind F_1	90° to non-uniform wind F_2	5° to wind F_1	5° to wind F_2	10° to wind F_1	10° to wind F_2
5	100	110	130	25	75	50	150
10	380	430	510	100	300	200	600
15	860	960	1150	225	675	450	1350
20	1530	1700	2040	400	1200	800	2400
Curve in Figure 154	A	B	C	D	E	F	G

twisting moment may be produced on a suspended panel, so that it will tend to twist about the suspension cable. This may be brought about in two ways, thus:

(a) the pressure on the panel is not uniform if the speed of the wind is non-uniform, for example in the vicinity of existing buildings.

(b) the pressure on the panel is not uniform if the wind is turbulent.

In addition, even if the wind is uniform,

(c) if the panel is almost lined up with the wind a 'lift' force acts so as to rotate the panel.

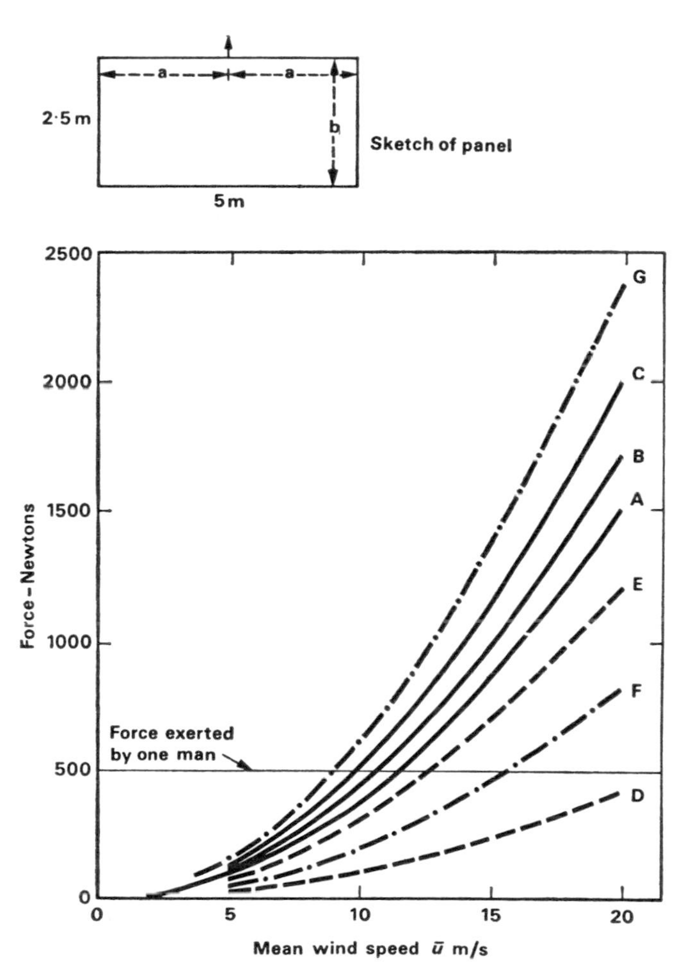

Figure 154 Forces needed to restrain in wind a panel 5 x 2·5 m freely suspended at mid-point of top edge. It is assumed that the panel is held by a man at each end. If there is a turning moment the force required is greater at one end than the other.
A: Panel perpendicular to wind, equal force each end
B, C: Panel perpendicular to wind, steady non-uniform wind
D, E: 5° to wind direction
F, G: Panel 10° to wind direction

Forces on a suspended panel in a non-uniform wind

If there are obstructions up-stream of a construction site, so that the wind is deflected from its unobstructed course, there will usually be variations in the speed of the wind from one point to another. For example, the air will be speeded up round the corner of an existing building, in such a way that the speed a metre or so away from the corner may be twice as high as that 10 or 20 metres away (see page 63). It follows that a panel supported across the general flow of the wind will in such a place experience a stronger wind at one end than at the other.

If this panel is the same size as that already considered, 5 m long by 2·5 m high, and if the gradient of wind speed is such that the mean speed on one half of the panel is $\bar{u}$ m/s, and on the other half 1·2 $\bar{u}$ m/s, the restraining forces at the two ends will be approximately as shown in Table 63 and by curves B and C in Figure 154. Because of the turning moment induced by the non-uniform wind field, the restraining force needed at one end is some 20 per cent higher than that needed at the other.

Forces on a suspended panel in a turbulent wind

The calculations in the preceding paragraph treated the wind as though it was perfectly steady at any point. In fact, as explained on page 162 (and Figure 153), the wind is usually turbulent, so that even though the mean speed is constant, the instantaneous speed is continually varying. Because of this, the forces on a panel held transverse to the mean wind are continually changing. It is probable that the mean force on one half of the panel is usually different from that on the other half, so that there is a moment tending to turn the panel one way or the other.

If at any instant the gust speeds were such that the mean speed over half the panel was 20 per cent greater than over the other half, the restraining forces would be the same as those for a non-uniform steady wind as described above. In very turbulent conditions the differences between the speeds acting on the two halves may be much greater. However, it is likely that, because the gust forces acting at any point are so short-lived, the inertia of the panels does not permit turning movements to build up to a significant degree.

Forces on a suspended panel aligned with the wind

If a panel is aligned precisely parallel to a perfectly uniform airstream, there will be a drag-force upon it, but no turning forces. In practice, the panel will rarely if ever be exactly parallel to the wind flow, if only because the direction of the wind is continually fluctuating. When the plate is deflected from the mean direction of airflow by a small angle α, the resulting disturbance in the airflow causes a force L to arise, which acts perpendicularly to the direction of airflow

[Figure 153(e)]. Depending on the angle α, the restraining forces required to stop the panel turning are comparable with those required with a panel normal to the wind (Figure 154).

Assume that the panel acts like a section of a flat plate of infinite aspect ratio, and chord length 2a. For a thin flat plate the centre of pressure will be at the quarter-chord point. The turning moment about the point of suspension will be

$$M = L \times \tfrac{a}{2} \times \cos \alpha.$$

$$\text{Now } L = C_L \times \tfrac{1}{2}\rho\, \bar{u}^2 \times 2a \times b$$

where b is the width of the plate. From aerofoil theory

$$\frac{dC_L}{d\alpha} = 2\pi \text{ (for small } \alpha).$$

Hence $C_L = 2\pi\alpha$

and $M = 2\pi\alpha \times \tfrac{1}{2}\rho\, \bar{u}^2 \times 2ab \times \tfrac{a}{2} \cos \alpha,$

so that, for small values of α,

$$M = \pi a^2 b \times \rho\, \bar{u}^2 \times \alpha.$$

Now in the present example, a = 2·5 m, b = 2·5 m, and ρ, the density of air, = 1·2 kg/m³.

Therefore, M is approximately equal to $\bar{u}^2\alpha$ N m, when $\bar{u}$ is in metres per second, and the angle α in degrees.

The restraining forces to be applied at the ends of the panel to stop it turning are F_1 and F_2 [Figure 153(e)]:

$$F_2 \times 2a = L \times \tfrac{3a}{2}; \quad F_1 \times 2a = L \times \tfrac{a}{2}$$

so that

$$F_2 = \tfrac{3}{4} L \text{ and } F_1 = \tfrac{1}{4} L$$

and $L = M/\tfrac{a}{2} \simeq \tfrac{4}{5}\bar{u}^2 \alpha,$

therefore $F_1 = \tfrac{1}{5}\bar{u}^2 \alpha$ ....(39)

and $F_2 = \tfrac{3}{5}\bar{u}^2 \alpha$ ....(40)

Values of the forces F_1 and F_2 at different wind speeds and for two angles, 5° and 10° to the wind direction, are given in Table 63. Because of the large turning moments, the force at the up-stream end is three times that at the down-stream end of the panel.

The force L is analogous to the lift-force experienced by a nearly-horizontal flat plate or by an aircraft wing. It increases rapidly at first as the angle of deflection α increases, so that a suspended plate or panel lined up with the wind is in unstable equilibrium. However, at angles above about 10° the deflecting force L decreases rapidly again, as the plate 'stalls'.

Observations of building operations roughly confirm these estimates, although no exact measurements are available. At the same time it was noted that with experience workers can allow a panel to swing enough to 'spill the wind' and thus reduce the force of the wind on the panel. However this is obviously made more difficult when the direction of the wind is also fluctuating, which is usually the case.

Interference with plastering and painting by condensation and high humidity

Painting on damp surfaces is often risky, although some paints are claimed to be formulated to permit their use in such circumstances. A surface may be damp because the material itself is wet (or because moisture is moving through the material), or it may become wetted by condensation of atmospheric water vapour upon it. Steel is commonly affected by the latter, plaster and masonry surfaces by both processes, which are often difficult to distinguish. Once plaster has set it should be helped to dry out so that it can be decorated as soon as possible. The drying out of new buildings is assisted more by ventilating than by merely heating, air movement being helpful even at high humidities (except during periods of rapidly rising dew-point, as explained below).

A surface becomes damp when its surface temperature falls below the dew-point of the surrounding air, so that moisture in the air condenses on the surface. This may occur for two reasons. Most commonly, in the case of surfaces out of doors, the surface becomes chilled at night when the sky is clear. The surface loses heat by radiation to the cold sky, and its temperature falls below that of the air, and even below the dew-point of the air, so that atmospheric moisture condenses on to it. Lightweight materials such as thin panels (especially if there is good insulation behind the surface), and metal structures have a low thermal capacity and in consequence their temperature can fall rapidly.

By dawn, therefore, such materials may be covered with a film of moisture (dew). If, as often happens on clear nights, mist or fog forms, the atmosphere in contact with the surface will be saturated and may remain in this condition for some hours after sunrise, in fact until the fog disperses. As long as the air remains saturated, the moisture on the surfaces cannot evaporate and they will remain wet (see also page 165).

Surfaces inside a building are unlikely to be affected in this way, even when unheated and with the windows unglazed, for the thermal capacity of the materials will usually be high enough to maintain their surface temperatures above the atmospheric dew-point. Furthermore they are sheltered from the cold sky and cannot lose heat so rapidly as exposed surfaces. However, high humidities outside will restrict the rate at which damp surfaces such as fresh plaster can lose their moisture (see page 115 for a note on the rate of loss of water from walls).

Rapid rise in dew-point

Surfaces may also become moistened by condensation of atmospheric moisture when certain changes of weather take place. An unheated building of heavy construction (masonry or concrete) will change in temperature rather slowly, so that at any time its temperature will tend to be only a degree or two above the mean temperature of the air during the previous 24 hours or so, especially if the windows are unglazed. If there is a change of air mass, with the original air being replaced by air which is warmer and more humid, the dew-point of the new air may be appreciably higher than the temperature of the building. Moisture will then condense on all parts of the building which are exposed to this humid air. Since the dew-point of the air may rise to some 5 deg or more above its original value within a few hours, the rate of condensation may be high, and sufficient to ruin recently applied decorations. Figure 20 illustrates the conditions on a day when dew-point rose by some 7 deg C in about 12 hours.

It will be obvious that under these conditions it is of no use to increase the ventilation of a building in an attempt to get rid of the moisture. This will in fact increase the rate of condensation, because it will only make a larger supply of humid air available. It is necessary to heat the building to overcome the risk of condensation from this cause, and the most economical procedure would probably be to arrange for forecasts to be provided whenever the risk occurs, so that heaters can be switched on in good time. The Meteorological Office could arrange for such a forecast to be despatched to the responsible authority whenever it was expected that the atmospheric humidity would increase to such an extent that there would be appreciable risk of condensation in unheated stores.

In Figure 155 statistics of mean air temperature and of atmospheric dew-point are presented in such a way as to demonstrate the likely frequency of the conditions which give rise to this form of condensation. The 24-hour mean air temperature at Garston is compared with the dew-point, measured at the end of the 24-hour period, for each day of the year. The winter month statistics (October–March) are presented separately from the summer ones (April–September). The two histograms represent the frequency with which the atmospheric dew-point at 09 00 GMT can be expected to be below or above the mean air temperature for the preceding 24 hours. If we assume that a surface takes up this 24-hour mean temperature, the diagram shows that on about 20 mornings in an average winter, the dew-point is just equal to the 24-hour mean air temperature, so that condensation will just not occur. On about 28 mornings the dew-point is above the 24-hour temperature, so that condensation is possible. However, on about 23 of these mornings the excess is no more than 2 deg, so that condensation is unlikely unless the building has become exceptionally chilled. On a few mornings the excess of the dew-point over the mean air temperature may be as high as 6 deg, and copious condensation may be expected. Indeed, on such a morning even the wooden rafters in a pitched roof have been found to be so damp as to arouse suspicion of rain penetration.

Although the frequency of such severe condensation is not great, it is worth bearing in mind that, whenever there is a cold spell lasting two or three days or more, it may well be ended by an incursion of warm moist air. Even in the summer half of the year, these conditions may occur, though much less commonly. Usually the only noticeable effect in summer is condensation on exposed cold-water pipes.

These remarks are concerned with the effects of condensation on new construction, but it will be obvious that they may be taken to apply to any unheated building. In an unheated, ventilated store, for example, condensation from this cause may be expected to occur on the stored contents. Unventilated buildings may also be affected, for example opening an unventilated store to warm spring air after a cold winter may have the opposite effect to that intended.

It may be noted that a sunny day following a cold night will not be expected to cause condensation because the warmth of the day arises from the sunshine, and the humidity of the air will remain more or less constant if there is no change of air-mass.

Length of time for which surfaces remain wet

Shellard (1962) has quoted some results obtained with a surface-wetness recorder, which consists of an expanded-polystyrene block, exposed so that it can become wetted by dew or by rain, mechanically coupled to a pen recorder so that in effect the block is continuously weighed. The length of time for which the block was wet was found to be highly correlated (correlation coefficient = 0·97) with the number of hours during which the relative humidity of the air was 90 per cent or more. This instrument was designed to behave, in relation to wetness, in the same way as leaves of crops. It might therefore be expected to behave in this respect in the same way as lightweight building components, and especially thin sheets.

Figure 18 (page 22) shows the proportion of time during the summer for which the relative humidity is 90 per cent or more in various parts of Britain. Figure 19 shows more detailed information for one place only. It demonstrates how the proportion of the time for which the relative humidity is *below* 90 per cent varies, both with time of day and with time of year.

As will be expected, the frequency of high relative humidity is much the lowest in the afternoon, and especially in the summer months. Indeed it appears that in May to August there is only about a 1 in 20 chance of the relative humidity being 90 per cent or more in the afternoon at Heathrow. However the chance of the relative humidity being 90 per cent or more has risen to about 1 in 10 by sunset and about 1 in 4 by midnight (GMT).

In the morning, the 50 per cent line (ie that which represents the evens chance of the relative humidity being below 90 per cent) is roughly parallel to the sunrise line. It is about 1 to

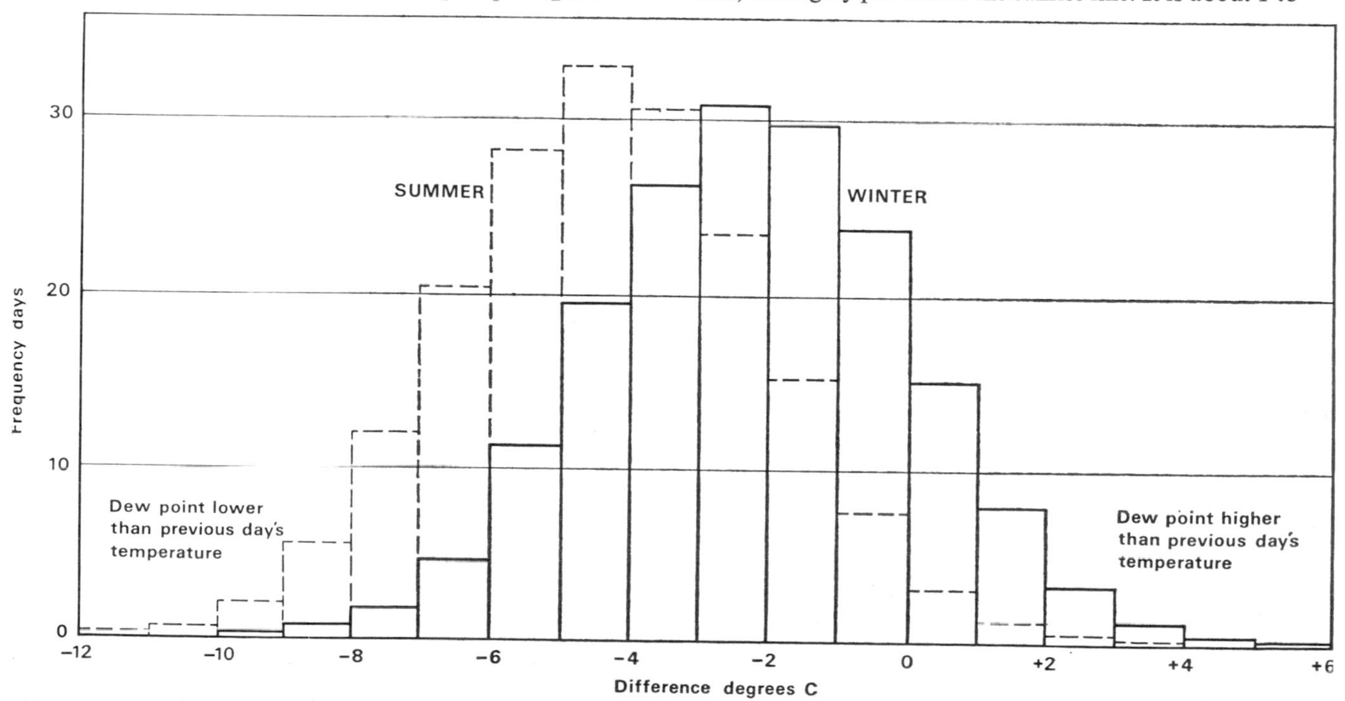

Figure 155 Average frequency of conditions likely to cause condensation when there is a rapid rise of dew point. Frequency, in days per 6-month season, of difference between 09 00 GMT dew point and mean air temperature during preceding 24 hours, at Garston

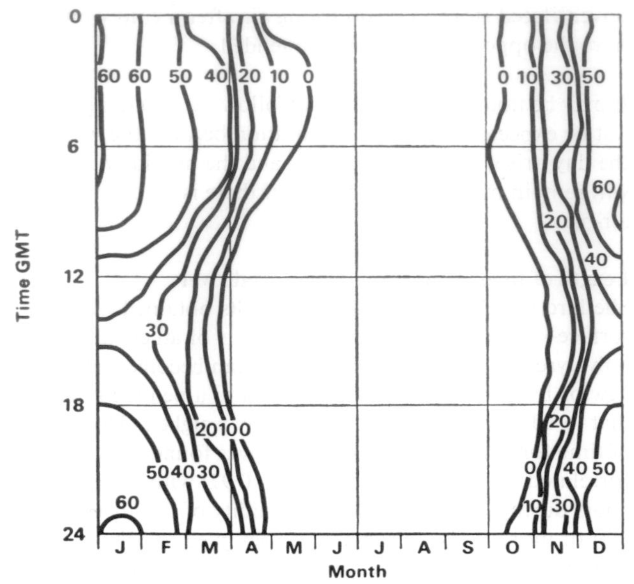

Figure 156 Percentage of the time during which the air temperature is less than 4°C at Heathrow. From hourly observations during the 10 years 1957–66

1½ hours later than sunrise during January to March, and about 2 hours later during the rest of the year. But it may be noted that during May and June there is just under an even chance that the relative humidity will remain below 90 per cent all night.

In winter, on the other hand, on only about two occasions in three will the relative humidity in the middle of the day be low enough to permit painting, although on half of these the temperature may be too low (Figure 156).

If the surfaces of materials which have to be painted behave in regard to wetness like the wetness-recorder, then Figure 19 gives a guide as to the best time for painting, especially if used in conjunction with Figure 156, which shows the proportion of time with air temperatures below 4°C.

Low temperatures and their effects on painting

The drying of paints, whether water-thinned or oil-based, is retarded by low temperature and high humidity. Emulsion paints may fail to coalesce below 5°C and thus not form a cohesive film. A subsequent rise in temperature may not prevent failure of the film. Oil-based paints will remain tacky for an excessive time at lower temperatures, while chemically-cured finishes are also slow to dry and should not be used in these conditions.

The most conveniently available statistics of temperature are those produced by the Meteorological Office from hourly data. These will give frequencies, as monthly or annual means, of temperatures below 4°C for each hour of the day (that is, temperatures of 3·9° or below). They are displayed diagrammatically in Figure 156. In midwinter, temperatures at Heathrow (which may be taken to be typical for much of south-east England at altitudes below about 100 m) are below 4°C for more than half the time. Even in the daytime these low temperatures persist for more than 25 per cent of the time in December, January and February.

A diagram like this gives a guide as to the average probability of experiencing temperatures unsuitable for painting at any time of the year. However, as already noted on page 164, damp conditions may also hinder painting. Table 64 is taken from a bi-modal analysis of hourly values of temperature and relative humidity at London Airport, Heathrow. The values are annual averages, during the 12 years 1957–68. The second column of the Table shows that, on average, the air temperature is 5·0°C or below for 17 per cent of the time in the daytime (08 00 to 18 00 GMT). During half or more of the night-time hours when the temperature is 5·0°C or below, the relative humidity is 90 per cent or more, but in the daytime only about a quarter of the cold hours are as humid as this. The right-hand column of Table 64 gives the detailed figures. On such occasions the conditions are doubly unsuitable, for not only is the temperature too low for satisfactory painting, but the surfaces are likely to be damp because of the high humidity, and the air too moist to allow water-thinned paints to dry.

Table 64 Frequency of occasions at each hour with temperature 5·0° C or below, and relative humidity greater than 90 per cent at Heathrow. Averages for 12 years 1957–68

Time GMT	Frequency of temperatures of 5°C or below %	Frequency when r.h. was greater than 90% at same time %
00 00	25·5	11·9
02 00	27·6	14·4
04 00	29·3	17·0
06 00	29·5	17·3
08 00	26·2	14·1
10 00	19·9	7·4
12 00	14·2	4·0
14 00	11·8	3·0
16 00	13·3	3·4
18 00	17·2	5·0
20 00	20·2	7·4
22 00	23·2	10·0
Mean	21·5	9·6

This analysis only takes into account the conditions at the time of painting. It is now also possible to state the frequency, at any time of the year, of the occurrence of spells during which conditions will be favourable both for applying a paint and for its subsequent drying or curing.

Wind-chill

The chilling effect of low temperatures on human beings is much increased if a wind is blowing, and a formula intended to give a measure of this effect was devised by Siple and Passel (1945). The formula gives the rate of heat loss from a dry, heated surface, and so takes no account of evaporation. The formula contains a factor F which varies with wind speed (Figure 157) and a temperature difference $(33 - t)$, where 33°C is the temperature of the dry skin, and t is the air temperature to which the skin is exposed. The wind-chill index is $F(33 - t)$ W/m², where the factor F varies with wind speed. The nomogram in Figure 158 is a convenient device for calculating the wind-chill index under any given conditions of air temperature and wind speed.

The index, although quite empirical in nature, has been found to correlate well with the sensation of bare, dry skin. The scale of wind-chill index in Figure 158 carries an indication of the

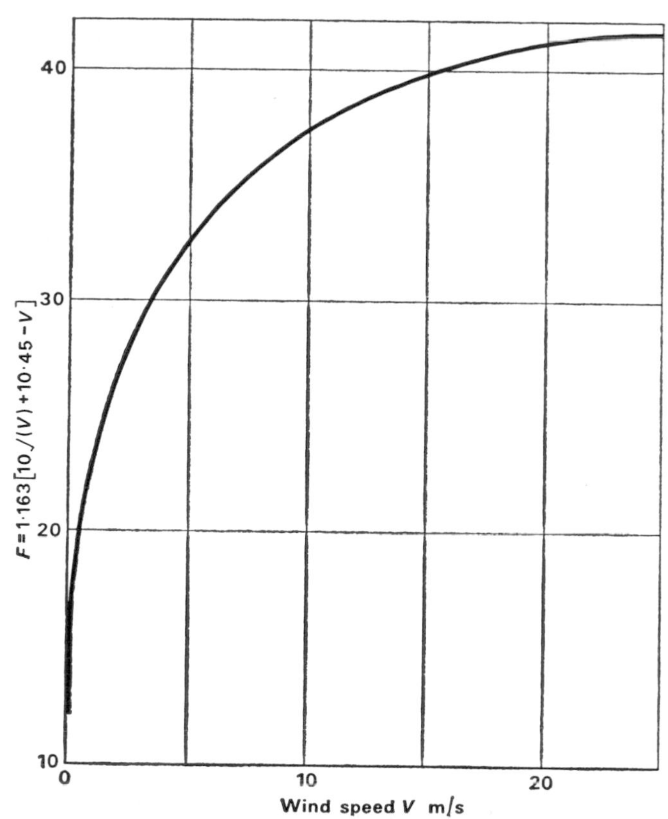

Figure 157 Variation of wind factor F in equation for wind-chill index, with change of wind speed

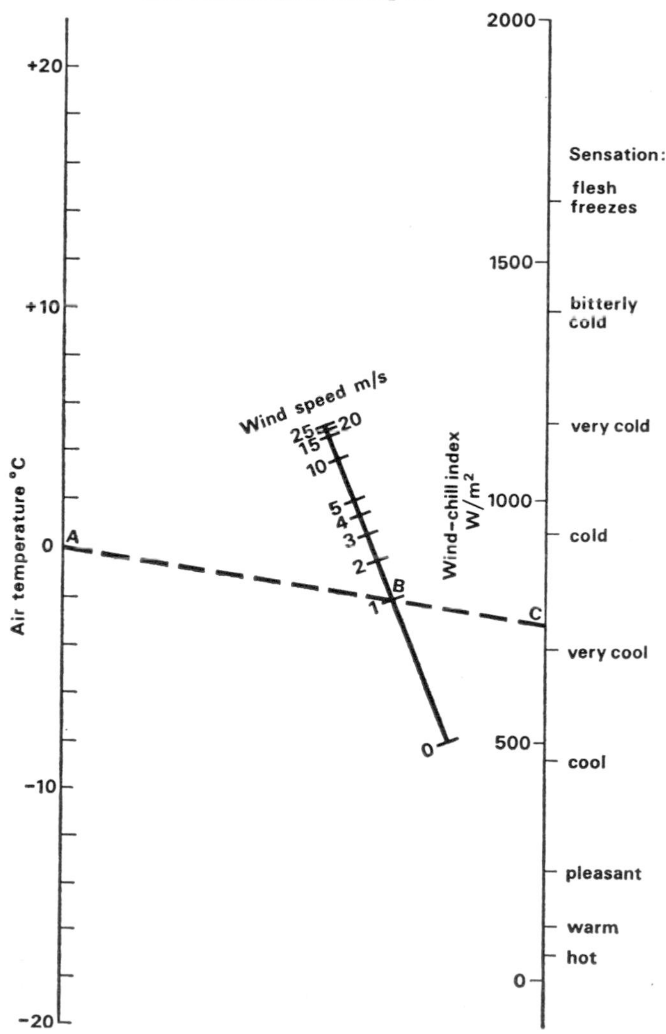

Figure 158 Nomogram for calculating wind-chill index. To use the nomogram, lay straight-edge through temperature and wind-speed scales and read index from right-hand scale, eg, line ABC through 0°C and 1 m/s gives index of 740 W/m²

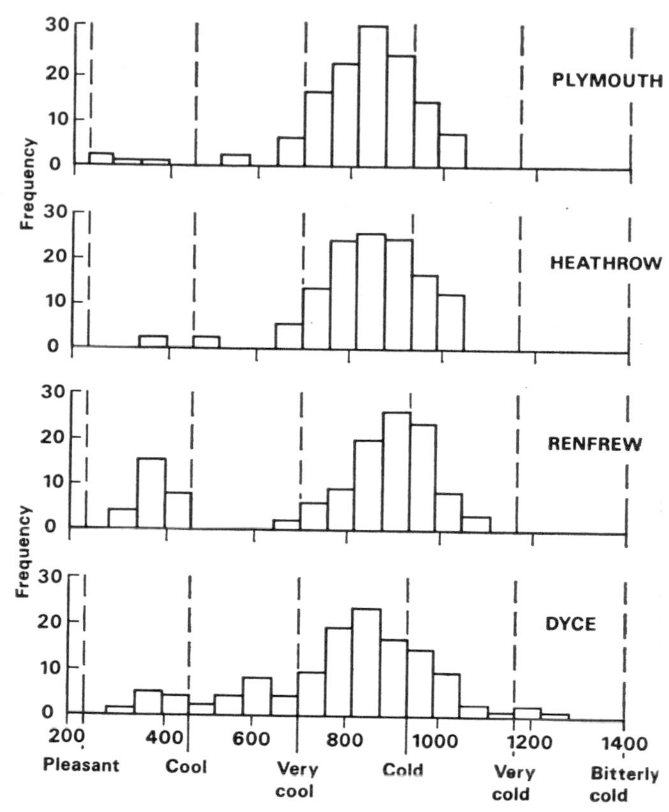

Figure 159 Frequencies of wind-chill index in January 1961, from wind and temperature at 00 00, 06 00, 12 00 and 18 00 GMT

sensation appropriate to each range of index, from hot to bitterly cold, and finally an indication that exposed flesh is liable to freeze when the index exceeds about 1600 W/m². It must be emphasized that the index does not give an indication of the sensation of a clothed body, but only that of exposed parts such as hands and face. It may therefore be a useful index for occasions when there is prolonged exposure and a job involves more or less delicate operations with ungloved hands, for example the erection of steelwork.

If the exposed skin is wet, the sensation of cold may be much greater than is suggested by the index, for a wetted skin may be cooled to the wet-bulb temperature. Unless the air is saturated, that is to say, unless the relative humidity is 100 per cent, the wet-bulb temperature will be some degrees lower than the air temperature (see page 15).

It has not been possible to prepare a map showing the variation of wind-chill index over the British Isles. Such a map would have to take into account the many local variations of wind speed and temperature. It will be expected that on average the index is highest in the windiest parts of the country, in spite of the fact that they are on the whole rather milder than the Midlands and south-east of England.

There may be great variations in the wind-chill index during any month. Figure 159 comprises frequency diagrams showing how wide a range of index may be experienced during one January, at four places representing south-west England, south-east England, west Scotland and east Scotland, respectively. The individual values were calculated from the temperatures and wind speeds given in the *Daily Weather Report* for four times each day, at 00 00, 06 00, 12 00 and 18 00 GMT. The calculated wind-chill index varies from about 200 W/m² (pleasant) on two occasions at Plymouth, to about 1250 (very cold to bitterly cold) on one occasion at Dyce, near Aberdeen. On exceptional days, as for example in the 1962–63 winter, an index of 1100 or 1200 may be experienced over a considerable area of the country.

The calculated values refer to wind speeds measured at a height of about 10 m above ground in open country. In built-up areas the wind speed near the ground may be lower than the assumed values, but on high buildings, at 50 or 100 m above ground, the wind speed may be even higher than that at 10 m in the open. Thus the wind-chill index may be higher than the 'air-field' value, although not by a great amount, for as can be seen by the shape of the curve of *F* against wind speed (Figure 157), at higher wind speeds the change in index with wind speed is small. This is because the surface of the skin has been cooled to nearly the same temperature as the air, and obviously can cool no more as long as it remains dry.

Plastic cracking of concrete

From time to time trouble is experienced with the cracking of large areas of concrete shortly after laying, the cracks sometimes extending through the greater part of the thickness of the slab. The phenomenon is observed on occasions when the relative humidity of the air is low, and is generally supposed to be caused by drying shrinkage of the surface layer of the slab, resulting from rapid evaporation of the water on the surface.

However it is unlikely that sufficient water could be evaporated in a period of only a few minutes as the rate of evaporation would probably not exceed about 1 mm/h. Moreover when the cracking takes place the concrete is still moist. It seems more likely that the cooling of the surface layer resulting from the high rate of evaporation causes thermal contraction. It is this shrinkage which causes the cracks to form in the still-weak material.

In any case the obvious way to avoid the trouble is to shield the freshly-laid concrete as quickly as possible from the dry air, especially from wind which will greatly increase the rate of evaporation. The higher the temperature of the concrete relative to that of the air, the more likely the trouble is to occur, for the higher the temperature difference the greater will be the rate of evaporation and hence the rate of cooling and shrinking.

No detailed statistics of combined values of relative humidity, and wind strength are at present available. As a general guide, it may be said that low humidities are most likely to be experienced in the afternoon, and especially at times when the sky is clear. Days with bright weather and a good breeze are the ones most conducive to a high rate of evaporation.

The conversion of high alumina cement by high temperature and humidity

High alumina cement is sometimes used in structural concrete when it is desired to attain high strength quickly, so as to permit prestressing and removal of shuttering within a day or so after casting and thus speed up the manufacturing process. It is well known that this cement will, when exposed to a high enough temperature after setting, convert rapidly to a different form with a reduced strength. It is usually considered that for practical purposes atmospheric temperatures exceeding about 30°C are the only ones that need consideration, and that because such high temperatures are rare in Britain there is no restriction on the use of high alumina cement in ordinary structures. However, the concrete may reach a high temperature, perhaps as high as 40°C, if cast in the open on a hot, sunny day, when the heat of reaction during setting is supplemented by solar heat absorbed by the dark-coloured material. Even a few hours at this elevated temperature can cause appreciable conversion.

Recent studies have shown that in fact, although the rate of conversion at lower temperatures, say around 20°, is very low indeed, even this low rate may be sufficient over the life of a building to reduce strength to an unacceptable degree. Furthermore, concrete beams in a roof structure may be warmer than is generally realized. Exact values will vary with the details of the structure: thermal insulation, amount of glazing in the walls, orientation and ventilation among other factors, but in a typical modern flat-roofed building covered with asphalt and new white chippings, the roof beams have been found during summer to have a daily average temperature about 2 or 3 deg C above that of the outside air. Therefore it would be expected that the roof beams in such a building in London would have a temperature of 25°C or more for about half per cent of their life. This value is about what would be expected from a study of the degree of conversion which has occurred in beams in a building about 17 years old. In wintertime the beams will be a little cooler than the rooms below, with temperatures averaging perhaps in the range 15 to 18°C, although electrical equipment or heating ducts may give higher temperatures locally.

High atmospheric humidity may sometimes be necessary for the conversion reaction to proceed, but there is usually enough free moisture in the concrete for this purpose.

Wet ground and its effect on construction

Building work may be hindered if the ground is wet. This is particularly troublesome in areas with a clay soil, not only because wet clay is difficult to dig, and the going underfoot is exceptionally unpleasant, but because clay drains slowly and the bad conditions tend to persist. In winter a clay soil may be more or less saturated for months on end, but in spring, summer and autumn, evaporation from the surface of the soil and transpiration of moisture by plants help to dry the ground. Calculations of the moisture-content of the ground may be made using information on the rainfall and other meteorological parameters, and an estimate obtained in terms of a moisture deficit. This is the quantity of rainfall required to restore the soil to field-capacity. While this may not be an exact measure of the state of a soil in the context of interference to digging or of movement over the ground, it may be of some help in planning operations.

In an area where there is no deficit of soil-moisture, the rainfall over a period (usually about 2 weeks) has equalled or exceeded the calculated transpiration from the ground. Thus the ground is at field capacity, which is to say that it contains sufficient water to satisfy the needs of plants, although it may not be completely saturated with water. When soils are in this state moving and digging will be difficult – it is typical of the conditions found in winter.

When there is a soil-moisture deficit, the surface of the ground will have dried to a greater or lesser degree, depending on the magnitude and duration of the deficit. At depth, the degree of drying will depend both on the deficit and the kind of soil. An open-pored soil will dry out more than a clayey one. Clay soils tend to shrink and crack, the blocks of soil between the cracks remaining in a wet state. A further complication is that clay soils can readily become puddled, so that drainage of rainwater is impeded.

Soil-moisture deficits

During the period of the year when there is most likely to be a deficit of moisture in the soil, from about April or May until October or November, statistics and maps of the estimated soil-moisture deficits are issued about once a fortnight by the Meteorological Office. Figure 160 is a copy of one of these

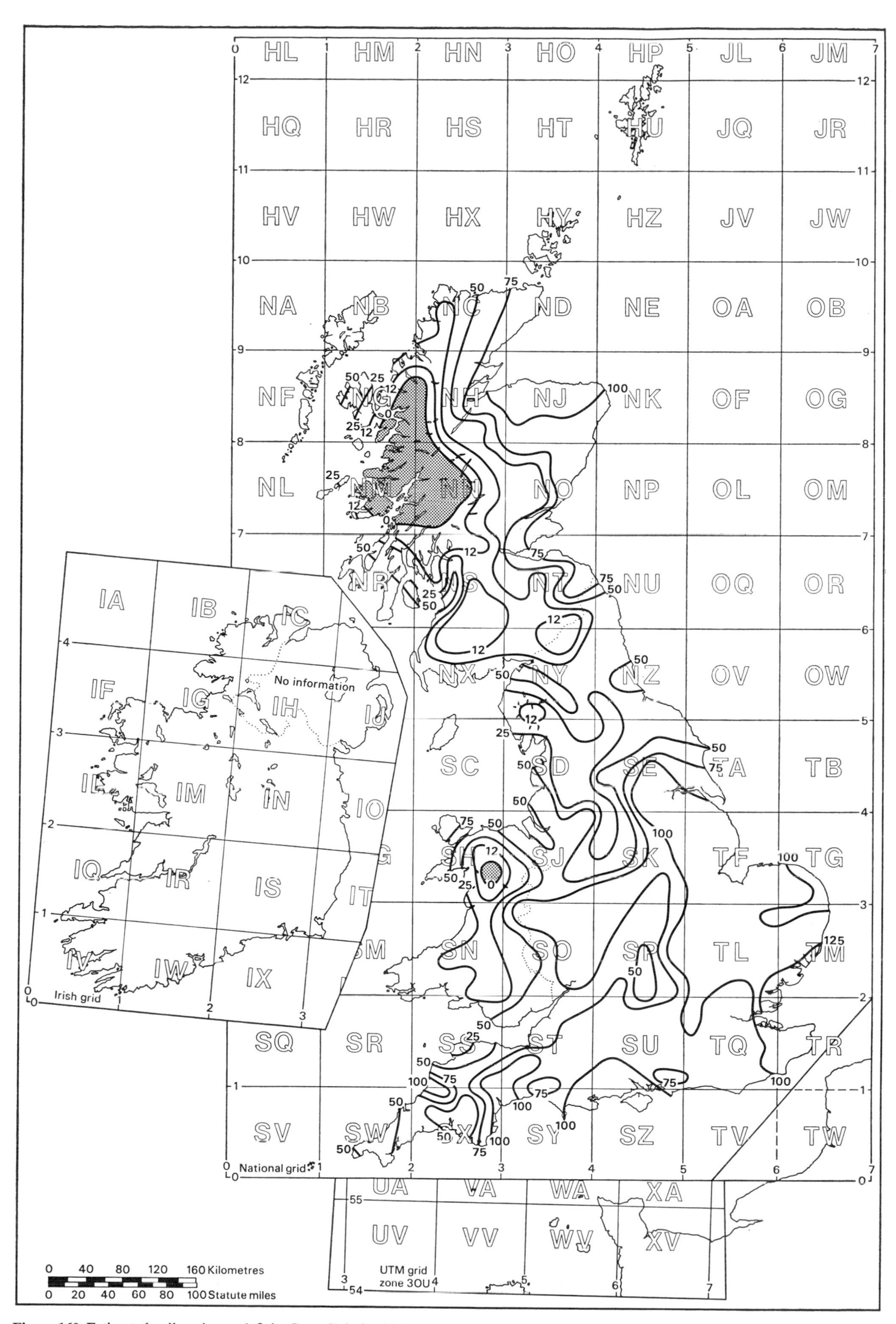

Figure 160 Estimated soil moisture deficit, Great Britain, 15 September 1971. Areas with no soil moisture deficit are shaded. Remaining areas are bounded by 0, 12, 25, 75, 100 and 125 mm lines

maps, for 15 September 1971, showing that in parts of western Britain there was no deficit, while over the rest of the country there was a greater or lesser deficit ranging up to over 125 mm in the coastal areas of Essex and Suffolk.

These data may be useful in assessing the likely delays in work caused by water-logged ground. They are calculated on the assumption that the ground is covered with vegetation which is evaporating moisture. An extensive site which has been stripped of all vegetation may therefore be wetter than is indicated by the map, because, unless the surface of the soil is wet, the rate of evaporation will be less than that from vegetation.

On the other hand, the maps only refer to the conditions of concern to crops. The ground at a depth of a metre or more may be wetter or drier than is indicated by the map. In southern Britain especially, the ground at depth may be appreciably drier than the surface soil where trees are growing or where they have recently been cut down. Because it takes so long to replenish the water in this dried-out ground, clay soils may continue swelling for some years after the trees have been removed. This is a not uncommon cause of damage to buildings.

The maps and statistics may be obtained on subscription as they are issued by the Meteorological Office. A paper describing how the maps are produced was published by Grindley (1967).

Construction delays caused by fog

The greatest delays to construction work as a result of fog will be those caused to road traffic, so that materials and workpeople will be delayed on their way to the site. This is discussed in Chapter 2. On the site itself, the interference is likely to be negligible except in the very thickest fogs, with the visibility less than 20 m. Table 65 shows the cumulative frequencies of fog of various densities at London Airport (Heathrow), which exemplifies conditions in the London area. The data are averages for the 3 months November, December and January, in the 10 years 1957–66. The visibility is 22 m or less for only 0·43 per cent of the daylight hours, on average, corresponding to about 3 hours in all during an average winter. A visibility of 40 m or less is experienced for 1·83 per cent of daylight hours on average but it is unlikely that visibilities of 23 to 40 m will cause difficulty on a normal construction site, except perhaps when tower cranes are being used on high buildings.

The visibilities quoted refer to observations in a horizontal plane. On some occasions the fog may be so shallow that upward visibility will be better than horizontal.

Table 65 Cumulative frequencies of visibility at London (Heathrow Airport). Daylight hours during November, December and January, averaged over the ten years 1957–66. Shown as percentage of all daylight hours

Visibility	Cumulative frequency percent of all hours of daylight
22 metres or less	0·43
40 metres or less	1·83
94 metres or less	5·70
200 metres or less	8·73
400 metres or less	12·86
1000 metres or less	24·79

As was remarked on page 76, on higher ground there may be many occasions when the clouds are so low that they reach the ground, so that visibility is as bad as in a fog at low level (for fog is really only a cloud with its base on the ground). Tall structures on high ground will be even more frequently enveloped in low cloud, perhaps for days on end. Even in Sheffield, at an elevation of some 150 m, it has been noted that a building 50 m high sometimes has the top in a cloud.

Thus the distribution of fog may be quite complex, being generally most frequent in inland valleys, decreasing in amount with increase in altitude until the occurrence of low cloud begins to become important. Local winds will also influence the distribution of fogs (see page 78).

Interference with construction by frost

If the ground becomes frozen to an appreciable depth, excavation may be hindered or, if sufficiently powerful machines are not available, may be stopped altogether. As recorded on page 126, the maximum depth to which the ground has been found to be frozen in lowland Britain is about 0·6 m. It appears that the depth of frost may be greater where the ground is traversed by water pipes leading from a cold reservoir, but this will be a very local effect.

As is explained on page 126, the depth to which the ground freezes depends on the kind of soil, the kind of cover (ie vegetation or snow) and on the weather conditions. Other things being equal, the depth of freezing depends approximately on the total frost-degree-days, which is the mean air temperature over a period multiplied by the number of days, in a spell when the mean temperature is continually below 0°C. The ground freezes to a depth of 0·3 m about once in 20 years on average in lowland Britain, although since 1940 the frequency has been rather greater.

It should be pointed out that in some circumstances frozen ground may cause less interference with work than wet ground, since frozen ground is easily traversed by men and by machines which could become bogged down in wet ground.

Frost is more of a nuisance when it freezes stocks of material, hinders the setting of mortar and concrete, and causes discomfort to operatives. These troubles can all be overcome by various means, though at added expense. Often the intelligent use of weather forecasts will enable work to be planned so as to reduce the risk of damage to new work. Some statistics of the frequency of days with minimum temperatures of 0°C or below are given on page 5, but the data required will depend on the particular problem. The question of discomfort is perhaps most likely to be linked with the concept of wind-chill (see page 166).

Figure 10 (page 14) shows that in winter the minimum temperatures at the surface of an extensive concrete slab average only about 1 deg C lower than screen minima (at 1·2 m above ground-level). These conditions are probably similar to those on an average building site, so that little error will be caused by assuming that the average number of frosts likely to damage fresh concrete or brickwork is the same as the number of air frosts, that is the occasions when the screen temperature falls to 0°C or below. It is likely that for all practical purposes the number of frosts to which an uncompleted building will be exposed will be the same as this up to heights of perhaps 100 m or more. Figure 9 (page 13) shows the annual mean number of days of air frost in England, Wales and Scotland, although it must be pointed out that there may be appreciable local differences in hilly districts.

Appendix I

CLIMEST–quick reply climate service for the construction industry

A service by the Meteorological Office

The following paragraphs are taken from a Meteorological Office circular describing a service started in 1969, CLIMEST, which is intended to give building contractors a guide to the amount of adverse weather which may be experienced during a contract for the period of the year when construction will be in progress.

Annex A to Met O 0/A6/1973 : CLIMEST – Quick Reply Climate Service for the Construction Industry

1. In preparing tenders contractors may need to allow for time lost on a contract because of adverse weather. To do this in the time available, which is often very limited, they need ready access to climatological information for the locality of the contract. The CLIMEST service provided by the Meteorological Office is designed to meet this need.

What the service offers

2. CLIMEST provides estimates for each month of the year and for the year as a whole of:

	Item	
Rain	1	Average total rainfall in millimetres (mm);
	2	Average number of hours in the period 0700 to 1700 GMT (0800 to 1800 BST) with rate of rainfall of 0·5 mm per hour or more at some time during the hour;
Temperature	3	Average number of days with air frost, ie the number of days when the air temperature measured at about four feet above ground level is below freezing point at any time of the day or night;
	4	Average number of hours (24-hour day) with air temperature below 0°C (32°F);
	5	Average number of hours (24-hour day) with air temperature below 2·2°C (36°F);
Wind	6	Average number of hours (24-hour day) with gusts of more than 38 mph (17 m/s) at 33 ft above ground level;
	7	Average number of hours (24-hour day) with gusts of more than 54 mph (24 m/s) at 33 ft above ground level

in the locality of the contract, anywhere in the United Kingdom. Enquiries will be dealt with promptly, usually within 24 hours.

How and where to apply

3. Write or telephone to one of these offices:

For sites in:	*Address*	*Telephone*
England and Wales	The Director-General Meteorological Office Met O 3b, London Road Bracknell Berks RG12 2SZ	Bracknell 20242 Ext 2279
Scotland	The Superintendent Meteorological Office 231 Corstorphine Road Edinburgh EH12 7BB	031–334 9721 Ext 524
Northern Ireland	The Senior Meteorological Officer Meteorological Office Tyrone House Ormeau Avenue Belfast BT2 8HH	Belfast 28457

Give — the position of the contract site, preferably its National Grid Reference plus distance and direction from nearest town centre
— the height of the site above sea level
— the client's order number (if appropriate)
— the address for reply.

How the reply will be sent

4. The information will be posted by first class mail, normally within 24 hours.

The cost

5. A standard charge is made for each CLIMEST. The bill is usually sent within a week of the written reply. If the estimates are required for a route (eg for roadworks or pipe laying) or an exceptionally extensive site it may be that representative averages cannot be given in a single CLIMEST; the route or area may have to be divided into two or more sections and a separate CLIMEST issued for each.

Other climatological information

6. Only those items specified in paragraph 2 are provided under the CLIMEST service. It may be possible to supply other climatological information on request but this is likely to take longer and the charge made will be based on the time taken to prepare it.

Appendix 2

Weather and the builder

This Appendix reproduces below the text of a DOE Advisory Leaflet which is intended to inform people on the building site what kind of weather information is available and how they can get it. It refers to both climatological information (on long-term weather) and to forecasts of weather in the next day or so. Builders can get either routine forecasts or special ones warning them of the impending occurrence of conditions which will be of particular interest to them.

Department of the Environment

DGD (Housing & Construction)

ADVISORY LEAFLET 40

The Weather Man can help you to carry on building

Building in Britain, particularly in winter, is often delayed by adverse weather conditions. The effects of bad weather can be reduced *without undue expenditure* by getting an early warning of its approach.

General weather forecasts for the next 24 hours, with an outlook for a further 24 to 48 hours, are given on radio, television and in the press. The forecasts are for large areas: forecasts for smaller areas are available on the G.P.O. automatic telephone weather service. General forecasts give an overall idea of the weather to be expected *but the conditions on a particular site may vary*. The severity of frost, for instance, may be influenced by the surrounding country. Cold air flows downhill and in a valley, with grassy slopes free of hedges, the temperature on a frosty night could be 3 to 6°C below that of a site on higher ground. Not only could the frost be more severe than that forecast but frost might occur when not mentioned in the general bulletin. *The Met. Office can prepare special forecasts for a particular building site. This leaflet tells you about these and other weather services of use to builders.*

Further Information

DOE Advisory Leaflets:

7 Concreting in cold weather

8 Bricklaying in cold weather

71 Site lighting

74 Protective screens and enclosures

78 Heating and drying out

DOE booklet:

Winter building

BRE report:

Driving-rain index

All above items obtainable from HMSO and other booksellers

Monthly Forecasts

The type of weather for the coming month is given in *Monthly Weather Survey and Prospects* published at the beginning of each month. There is also a statement on the prospects for the next thirty days (published about the middle of each month) which brings up to date the forecast made at the beginning of the month.

The forecast tells the builder the expected mean temperature, the total rainfall in relation to long-term averages and the sort of weather to expect, eg 'frost is likely to be more frequent than usual for November'. The prospects apply to the country as a whole, but it is sometimes possible to differentiate between England/Wales and Scotland/Northern Ireland or between eastern and western districts.

The *Monthly Weather Survey and Prospects* also contains a table and maps giving data on climatology for places throughout the country, a record of the weather in the month just past and a general survey. The yearly subscription rate for this publication (including the mid-monthly statement) can be obtained from the Secretary, Meteorological Office, London Road, Bracknell, Berks RG12 2SZ.

Short Period Forecasts

These are compiled specially for a particular building site. They help the builder to anticipate how the weather will affect the *day to day* work on site. He can then plan to avoid delays due to bad weather.

The forecast can be ordered to give details of winds (at the surface and other working levels), temperatures – *with special emphasis on frosts* and the general weather. A further advantage in having a short period forecast at specified times is that the office issuing the forecast will also issue amendments as and when necessary.

Climatology Services

Climatology is mainly related to the planning stages of the project; it is the study of *average variations* in the weather as distinct from the *actual* weather at a specified time. Climatology can help the builder:

a. To decide if a building practice used in one area is likely to be successful elsewhere – the average annual total amount of driving rain in parts of Scotland is twenty times greater than in the London area.

b. To assess how outdoor work may be interrupted by frost, heavy rain or gales. Their frequency and severity can be given for different parts of the country. Information is available on the average frequency of wind *gusts* above the critical speed for safe working – this is especially useful when thinking of using tower cranes.

c. To plan excavation work. Data on the frequency of heavy rains and probable rates of evaporation, together with details of the type of ground, can indicate the working conditions to be expected.

Charges for this service are based on the type of data required and the time spent on the work.

For construction sites the Met. Office provides a special quick-reply service, CLIMEST. This gives estimates of the monthly and annual averages of frost and low temperatures, rainfall and wind gusts above certain values, in the locality of the contract.

Effects of Weather

Phenomenon	In conjunction with	Effect
Rain		1. Affects site access and movement 2. Spoils newly finished surfaces 3. Delays drying out of buildings 4. Damages excavations 5. Delays concreting, bricklaying and all external trades 6. Damages unprotected materials 7. Causes discomfort to personnel 8. Increases site hazards
	High wind	1. Increases rain penetration 2. Reduces protection offered by horizontal covers 3. Increases site hazards
Low and sub-zero temperatures		1. Damages mortar, concrete, brickwork, etc. 2. Slows or stops development of concrete strength 3. Freezes ground and prevents subsequent work in contact with it, eg concreting 4. Slows down excavation 5. Delays painting, plastering, etc. 6. Causes delay or failure in starting of mechanical plant 7. Freezes unlagged water pipes and may affect other services 8. Freezes material stockpiles 9. Disrupts supplies of materials 10. Increases transportation hazards 11. Creates discomfort and danger for site personnel 12. Deposits frost film on formwork, steel reinforcement and partially completed structures
	High wind	Increases probability of freezing and aggravates effects of 1–12 above
Snow		1. Impedes movement of labour, plant and material 2. Blankets externally stored materials 3. Increases hazards and discomfort for personnel 4. Impedes all external operations 5. Creates additional weight on horizontal surfaces
	High wind	Causes drifting which may disrupt external communications
High wind		1. Makes steel erection, roofing, wall sheeting, scaffolding and similar operations hazardous 2. Limits or prevents operation of tall cranes and cradles, etc. 3. Damages untied walls, partially fixed cladding and incomplete structures 4. Scatters loose materials and components 5. Endangers temporary enclosures

Warnings

A warning service can be provided as an alternative to short period forecasts. Standing arrangements can be made for the builder to be told when special weather conditions are likely to occur, eg local warnings of frost, gales, fog or sunshine and dry spells. This enables the builder to plan for bad (or beneficial) conditions.

Pre-arranged *short period forecasts* or a *warning service* have considerable advantages over random telephone calls to a forecast office because they give plenty of warning about important changes in the weather. The builder can also make his special requirements known to the meteorologist who will see that any peculiarities of the site are considered. Forecasts and warnings, 'tailor-made' to the requirements of the individual builder, are passed by telephone and enable the builder to discuss his problems directly with the forecaster.

Weather Information Centres

London
284-286 High Holborn, London WC1
Telephone 01–836 4311

Glasgow
118, Waterloo Street, Glasgow C2
Telephone 041–248 3451

Manchester
56, Royal Exchange, Manchester, M2 7DA
Telephone 061–832 6701

Southampton
160, High Street-below-Bar, Southampton, SO1 0BT
Telephone Southampton (0703) 28844

Newcastle upon Tyne
Newgate House, Newgate Street,
Newcastle upon Tyne, NE1 5UQ
Telephone Newcastle (0632) 26453

These centres are open for personal enquiries from Monday to Friday (mornings and afternoons) and on Saturday (morning only). Forecasters are available to deal with both climatology and forecasting enquiries.

Some of the Meteorological Services attract charges – for details it is best to apply to the Senior Meteorological Officer at any of the Offices shown on the map, alternatively reference may be made to one of the following addresses:

England and Wales
The Director-General,
Meteorological Office,
London Road,
Bracknell, Berks. RG12 2SZ
Telephone 0344 20242

Scotland
The Superintendent,
Meteorological Office,
231, Corstophine Road,
Edinburgh, EH12 7BB
Telephone 031-334 9721

Northern Ireland
The Senior Meteorological Officer,
Meteorological Office,
Tyrone House,
Ormeau Avenue,
Belfast, BT2 8HH
Telephone Belfast 28457

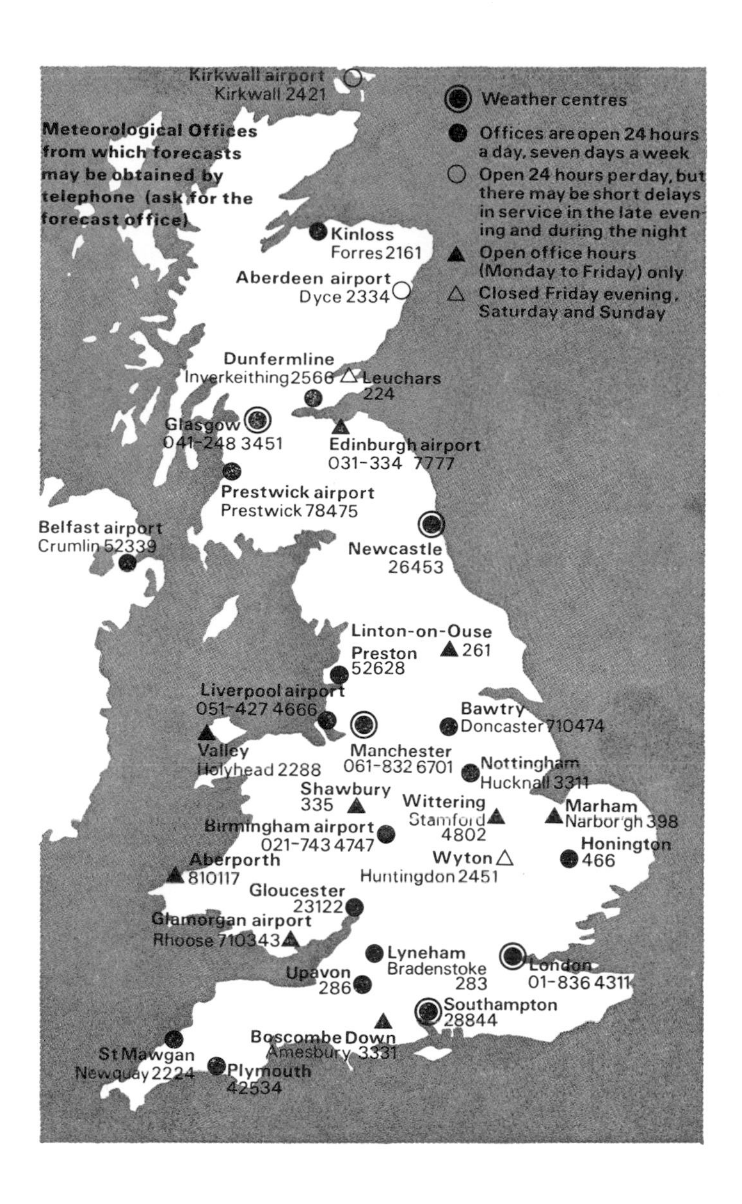

Professional Consultations

Professional consultations with Met. Office specialists can be arranged both within and outside the Met. Office – they are always ready to discuss your problems and requirements.

Terms used in forecasts

The meanings of terms used in forecasts for building are unlikely to raise doubts. In frost forecasting special descriptive terms are used because damage caused by frost depends on *wind speed* as well as air temperature.

Term	Corresponding air temperature (°C) — Wind Speed below 10 knots	10 knots or over
Slight frost	Below 0 to –3	Just below 0
Moderate frost	–4 to –6	–1 to –2
Severe frost	–7 to –11	–3 to –5
Very severe frost	–12 or below	–6 or below

A temperature of –1°C is often referred to as 'one degree of frost', –3°C as three degrees of frost, and so on

Term	Wind Force (Beaufort No.)	Wind Speed in knots	in miles/hour	in kilometres/hour	in metres/second
Calm	0	less than 1	less than 1	less than 1	less than 0.2
Light air	1	1–3	1–3	1–5	0 3 – 1.5
Light breeze	2	4–6	4–7	6–11	1.6 – 3.3
Gentle breeze	3	7–10	8–12	12–19	3.4 – 5.4
Moderate breeze	4	11–16	13–18	20–28	5.5 – 7.9
Fresh breeze	5	17–21	19–24	29–38	8.0–10.7
Strong breeze	6	22–27	25–31	39–49	10.8–13.8
Near Gale	7	28–33	32–38	50–61	13.9–17.1
Gale	8	34–40	39–46	62–74	17.2–20.7
Strong Gale	9	41–47	47–54	75–88	20.8–24.4
Storm	10	48–55	55–63	89–102	24.5–28.4
Violent Storm	11	56–63	64–72	103–117	28.5–32.6
Hurricane	12	over 63	over 72	over 117	over 32.7

References and Bibliography

See also list of useful publications at the end of Chapter 1, page 54

Aanensen, C J M. 1964. Gales in Yorkshire in February 1962. Meteorological Office, *Geophys Mem.*, No 108, HMSO, London

Addleson, L. 1972. *Materials for buildings* – Vol 2 and 3. Water and its effects, Iliffe, London

Atkins, J E. 1968. Changes in the visibility characteristics at Manchester/Ringway Airport. *Met. Mag.*, **97** (1151), pp 172–174

Atkinson, B W. 1966. Some synoptic aspects of thunder outbreaks over south-east England, 1951–60. *Weather*, **21** (6), pp 203–209

Balchin, W G V and **Pye, N.** 1947. A micro-climatological investigation of Bath. *Quart. J.R. Met. Soc.*, **73** (317–8), pp 297–319

Best, A C. 1950. The size distribution of raindrops. *Quart. J.R. Met. Soc*, **76,** pp 16–36

Best, A C, Knighting, E, Pedlow, R H and **Stormonth, K.** 1952. Temperature and humidity gradients in the first 100 m over south-east England. Meteorological Office, *Geophys. Mem.*, No 89. HMSO, London

Bilham, E G. 1936. Classification of heavy falls of rain in short periods. *British rainfall 1935.* HMSO, London pp 262–280

Bilham, E G. 1938. *The climate of the British Isles.* Macmillan, London

Bonacina, L C W. 1957. Snowfall in Great Britain during the decade 1946–55. *British rainfall 1955.* HMSO, London, pp 219–230

Boobyer, E H and **Brookes, A J.** 1967. Inclement weather, assessing its effects on building contracts. *Building*, **212** (6464), pp 133–142

Booth, R E. 1964. 1931–60 average monthly and annual maps of bright sunshine over Great Britain and Northern Ireland. Meteorological Office, *Clim. Mem.* No 42, Bracknell

Booth, R. E. 1969. 1931–60 monthly, seasonal and annual maps of mean daily maximum, mean daily minimum, mean temperature and mean temperature-range over the British Isles. Meteorological Office, *Clim. Mem.* No 43A, Bracknell

Brazell, J H. 1968. *London weather.* Meteorological Office, Met. O.783, HMSO, London

British Standards Institution, London. 1965. BS 1339: *Definitions, formulae and constants relating to the humidity of the air*

British Standards Institution. 1966. BS 2573: *Permissible stresses in cranes*

British Standards Institution. 1967. CP3: *Code of basic data for the design of buildings.* Chap. V, Loading: Part 1: Dead and imposed loads

British Standards Institution. 1972. CP3: Chap V, Part 2: Wind loads

British Standards Institution. 1973. CP121: *Walling*, Part 1: Brick and block masonry, p 18

Brooks, C E P and **Douglas, C K M.** 1956. The glazed frost of January 1940. Meteorological Office, *Geophys. Mem.* No 98. HMSO, London

Browning, K A and **Ludlam, F H.** 1962. Airflow in convective storms. *Quart. J. R. Met. Soc.*, **88** (376) pp 117–135

Brunt Committee. 1955. Basic design temperatures for space-heating. Ministry of Works, *Post-War Building Studies* No 33. HMSO, London

Building Research Establishment. 1970. The assessment of wind loads. *BRS Digest* 119, HMSO London

Building Research Establishment. 1971. An index of exposure to driving rain. *BRS Digest* 127, HMSO, London

Building Research Establishment. 1972. Wind environment around tall buildings. *BRS Digest* 141, HMSO, London

Building Research Establishment. 1973. Soakaways, *BRE Digest* 151, HMSO, London

Building Research Establishment. 1976. Roof drainage, parts 1 and 2. *BRE Digests* 188 and 189, HMSO, London

Butterworth, B. 1947. The rate of absorption of water by partly saturated bricks. *Trans. Brit. Ceram. Soc.* **46,** pp 72–76

Byers, H R and **Braham, R R.** 1949. *The thunderstorm.* US Dept of Commerce, Washington

Carlson, T N and **Ludlam, F H.** 1968. Conditions for the occurrence of severe local storms. *Tellus,* Stockholm, **20** (2), pp 203–226

Caspar, W and **Sandreczki, A.** 1964. Eisablagerungen aus meteorologischer Sicht. *Elektrotech. Z.* B, **16** (26), pp 763–767

Chalmers, J A. 1957. *Atmospheric electricity.* Pergamon, London. p 254

Chandler, T J. 1965. *The climate of London.* Hutchinson, London

Chandler, T J. 1965. Absolute and relative humidities in towns. *Bull. Am Met Soc.,* Lancaster, Pa, **48** (6), pp 394–9

Clapp, M. 1966. Weather conditions and productivity – detailed study of five building sites. *Building,* **211** (6439), pp 171–2, 175–6, 179–180

Conrad, V and **Pollak, L W.** 1950. *Methods in climatology.* Harvard University Press, Cambridge, Mass

Cooling, L F. 1930. The evaporation of water from brick. *Trans. Ceram. Soc.,* **29**(2), pp 39–54

Corfield, G A and **Newton, W G.** 1968. A recent change in visibility characteristics at Finningley. *Met. Mag.,* **97** (1152), pp 204–209

Cottis J G. 1960. Rain and/or low temperature as factors interrupting external building work in the London area. Meteorological Office, *Clim. Mem.* No 27

Craddock, J M. 1965. Domestic fuel consumption and winter temperatures in London. *Weather,* **20** (8), pp 257–258

Craxford, S R, Gooriah, B D and **Weatherley, M-L P M.** 1964. Air pollution and sunshine. Warren Spring Laboratory note SCCB 62/4 (unpublished)

Currie, G. 1955. The London blackout. *Weather,* **10** (4), p 136

Davis, A, Gordon, D and **Howell, G V.** 1973. Continuous world-wide uv monitoring programme. ERDE *Tech. Rep.* No 141, Waltham Abbey, Essex

Davenport, A G. 1965. The relationship of wind structure to wind loading. *Proc. Conference on Wind Effects on Buildings and Structures, 1963.* National Physical Laboratory. HMSO, London, pp 53–111

Davenport, A G. 1968. The dependence of wind loads on meteorological parameters. *Proc. International Research Seminar on Wind Effects on Buildings and Structures, 1967,* National Research Council of Canada, Ottawa. University of Toronto Press, Toronto, 1968, pp 19–82

Deacon, E L. 1955. Gust variation with height up to 150 m. *Quart. J. R. Met. Soc.,* **81** (350), pp 562–573

Department of the Environment, DGD (Housing & Construction). Advisory Leaflet 40. *Weather and the builder.* 6th edn 1975 HMSO, London

Dick, J B. 1950. The fundamentals of natural ventilation of houses. *Inst. Heat. and Vent. Eng. J.* **18** (179), pp 123–134

Evans, D C. 1957. A second report on fog at London Airport. *Met. Mag.,* **86,** pp 333–339

Eyre, S R. 1962. Fog in Yorkshire. *Weather,* London, **17** (4), pp 125–131

Gajzágó, L. 1973. Outdoor microclimates and human comfort. *CIB Colloquim 'Teaching the teachers',* Stockholm, 1972. State Institute for Building Research, Stockholm

Granum, H. 1965. Evaporation of moisture from wall surfaces. *RILEM/CIB Symposium on moisture problems in buildings,* Helsinki, paper 3–6. Otaniemi, Finland. RILEM, 1966

Grindley, J. 1967. Estimation of soil moisture deficits. *Met. Mag.,* **96,** pp 97–108

Harris, P B. 1973. The BRE ultra-violet sensor. *BRE Current Paper* CP 17/73. Building Research Establishment, Watford

Harrison, A A. 1967. Variations in night minimum temperatures peculiar to a valley in mid-Kent. *Met. Mag.* **96** (1142), pp 257–265

Harrison, A A. 1971. A discussion of the temperatures of inland Kent with particular reference to night minima in the lowlands. *Met. Mag.,* **100** (1185), pp 97–111

Hawke, E L. 1944. Thermal characteristics of a Hertfordshire frost hollow. *Quart. J.R. Met. Soc.,* **70,** pp 23–48

Helliwell, N C. 1971. Wind over London. Proc. 3rd Conference on wind effects on buildings and structures, Tokyo, pp 23–32

Hoblyn, T N. 1928. A statistical analysis of the daily observations of the maximum and minimum thermometers at Rothamsted. *Quart. J.R. Met. Soc.,* London, **54** (227), pp 183–202

Holland, D J. 1964. Rain intensity frequency relationships in Britain. Meteorological Office, *Hydrolog. Mem.* No 33, with Appendix (1968), Bracknell

Holland, D J. 1967a. Rain intensity frequency relationships in Britain. *British rainfall 1961,* part III, pp 43–51. HMSO. London

Holland, D J. 1967b. Evaporation. *British rainfall 1961,* part III, pp 5–34. HMSO, London

Illig, W. 1952. The amount of water transferred by diffusion at walls. *Gesundheitsingenieur,* **73,** p 124

Illuminating Engineering Society. 1962. Lighting during daylight hours. *IES Technical Digest* No 4, London

Ingard, U. 1953. The physics of outdoor sound. *Proc. 4th Annual Noise Abatement Symp.,* October 1953, Chicago, pp 11–25. National Noise Abatement Council, New York

Institution of Heating and Ventilating Engineers. 1971. *IHVE Guide 1970, Book A.* London

Jansson, I. 1965. Testing for rate of water absorption. *RILEM/CIB Symposium 1965,* Helsinki, paper 2–25. Otaniemi, Finland. RILEM, 1966

Kelly, T. 1963. A study of persistent and semi-persistent thick and dense fog in the London area during the decade 1947–56. *Met. Mag.,* **92** (1091), pp 177–183

Khlusov, Z. 1954. The protection of buildings from atmospheric precipitation. *Arkhitektura SSSR,* **11,** pp 43–44 (English translation, *BRS Library Comm.* No 797)

Köppen, W and **Geiger, R.** 1930. *Handbuch der Klimatologie.* Borntrager, Berlin

Kratzer, P A. 1956. *Das Stadtklima.* Vieweg, Braunschweig

Lacy, R E. 1951. Variations of the winter means of temperature, wind speed and sunshine, and their effect on the heating requirements of a house. *Met. Mag.,* **80,** pp 161–165

Lacy, R E. 1964. Some measurements of snow density. *Weather,* **19** (11), pp 353–356

Lacy, R E. 1965. Driving-rain maps and the onslaught of rain on buildings. *RILEM/CIB Symposium on Moisture Problems in Buildings,* Helsinki, Paper 3–4. (Also *BRS Current Paper, Research Series* 54, *Building Research Station, Watford*)

Lacy, R E. 1966. Frost patterns on brickwork. *Weather,* **21** (4), pp 134–137

Lacy, R E. 1968. Tornadoes in Britain 1963–66. *Weather,* **23** (4), pp 116–124

Lacy, R E. 1976. *Driving-rain index.* BRE Report, HMSO, London

Lacy, R E and **Shellard, H C.** 1962. An index of driving rain. *Met. Mag.,* **91,** pp 177–184

Lamb, H H. 1969. The new look of climatology. *Nature,* **223,** pp 1209–1215

Landsberg, H. 1962. City air – better or worse? In Symposium 'Air over cities'. R A Taft Sanitary Engineering Center. *SEC Tech. Rep.* 62–5, pp 1–22

Lawrence, E N. 1966a. Sunspots – a clue to bad smog? *Weather,* **21** (10), pp 367–370

Lawrence, E N. 1966b. An indication of wind directions potentially favourable for atmospheric pollution. *Met. Mag.* **95** (1129), pp 241–248

Laws, J O and **Parsons, D A.** 1943. Relation of raindrop size to intensity. *Amer. Geophys. Union Trans.,* **24,** part II, pp 453–460

Loudon, A G. 1964. In discussion on Lumb (1964). *Quart. J.R. Met. Soc.* **90** (386), pp 493–495

Loudon, A G. 1965. The interpretation of solar radiation measurements for building problems. CIE Conference on sunlight in buildings. *BRS Current Paper, Research Series* 73. Building Research Station, Watford

Loudon, A G. 1968. Summertime temperatures in buildings. IHVE/BRS Symposium on thermal environment in modern buildings. *BRS Current Paper* CP 47/68. Building Research Station, Watford

Loudon, A G and **Danter, E.** 1965. Investigations of summer overheating at the Building Research Station, England. *Building Science,* **1** (1), pp 89–94

Ludwig, F L. 1968. Urban temperature fields. World Meteorological Organisation, *Tech. Note* No 108, pp 80–107. Geneva

Lumb, F E. 1961. Seasonal variation of the sea surface temperature in coastal waters of the British Isles. Met. Office *Scientific Paper* No 6. HMSO, London

Lumb, F E. 1964. The influence of cloud on hourly amounts of total solar radiation at the sea surface. *Quart. J.R. Met. Soc.,* **90** (383), pp 43–56

Macklin, W C. 1962. The density and structure of ice formed by accretion. *Quart. J.R. Met. Soc.,* **88** (375), pp 30–50

Manley, G. 1939. On the occurrence of snow cover in Great Britain. *Quart. J.R. Met. Soc.,* **65,** pp 2–27

Meetham, A R. 1956. *Atmospheric pollution its origins and prevention.* 2nd edn. Pergamon, London

Meteorological Office. 1936. The great Northamptonshire hailstorm of 22nd September, 1935. *British Rainfall* 1935. M.O. 395. HMSO, pp 281–285

Meteorological Office. 1956a. Monthly and annual maps of average temperature over Great Britain and Northern Ireland 1921–50. *Clim. Mem.* No 3, Bracknell (Reprinted 1959)

Meteorological Office. 1956b. *Handbook of meteorological instruments,* part 1, Met. O. 577. HMSO, London

Meteorological Office. 1958a. *Tables of temperature, relative humidity and precipitation for the world,* part 3. Europe and the Atlantic Ocean. Met. O. 617c. HMSO, London

Meteorological Office. 1958b. Combined distribution of hourly values of dry-bulb and wet-bulb temperatures 1946–55. *Climatological Branch Memoranda* Nos 10 to 20. Meteorological Office, Bracknell

Meteorological Office. 1964. *Hygrometric tables:* Part II: Stevenson Screen readings, Degrees Celsius (Met. O. 265b); Part III: Aspirated Psychrometer readings, Degrees Celsius (Met O. 265c). HMSO, London

Meteorological Office. 1968a. *Tables of surface wind speed and direction over the United Kingdom.* Met. O. 792. HMSO, London

Meteorological Office. 1968b. *Averages of earth temperature at depths of* 30 *cm and* 122 *cm for the United Kingdom,* 1931–60. Met. O. 794. HMSO, London

Meteorological Office. 1972. *Meteorological glossary.* Met. O. 842. HMSO, London

Meteorological Office. 1975. *Flood studies report,* Volume II Meteorological studies, Volume V Maps. Natural Environment Research Council, London

Meteorological Office. *Observatories' Year Books* for 1922 to 1942. HMSO, London

Monteith, J L. 1961. The long-wave radiation balance of the British Isles. *Quart. J.R. Met. Soc.,* **87** (372), pp 171–179.

Monteith, J L. 1962. Attenuation of solar radiation: a climatological study. *Quart. J.R. Met. Soc.,* **88,** pp 508–521

Monteith, J L. 1966. Local differences in the attenuation of solar radiation over Britain. *Quart. J.R. Met. Soc.,* **92,** (392), pp 254–262

Monteith, J L and **Sceicz, G** 1962. Radiative temperature in the heat balance of natural surfaces. *Quart. J.R. Met. Soc.* **88** (378), pp 496–507

Murphy E J. 1973. Distribution of driving rain in Ireland. *Climatological Note* No 3, Department of Transport and Power, Meteorological Service, Dublin

Newberry, C W and **Eaton, K J.** 1974. *Wind loading handbook.* BRE Report, HMSO, London

Oke, T R. 1972. City size and the urban heat island. *Amer. Met. Soc. Conference on Urban Environment,* Philadelphia, Pa.

Page, J K. 1969. Heavy glaze in Yorkshire – March 1969. *Weather,* **24,** (12), pp 486–495

Pedgley, D E. 1962. A meso-synoptic analysis of the thunderstorms on 28 August 1958. Meteorological Office, *Geophys. Mem.* No 106. HMSO, London

Pedgley, D E. 1967. Why so much rain? *Weather,* **22** (12), pp 478–482

Penman, H L. 1940. Meteorological and soil factors affecting evaporation from fallow soil. *Quart. J.R. Met. Soc.,* **66** (287), pp 401–410

Penwarden, A D. 1973. Acceptable wind speeds in towns. *Building Science,* **8,** pp 259–267

Penwarden, A D and **Wise, A F E.** 1975. *Wind environment around buildings.* BRE Report, HMSO, London

Petherbridge, P. 1969. *Sunpath diagrams and overlays for solar heat gain calculations.* HMSO, London

Reuben, B. 1968. An unusual and expensive sunshine recorder. *Weather,* **23** (3), pp 125–126

Reynolds, G. 1954. Snowfall probabilities on Merseyside. *Quart. J.R. Met. Soc.,* **80** (345), pp 444–447

Rowsell, E H. 1956. Damaging hailstorms. *Met. Mag.* **85,** pp 344–346

Scholes, W E and **Parkin, P H.** 1967. The effect of small changes in source height on the propagation of sound over grassland. *J. Sound Vib.,* **6** (3), pp 424–442

Scorer, R S. 1968. Air pollution problems at a proposed Merseyside chemical fertilizer plant. *Atmos. Environment, Oxford,* **2** (1), pp 35–48

Scorer, R S and others. 1968. Round table on plume rise and atmospheric dispersion. *Atmos. Environment, Oxford.* **2** (3), pp 193–196

Shellard, H C. 1956a. Maps of standard deviation of monthly mean temperature, 1921–50. Meteorological Office, *Clim. Mem.* No 2 Bracknell

Shellard, H C. 1956b. Monthly averages of accumulated temperature above and below various base temperatures for stations in Great Britain and Northern Ireland 1921–50. Meteorological Office, *Clim. Mem.* No 5, Bracknell

Shellard, H C. 1962. Can the meteorologist help you? *J. Oil Colour Chem. Assm,* **45** (3), pp 200–217

Shellard, H C. 1965. Extreme wind speeds over the United Kingdom for periods ending 1963. Meteorological Office, *Clim. Mem.* No 50, Bracknell

Shellard, H C. 1967. Results of some special measurements in the United Kingdom relevant to wind loading problems. *Proc. International Research Seminar on Wind Effects on Buildings and Structures,* National Research Council of Canada, Ottawa. Univ. of Toronto Press, Toronto, 1968, pp 525–533

Simpson, J E. 1967. Aerial and radar observations of some sea breeze fronts. *Weather,* **22** (8), pp 306–316, 325–327

Siple, P A and **Passel, C F.** 1945. Measurements of dry atmospheric cooling in sub-freezing temperatures. *Proc. Am. Philos. Soc.,* **89** (1)

Steinhauser, F, Eckel, O and **Sauberer, F.** 1955–59. *Klima und Bioklima von Wien,* parts I to III. Österr. Gesellschaft für Meteorologie, Vienna XIX

Stevenson, C M. 1969. The dust fall and severe storms of 1 July 1968. *Weather,* **24** (4), pp 126–132

Thomson, A B. 1963. Water yield from snow. *Met. Mag.,* **92,** pp 332–335

Tunnell, G A. 1958. World distribution of atmospheric water vapour pressure. Meteorological Office, *Geophys. Memo.* No 100. HMSO, London

Watkins, L H. 1963. Research on surface-water drainage. *Proc. Inst. Civ. Engrs.,* **24** (March), pp 305–330

World Meteorological Organisation. 1971. Guide to meteorological instrument and observing practices, Table 6.2, *WMO-No* 8, *TP* 3, 4th edn. Geneva

Index

Printed in England for Her Majesty's Stationery Office
by Bemrose & Sons Ltd., Derby

Dd 496366 K 12 6/77